INTRODUCTION TO ESTIMATING, PLAN READING AND CONSTRUCTION TECHNIQUES

To understand Construction Estimating one must also understand plan reading and construction techniques. This book is designed to teach the construction student these three core skills in equal measure. Using hundreds of plans, sketches, and photos, the book builds case studies of the major construction divisions including concrete, masonry, carpentry, and more. Over forty cases are divided into sections following a specially designed format:

Plans: Scale drawings of floor plans, sections, or elevations.
Plan Interpretation: The drawings are explained with comments.
Scope of the Work: A written description of the boundaries of the work is given for each section.
Construction Techniques: The construction processes and their sequence are explained.
The Takeoff: A takeoff is shown at the end of each section.

This approach helps foster confidence in plan reading, building methods, arithmetic, takeoffs, and estimates. The various products and terms used in the industries of structural steel, doors and hardware, and roofing are defined. The shop drawing process is explained, which is so important in many industries, as are the role of and difference between manufacturers, fabricators, and suppliers/distributors. The book ends with a study of "front-end" documents, including Division 00 General Conditions, AIA 201, and Division 01 General Requirements, and a chapter on ethics. This textbook can be used to teach a variety of classes including plan reading, construction techniques, and estimating 1 and 2 (takeoffs and pricing).

Gary Anglin is a construction firm owner with over 35 years' experience and former Adjunct Professor at the Rinker School of Building Construction, University of Florida, USA.

INTRODUCTION TO ESTIMATING, PLAN READING AND CONSTRUCTION TECHNIQUES

Gary Anglin

Routledge
Taylor & Francis Group
LONDON AND NEW YORK

First published 2020
by Routledge
2 Park Square, Milton Park, Abingdon, Oxon OX14 4RN

and by Routledge
52 Vanderbilt Avenue, New York, NY 10017

Routledge is an imprint of the Taylor & Francis Group, an informa business

British Library Cataloguing-in-Publication Data
A catalogue record for this book is available from the British Library

Library of Congress Cataloging-in-Publication Data
A catalog record for this book has been requested

ISBN: 978-0-367-20903-2 (hbk)
ISBN: 978-0-429-26405-4 (ebk)

Typeset in Bembo
by Apex CoVantage, LLC

Visit the eResources: www.routledge.com/9780367209032

FSC www.fsc.org MIX Paper from responsible sources FSC™ C013985

Printed in the United Kingdom
by Henry Ling Limited

CONTENTS

The "Parts" of this textbook, from number 3 to 10, match Divisions of work used in standard specifications found in project manuals. The Construction Specifications Institute (CSI) is responsible for the nomenclature of these "Divisions" of work used in the construction industry. Division 3 is concrete, Division 4 is Masonry, and so on, while in this textbook, Part 3 is concrete, Part 4 is masonry, and so on up through Part 10.

Part 1 covers Plans and Specifications, Part 2 is about Estimating, Part 11 is Construction Documents, and Part 12 concludes with bidding ethics.

ACKNOWLEDGMENTS

Several individuals and organizations contributed to and helped with this textbook. Professor Michael J. Cook, University of Florida, encouraged me to write it and proofread several chapters. He has a law degree, appears as an expert witness, and was employed for many years as an estimator. We had many beneficial discussions about contract law and ethics.

Several industry trade institutes were helpful in providing fact checking and other resources. These groups publish industry standards about their materials, including installation and nomenclature. Their publications were instrumental in writing several chapters. These groups were:

The Steel Joist Institute
The Steel Deck Institute
National Roofing Contractors Association
The Gypsum Association

Mr. Atila Bodo, PE, ME, structural engineer, reviewed the chapter about structural steel. He has taught at the University of Florida, has been president of his consulting firm for over 40 years, and frequently serves as an expert witness.

Mr. Armin Goldberg, AHC (architectural hardware consultant), contributed greatly to the door and hardware chapters. For over 23 years he has been bidding and specifying these often-complicated products that are installed in hospitals, courthouses with detention, assisted living facilities, clean rooms with electronic interlocks, and colleges and schools.

Mr. Jason Warren, BS Civil Engineering, Structural Emphasis. He is the Engineering Manager for SCAFCO Steel Stud Company, member of the Steel Stud Manufacture's Association (SSMA), and the Supreme Steel Framing System Association (SSFSA). He is also a member of the Structural Engineers Association (SEA) and the Cold-Formed Steel Engineers Institute (CFSEI). Mr. Warren helped greatly with the chapter on steel studs.

Mr. Alex Poirier, with a master's degree in Construction Management, and Mr. Jonathan Sheets, BAS in Organizational Management, are two young men that work for me. They helped do everything from arranging the photos and drawings, checking the arithmetic, proofreading, and completing the index.

INTRODUCTION

Section 1 How Parts 3, 4, and 6 are covered in this textbook
Section 2 Plans and sketches for Parts 3, 4, and 6
Section 3 Plan interpretation of Parts 3, 4, and 6
Section 4 Scopes of work for Parts 3, 4, and 6
Section 5 Construction techniques for Parts 3, 4, and 6
Section 6 Takeoff formats of Parts 3, 4, and 6
Section 7 How Parts 5, 7, 8, 9, and 10 are covered in this textbook
Section 8 Terminology
Section 9 Parts 11 and 12, contract documents and ethics

Section 1 How Parts 3, 4, and 6 are covered in this textbook

Estimating should be learned with equal measures of plan reading and construction techniques. All three of these subjects are better understood by reviewing plans, lots of them. This book has hundreds of drawings, plus sketches to explain construction techniques. There is a good reason that the industry uses the term "interpretation" instead of plan reading, and the best way to learn this subject is to see a great many examples. Construction techniques, which here is assumed to mean "how to build", is explained for these projects.

Estimators use a two-step process to price a construction project. The first one, describing and quantifying labor and materials, is called "taking off" the work, a reference to collecting or quantifying the information from the plans. This involves measuring, adding, and multiplying various units of measure – lengths, widths, heights – and working left to right on a "takeoff". The result in the right-hand column *is the appropriate unit of measure needed for pricing* (cubic yards of concrete, linear feet of wood baseboard). Only these ending quantities are forwarded to Step 2, the pricing sheets, or estimate.

The quantities found in Step 1 are sent to an "estimate", the pricing sheet, where labor cost and material costs are figured for them. A portion of the pricing estimate, a quantity column, is devoted to receipt of the quantities just figured on the takeoff. Of course, individual costs for concrete footings or wood baseboard would not be counted on a takeoff and priced on an estimate unless these individual costs are needed. If subcontractors are doing this work, the estimator does not need to count them. They do need to be counted, however, if either the work is being done in-house or an estimator is preparing a budget for an owner and perhaps does not have a subcontractor involved in pricing at the time.

Three trades that require a lot of takeoff arithmetic include concrete, masonry, and carpentry. This book contains dozens of case studies for them that all end with a takeoff. A unique takeoff format (column arrangement) for each trade is used, which provides a head start for the student. This structure focuses attention on the important metrics that must be determined from the plans. The use of formatted takeoffs greatly helps with the organization of a great many numbers and prevents some of the confusion that a lot of data collecting presents.

Each case example has its own short set of plans and details to scale; this variety and range is needed to foster confidence in plan reading. Working conditions and "how to build" are discussed in the paragraphs about construction techniques and scopes (inclusions and exclusions) of the work. Teaching time for counting quantities can be divided into thirds: one third for explaining how to read the plans, another third for explaining how to build, and the last third for the geometry and arithmetic of counting quantities.

Each of the three trades of concrete, masonry, and carpentry have their own geometrics and methods of measurement and completely different construction techniques and plan details, and their case studies present a variety of conditions. However, their commonality is that the measurements and quantities of measure of these ordinary and relatively inexpensive materials (compared to the products and equipment in later divisions of work) take so much arithmetic that it is done on quantity or takeoff sheets, not the estimate where the pricing takes place.

The chapters within Part 3 Concrete, Part 4 Masonry, and Part 6 Carpentry are all divided into sections that follow the same format emphasizing plan reading, construction techniques, and takeoffs.

Section 2 Plans and sketches for Parts 3, 4, and 6

Drawings (floor plans, sections, elevations) are shown at the beginning of each section. There are hundreds of these throughout the book. These partial sets are drawn to illustrate the specific condition being studied within a chapter. Many of them are similar to plans and details the author has used. The Livingston Square Condominiums is a completed project designed and built by the author. The plans for the material bins, basketball courts, library, and others are parts of public projects. Other plans are whimsical, such as the Mad Hatter's residence and Escher's elevator shaft, and are provided to show a variety of conditions.

Estimators review many sets of plans and are witness to a wide range of how plan instructions are conveyed. A project can be illustrated in various ways depending on the style of the architect or engineer, and the plans in this book convey some differences in presentation. These plans were developed to be easy to read, as this is an introductory textbook, but they still illustrate how the same information can be shown to the contractor in a number of ways. Hence the term "interpretation".

Sketches are provided for illustration. When a drawing is labeled as a "Sketch", it is not a part of the plans and is provided to explain a construction technique or an estimating aid.

Section 3 Plan interpretation of Parts 3, 4, and 6

In every chapter section the drawings are explained, or "interpreted". Interpretation is a term used to describe "reading plans" that means details and notes on one sheet are associated with other details and more notes on other sheets. In addition, the written specifications add technical information such as compressive strength of concrete and block, installation tolerances, names of approved manufacturers, etc. The specifications are usually a written document separate from the plans, but it is common for the civil and structural engineer to place an entire sheet or two of specifications into their portion of the plan set.

The "plan interpretations" are lists of observations about the given plans or details. For example, a comment about the plans might be, "Footing A extends around the exterior at 2′ below grade" or "The concrete pads in Section B are 6′ square and at 5′ below grade". The plan interpretations in the text are meant to be a statement of fact about the drawings.

Section 4 Scopes of work for Parts 3, 4, and 6

A written description of the boundaries of the work is given for the case studies. The word "scope" is an important term in estimating, used to define what items are included or excluded. These scopes are short versions of instructions that estimators and project managers write for subcontractors.

Scopes of work are used to divide the work up in a logical manner and to keep from "doubling up". For example, the scope of the concrete trade might include the concrete filling of block cells and include the purchase of rebar cast into concrete and within the block cells, but exclude the labor to place rebar inside blocks. This is left for the mason, who does not install rebar in the footings, slabs, and beams but does place rebar inside the blocks. It is best for all of the rebar to be purchased by one trade, and out of the two that install it, concrete and masonry, the material cost for rebar is usually in the scope of the concrete trade.

Reading scopes of the work helps to understand construction techniques and the division of work between the trades.

Section 5 Construction techniques for Parts 3, 4, and 6

Explanations are given for the specific scopes of work. An estimate cannot be made without understanding "how to build" and sequencing of events. General techniques required to accomplish the work are explained within each chapter. For example, observation is made about where edgeforms are required, that anchor bolt template forms will have to be placed at the column footings, concrete will be pumped, and scaffolding will be used over there, etc. These explanations describe how the work is going to be accomplished and helps the student envision the work.

Section 6 Takeoff formats of Parts 3, 4, and 6

A takeoff is shown for the scope of work at the end of each chapter section. *The column headings of takeoffs are arranged in a standard format in this text.* The appropriate units of measure (L, W, Ht, etc.) are arranged left to right so that they give an answer in the right-hand column that is needed on the estimate. This layout shows the student where to start and end.

Several formats are used; for example, the concrete takeoff headings are different from the carpentry headings, but for each trade there is one takeoff form. To complete them requires bouncing around from plans to a takeoff, then to the specifications, then over to the bid form (to see what alternates and unit prices are needed, all requiring mini-estimates), and back again. It requires an understanding of plan reading and construction techniques. That's how these chapters are arranged; they move from a plan drawing to a discussion of techniques and end with a spreadsheet of quantities.

Section 7 How Parts 5, 7, 8, 9, and 10 are covered in this textbook

Some chapters are about topics that are easy to count, or at least they do not involve lengths, widths, and heights. For example, the unit of measure for doors is simply "each", as well as for toilet accessories and marker boards. These chapters are presented differently from those on concrete, masonry, and carpentry.

Lump sum product quotes (metal and wood trusses, doors and hardware, toilet partitions, etc.), and lump sum subcontract prices (painting, roofing, electrical, etc.), are entered straight onto the estimate. There are many prices that an estimator receives that never appear on a takeoff – takeoffs are for calculating quantities!

The introduction to many trades is made by first understanding what the products are, their names, and the nomenclature of their industries. Plan details are shown to help understand the connections of structural steel, the anchorage of door frames, and the blocking for bathroom accessories. Door schedules, legends, and enlarged floor plans are presented. Shop drawings, and who does what in this process (the role of suppliers and manufacturers varies depending on the trade), is an important part of the introduction to these divisions of work. Fabrication is a huge topic and includes structural and miscellaneous steel, the prepping of doors, and wood and steel trusses.

A different format is used in this part of the book from that used for concrete, masonry, and carpentry. A shift in emphasis (less about takeoffs) is made, because how to count *products* is easier than counting the materials in the early divisions.

However, the products are expensive, and there is much to learn before they can be properly accounted for on an estimate or purchased by a project manager. The role of suppliers, manufacturers, and fabricators takes on great significance in these divisions of work. While form lumber can be purchased at the local lumber yard but require a lot of arithmetic to count, toilet partitions and doors are easier to count but require a complicated pricing and shop drawing process through suppliers and manufacturers.

It does not take a complicated takeoff to count 100 each hollow metal doors; this is a quantity that can be placed *directly on the estimate without using a takeoff.* The count still must be made and the estimator must understand the plans, the industry, and its products, but the quantification is easier because the unit of measure is "each" or "lump sum". There is no "length times width times height equals . . ." required.

A takeoff is not concerned with subcontract pricing or with product quotes, such as the price of the roof hatch that might cost a thousand dollars or so, plus the labor to install it. *These types of costs can be handled directly on the estimate and do not appear on the takeoff.* An estimate format is presented in this book that efficiently takes both (a) quantities from takeoffs, and (b) pricing from quotes and subcontracts onto a spreadsheet to arrive at a final price in an organized way.

Section 8 Terminology

One note about language. Consider some of the construction terms used in this textbook, such as earthform, that in normal usage is separated into two words. That is fine if the word is used as a noun, a "thing". But, in construction language there are many such terms used as verbs, such as "Bob's crew is going to edgeform the sidewalks today". In this text, words such as edgeform, earthform, formwork, and overexcavate are used as one word.

Section 9 Parts 11 and 12, contract documents and ethics

After Part 10 Specialties, Part 11 is a study of "front-end" documents, so called because they are placed at the beginning of project manuals. They are also termed "nontechnical specifications". Front-end documents include Division 00 General Conditions (usually AIA 201) and Division 01 General Requirements (owner requirements). In contracting language, the first might be said to be construction law, while the second concerns a number of owner issues (dumpsters, fencing, insurance, etc.), many of which are a cost to the contractor and fall into the category of job overhead. The majority of front-end documents become "Contract Documents".

Part 12 concerns the subject of bidding ethics. It might generate some disagreement, as the industry has so many kinds of bidding environments, which results in varying relationships between owners and contractors and subcontractors. However, with some qualification and limitation on the kinds of bids being discussed, some recurring bidding situations are presented that illustrate choices between questionable and ethical behavior.

PART 1

Plans and specifications

1

READING PLANS

Section 1 Drawing illustrations

Plans are diagrammatic

Architectural plans are diagrammatic, dimensions are rounded off, and the entire "design concept" is only understood after reading the "whole" set of plans. Individual building parts and pieces on plans are not shown to a scale such that all of them are shown, hence the word "diagrammatic". In the floor framing plan below, at a scale of 1/8″ = 1′-0″, a single 2 × 12 is shown as a single line.

See the online resources for diagram 111.1

In Sections A and B, double joists are shown under the exterior walls at the perimeter. The framing plan could have two lines side by side to indicate double joists, but it doesn't. Perhaps it would if it were drawn to a larger scale. It doesn't matter, it's diagrammatic, only an outline. If two or three joists are meant to be there, instead of one, that's left for details to show or a note to explain. Whether the lines in this floor plan are joists, trusses, or ceiling rafters, they could be depicted the same way. Architectural drawings are not as precise as machine drawings!

Architectural plans are read and interpreted; they are not exact images of construction like a machine bolt drawing. A four-inch frame wall is not four inches wide, and an eight-inch block wall is not eight inches. Yet architectural drawings routinely use these dimensions. For actual dimensioning, the contractor carries the ball to completion. Some of this is through superintendent layout, and some through the submittal process and shop drawings. This is covered in detail later in the text.

Hidden lines can complicate the understanding of plan reading. Deciphering layers of construction shown with dashed lines is a part of plan interpretation. Sometimes only one of several edges, which is what a line is, an edge, is chosen by the drafter to be shown, and several others are left out "for clarity". That's fine, the designer is trying to paint as neat a picture as possible and avoid having confusing lines all over the place.

A plan with hidden lines is only pictorial and need not show them all. Indeed, the plan may show some hidden lines intersecting each other with a ninety-degree turn when one or both lines actually continue. This may not matter to a casual reader of the plans, but an estimator counting concrete or a superintendent building the structure must understand where various levels (elevations) of concrete begin and end.

Hidden lines are a subject of the quizzes at the end of this chapter.

Multi-dimensioning

Sometimes a sketch does double duty with multiple dimensions. A single drawing of a column can be used to note several column heights used at various locations. See the left side of the sketch of columns below. The same principle can be used to dimension lengths and widths with repeating products or equipment; see the picnic shelters below on the right.

See the online resources for diagram 111.2

Lines and symbols

A solid line is an "object line". An object line is directly in front of the eye and is an edge. A drawing is a snapshot directly above an object (a floor plan, a roof plan, etc.) or facing it (an elevation). A dashed line is drawn to represent an edge that is either below (in plan) or beyond (in elevation) and sometimes even above (an upper cabinet). The layering of dotted hidden lines, all occurring at different depths, can cause confusion.

Typical lettering is drawn at a height of 3/32″. Lettering should remain consistent – the paragraphs of written instructions should contain the same lettering, the larger size of letters used to name sections and details should be the same. There is convention, yes, but no rules. If the drafter doesn't have the space to add a few needed words at the same size that's already there, it will be entered with tiny lettering.

The following are some examples of drawing conventions found on plans.

See the online resources for diagrams 111.3, 111.4, 111.5, & 111.6

Architectural and engineering scales

The scale (size) of each drawing illustration (detail or section) within a set of plans and the number of them is up to the designer. The overall level of detail can vary significantly depending on the drafter. First, it is helpful to study the size of different kinds of drawings.

Architectural plans, as well as structural plans, are typically drawn to a scale of 1/8″ = 1′, or 3/16″, 1/4″, or sometimes 3/8″ = 1′. This is for "plans", which are a view from above. Unless otherwise noted (this is an often-used term found in notes on plans, and is usually abbreviated U.O.N.), the viewer is four feet above the plan of the floor looking down. Sometimes a bathroom or kitchen or other busy room is shown separately in an "enlarged plan". This might be to a scale of 1/2″, which allows for more detailed viewing.

"Details and sections" are shown to a larger scale than plan views, 1/2″ = 1′ up to 3″ = 1′. Note that a scale of 3″ =1′ is a scale of one to four. The smallest architectural scale used, 1/16″ = 1′, is a scale of one to 192.

The scales used to draw architectural work represent quite a range, and all of the scales, each of them equaling one foot, shown on an architectural scale used in the United States are:

1/16″	1/2″
1/8″	3/4″
3/32″	1″
3/16″	1-1/2″
1/4″	3″
3/8″	

This is eleven different scales, with a range from 1/16″ to 3″.

Civil engineering plans are used for long distances, hundreds of feet and longer. An often-used scale is 1″ = 20′. The actual measure of one inch is used to represent twenty feet. *The range of scales used in the United States shown on an engineering scale, all of them equaling one inch, are:*

> 10′, 20′, 30′, 40′, 50′, and 60′. This is six different scales, with a range of one to six, the largest scale being one inch equals sixty feet. Different scales are sometimes used to represent long distances, for example 1″ = 100′, but the typical engineering scale has but six scales.

Section 2 The interpretation of plans

Slab and footing plan study

Review the following slab and foot plan before some interpretations are made. Study the drawings before reading the interpretations.

See the online resources for diagrams 112.1, 112.2, 112.3, & 112.4

The lowest concrete, the twelve each 4′ × 4′ × 18″ footing pads, would be poured first. A note on the plan points to two of these pads, and without even looking at Section A, the plan interpretation would be that all of the dotted squares are the same thing. The position of these pads are shown to be underneath the mono slab edge in Section A. The elevation of the slab is shown as zero, which is the "assumed" elevation of many slabs on many drawings. Measure down 18″; see Elevation A, for the top of the footing pads (TOF).

There are anchor bolts embedded in the twelve pads, with steel columns above. Before concrete is poured, these bolts will have to be "suspended" in place with formwork called an anchor bolt template, precisely placed so that the base plates of the columns will fit just right.

The edge of the slab is 18″ thick, and is noted as a "mono edge", a feature of monolithic slabs. This outside edge has a brick ledge at minus 6″, which is an indentation for a brick veneer wall to rest on. The inside bottom edge of the footing pads aligns with the bottom 2′-8″ width of the mono edge; see the single dashed line noted as "D" in plan (shown for illustration).

The inside edge of the monolithic slab has two edges. Besides the dashed line shown at D, see the edge E shown in Section A. The slab plan doesn't show it, but edge E, 12″ inside line D, travels around the perimeter of the slab.

Review Section B, which has a thickened slab at 2′ wide at its bottom, but widening six inches on each side. There are four edges. Only two of these edges are shown, the ones 2′ apart at the lower elevation. A review of Section C shows that its outer edge is also not shown in plan.

Estimators, superintendents, and reinforcing bar detailers (completing shop drawings) are some of the people who must interpret the exact shape and dimensioning of the concrete and rebar.

The geometric shape of the slab is shown in the cross-section below.

See the online resources for diagram 112.5

Quiz # 1 Interpreting the hidden lines of floor plan T

See plan T and details A and B for more about hidden lines.

Floor plan T has a "shortened" version of the full set of hidden lines. There is nothing wrong with this; showing all of them could be very confusing. Realize that what is drawn may just be an outline and may not even be correct.

The dashed hidden lines in floor plan T evidently intend to show the footing under the exterior wall and the thickened slab under wall B. So far, so good. However, look at the ninety-degree intersection of the dashed lines. Does this really occur? The continuous footing of the exterior wall is far lower in elevation than the thickened slab of the interior wall, so these two pieces of concrete do not intersect. However, with the full set of hidden lines being abbreviated, minor "mistakes" like this often occur.

Note that the thickened slab edge shown in Section A does not appear in plan. The hidden lines on the plan are simply the best way the drafter has chosen to depict the conditions. The plan is diagrammatic, not a true representation.

For a look at floor plan T with all of the hidden lines shown, see the quiz answers at the end of the chapter. Before looking, study Sections A and B and envision all the hidden lines.

See the online resources for diagram 112.6

Quiz # 2 Visualize the shape of the garage plan

Review the following garage plan and Section A. What do you make of the opening in the concrete wall that is 3′-4″ wide? *What is this picture, beyond exactly what is drawn? What does Section A-1 look like?*

Consider the clues and see how this drawing can be interpreted. What would the section look like if drawn through the opening? Go to the quiz answers for the garage perspective 112.11 and Section A-1, 112.12.

See the online resources for diagram 112.7

Quiz # 3 Wall sections of the two rectangle plan

How many "kinds" of block walls are there in the two rectangle plan? Where do they start and stop? By definition, a wall type has a consistent height, but it can be continuous through doors and windows. Identify each wall and its length.

See the online resources for diagrams 112.8 & 112.9

Answer to quiz # 1 Floor plan T

The following observations can be made about the plan and two sections:

1 The exterior 8″ wall, Section A, has a continuous footing under it (two hidden lines).
2 The interior 8″ wall, Section B, has a continuous thickened slab under it (four hidden lines).
3 The concrete slab, Section A, has a thickened edge where it abuts the exterior block wall (one hidden line).

For a look at *all* the edges shown in Sections A and B, see below.

See the online resources for diagram 112.10

Answer to quiz # 2 Garage plan showing all hidden lines

Note that a 3′ door with two 2″ frames requires 3′-4″ in width. Anytime a contractor sees this dimension used for an opening, the first impression is that it is a door width.

See the perspective drawing below that fits the information given in the plan and elevation. The low concrete wall, called a stem wall, is shorter under the garage slab at the door opening, and the slab extends through the opening.

See the online resources for diagrams 112.11 & 112.12

Answer to quiz # 3 Wall sections of the two rectangle plan

The quiz is to find the wall types and identify where they start and end for the two rectangle plan. For the answers, see the following:

See the online resources for diagram 112.13

2

PLAN TYPES

Section 6 Structural plan sets

Structural foundation plan
Structural slab plans, slab and footing plans
Structural wood or steel floor framing, ceiling framing, roof framing

Section 7 Mechanical plan sets

Plumbing plans
HVAC plans
Electrical plans

Section 8 Plan revisions

Addendums
Plan changes
Record drawings

Section 1 Introduction

Architects and engineers (civil, structural, mechanical, and electrical engineers) produce plans for the purpose of building construction. These professions require different talents and Section 2 discusses the range of drafting techniques the constructor will see used by them, as they often convey information differently. So, before plan sheets and plan sets are defined, some differences in how drawing instructions are conveyed are discussed. The instructions to build a structure can differ while still producing more or less the same product. Whether it's the short instructions to build a kid's swing set or a big set of plans to build a building, illustrations and notes can vary depending on the drafter. Each architect and each engineer is going to draw a different set of plans given the same building. The information can be clear and concise or muddled and ambiguous, while still mostly getting there at the end. That's why plans are not simply read – they are "interpreted".

What the architect thought was drawn can actually have a different legal interpretation. The preponderance of notes, of large verses small, of written verses drawn, all go into shaping the outcome. The architect, while discharged with the authority by the owner as the final interpreter of the plans, especially when it comes to aesthetics, may be successfully challenged by experienced constructors. An estimator, studying in detail hundreds if not thousands of sets of plans during a career, experiences a wide range in the presentation of plan sets.

Plan completion goes through many iterations. Owners commence a project, then halt when the bank says no, or delay the project when a preliminary budget reveals the need for value engineering or downsizing. For many reasons, the lengthy task of completing a full set of plans for permitting and construction go through many versions. Each one has a name, and means that a certain phase of the plans has been reached. The terms in Section 3 introduce the student to language that the architectural profession uses to describe these iterations of the plans.

These various plan sets are discussed in Sections 4 through 7. Individual plan sheets are also defined, from architectural reflected ceiling plans to structural framing. Also, sections and details, the parts and pieces of plan sheets, are defined.

There is a history and logic to how plans are arranged. However, the content of a set of plans is individualistic. One person may show a section here, another drafter may choose the section to be "cut" elsewhere. Different things can be emphasized – those window sills may be prominently displayed by one drafter and only mentioned in a note by another. However, when notes and plans are viewed in their entirety, the entire structure should be adequately "defined".

Once a set of plans is completed (well, almost) and is issued to contractors for bids, changes to plans are called revisions, and are known by their date. Revisions, and the addendums that announce them, are covered in Section 8. This section also covers "record" drawings, which are done by the contractor at the end of construction.

The examples in this book are about interpreting plans; it is not about how to draw them. There is some practice about reading hidden lines, some study about determining the length of wall that a wall section applies to, and other interpretations. The way for a student to gain confidence in reading plans is to review dozens and dozens of various plan sets or parts of plan sets as shown throughout this textbook.

The eye is drawn to dark and bold lines first, to light and dashed lines second. It helps the picture if the lettering is neat and consistently styled so the instructions are easily read. The large lettering at the bottom of the sheet – that's obviously the name of the plan above it. If details and sections are too close to each other, or actually overlap, it is confusing.

Architectural drawings have a long history in presentation. In the days of pencil and paper, drawing "conventions" were developed, which means following the order of the past. Experience allows a degree of familiarity in how to follow plan instructions, and one begins to be able to "interpret" the plans. This means not only understanding them as drawn, *but also applying reasoning that allows inferences beyond what is specifically shown.*

Because various plan details and the written specifications all contribute to the description of a material or product or installation, the plans and specs are supposed to be "read as a whole" so they can be properly interpreted. Only with time and experience can an estimator judge overall plan intent. "Intent" is a big word in architectural language and construction contracts. So is "interpretation", which means using all the drawings and specifications to arrive at their correct intent, even when the note on Section 4 of the plans conflicts with what it says on sheet 6 or what is in the specs!

A "conflict" in the plans means that one instruction is inconsistent with another. Minor plan conflicts, where a small mistake on sheet 3 of the plans is contradicted by a large detail on sheet 5, is easily interpreted. Just as the written specifications are usually more detailed and take precedence over the plans, the law is that repetition and large-scale drawings govern over short notes or smaller drawn details.

The legalities of plan interpretation, intent, and conflicts are covered at the end of the text in Part 11, Chapter 2, Division 00 General Conditions. This chapter about plan reading concerns:

- The illustration of plans.
- The nomenclature of plans sets.
- The professions that draws plans.
- The kinds of plan sheets.

The contractor uses the drawings to count quantities. The specifications are used to define the quality of products and materials.

Section 2 Architects and engineers

The range of plan types and differences between architects and engineers

The architect is usually hired by an owner for the full services of architecture, structural engineering, civil engineering, and mechanical engineering (plumbing/electrical/air conditioning/fire protection). The architect hires and leads the team, bringing them into the project at the appropriate time. If it is a new site, the civil engineer will begin early. The structural and MEP (mechanical, electrical, plumbing) engineers begin after the architect provides beginning floor plans.

An exception to the above is that the owner often hires the surveyor, not the architect.

There are many, many more building projects that are additions or renovations than construction of new buildings. A wing of a hospital might be renovated, or another floor added to the whole building. Even a simple job like reroofing can involve lots of plan sheets because the owner decides that associated work should be done concurrently with the reroofing. Old air conditioning units are replaced with new ones installed on new curbs, and the old penthouse, which has a wood-structured roof, is being replaced with structural steel. Jobs like this happen across a campus or a downtown far more often than a new building is constructed. All of this creates the need for a great many sets of plans.

Institutional and major owners store project plans on their buildings over the years, and these are what architects and engineers use to work from on current projects. They do not go to the site and look around in attics and basements and measure the building. Their work begins with the use of prior plans, and they do not investigate above acoustical ceilings or within walls to verify the existing conditions. This is why contractors often find unforeseen conditions during the first weeks of a job.

An architect is not licensed to do engineering work and the engineer is not licensed to practice architecture. However, each one can do the other trade as long as it is *minor and incidental to the main project* (according to chapters 471 and 481 of the Florida Statutes). In other words, if an architect designs a building that has a couple of steel columns and a steel beam, the architect can design the size of the columns and beam because the amount of structural steel is minor. Conversely, if a mechanical engineer designs a new air conditioning system for a building and adds a few doors, the doors are incidental and an architect does not have to be hired.

Plans can be conveyed to the contractor in various formats depending on the drafter. There is a wide range in the approach that designers can take in presenting the work to the contractor. The scale (size) of each drawing within a set of plans and the number of sections and details is up to the designer. It is helpful to consider the range of drawings that can describe a project by comparing architectural drawings and engineering drawings, and architects and engineers.

The following case studies help conceptualize that what the contractor views in a set of plans is but one version of illustrations and notes, and that a different drafter would present different illustrations and notes but still end up giving the owner the same result. Well, close.

Both of the following jobs were completed by the author. These examples explain how projects can be described in polar opposite ways. The first project was a large mechanical job drawn by mechanical engineers. It required a lot of architectural work, but no architect was involved. The second job was a streetscape project (bricks and lights and trees) completed in a downtown with plans drawn by an architect. No engineers were involved.

These two plans are contrasted to show the range of plan instructions that a contractor must interpret.

Campus mechanical project

The purpose of this campus job was to remove and replace air conditioning, chillers, and other equipment within mechanical rooms on multiple floors in seven buildings on a campus. Each mechanical room was about the size of a large garage, one being on the first floor, one in a corridor on the third floor, another in an attic, etc. The existing equipment being removed and the new equipment being moved in was far larger than any of the entering doors or wall louvers. In only one location were a series of exterior windows used to create passage in and out. The walls must have been built around the equipment!

Of course the mechanical engineers that designed the new equipment went to great lengths to draw and describe it. However, explaining how the old equipment was going to be removed from the building and access gained for getting the new inside was kept simple. The instructions were few: "Remove all existing equipment. Patch and replace walls, floors, ceilings and roofs as necessary to restore to original condition. Means and method by the contractor." In other words, "You're on your own to figure out how to get this stuff in and out of here, but put the structure back just like you found it."

These simple instructions involved a million dollars of architectural and structural work! It included the demolition and replacement of entire walls and ceilings, shoring of floors above, new doors and hardware, concrete, roofing, and painting. In one building, where the mechanical room was on the top floor, an opening about 15′ square was cut into the roof. A curb about a foot high was built around the edge of the roof opening, and a temporary shingled roof hatch was built to sit on the curb and cover the opening at night. It was lifted off every morning with a crane and put back at the end of the day. This was done daily for a couple of weeks while the old equipment was taken out and the new equipment craned in. The roof structure and roofing were then repaired like new without any structural or architectural drawings, just the note about restoring to original condition!

This may be an extreme example in the range of plan instructions (or lack thereof), but it was primarily a mechanical project and by comparison a minor architectural job. The mechanical engineer sees the importance of the equipment and may not think the architectural work is significant or important. The author was the general contractor hired by a mechanical prime contractor on a project just like this.

Downtown streetscape project

One side of a downtown sidewalk is shown here; the original plans were of a dozen blocks, scaled at 1/4″ = 1′, easy to read. This drawing is not to scale; the actual sidewalks averaged 10′ to 12′ from curb to building. Both the streetscape plan, and the following legends and notes, are used together to read the plans. For their interpretation, start by focusing on the two areas where note 3 is pointing. The note is found to be about a concrete band at the ground level. The two different landscape symbols are found in the tree legend. At both corners of the blocks, a check of the symbols shown on two different legends verifies that these are streetlights.

See the online resources for diagram 122.1

The legends and notes were not in one place, as they are here; they were scattered around on different sheets of the plans. Legends are an organizer for the designer, but depending on how they are presented, they can be helpful or contribute to confusion! The architect is trained to use these drawing aids and sometimes lets this organization (it worked on other jobs like this!) get out of hand, causing confusion.

The light pole symbol at one end of the block is darkened and the other is not, and a review of the light pole legend shows that they represent two heights, one 10′ and one 12′ tall. For the lighting, another schedule must be reviewed to see how many "globes" (the globes housed the light bulbs), on the lighting fixture itself, are at the top of the steel post.

See the online resources for diagram 122.2

The actual project had more heights of light poles and various quantities of globes. It would have been convenient for each of these to have been part of the same legend, however there was a legend for poles and a legend for lighting and they were separated by several sheets. There were about four light poles per block, over forty total light poles. For each one, at least three different sheets of the plans had to be reviewed to figure out simple instructions about pole height and quantity of globes. It was confusing, and confusion in plans can cause mistakes.

The tree legend also added to the work that had to be accomplished to count the trees. The height of the light poles was on the plans, why wasn't the tree height? Instead, the tree heights were in the specifications. Again, some consolidation by the drafter would have helped. But, that is the purpose of this example, to show how sometimes one has to "work" at getting the information, and that work is called interpretation. The contractor must get used to information being presented in all sorts of ways. The information here was simple but made harder by having to chase all around the plans and specifications.

At least the streetscape job had plans, unlike the mechanical job that only had one note pertaining to architectural work! On one job, no plans, the other, too many plans and notes. Contractors have to read and interpret the work regardless of how it is shown. First case, no architectural plans by the engineers, second case, too much organizing by an architect. These two examples are drawing opposites.

Now that this range in drawing technique has been explained, consider another version of the streetscape plans. Organized differently, the following plan takes the separate notes and legend information and places it directly on the plan. (The original plans were to a large scale that could have easily accommodated this.)

See the online resources for diagram 122.3

The information on this plan is the same as that presented above in the plan and legends. This second plan does not need legends; the notes are on the plan, which makes it easier and faster to understand. Isn't this version simpler and easier to read? However, it is not that legends always make plans harder to understand, it depends on the contents of the notes and how the drafter organizes the information.

Here is an example about notes. A series of notes often appear together, a string of twenty or more of them down the right side of the plan sheet. This is often helpful. An architectural sheet may contain several wall sections cutting through doors, windows, and bearing and nonbearing conditions. These sections may have repeating materials such as gypsum board and wood studs. Consider the following two notes:

Notes

1. 2 × 4 studs 16″ o.c. to underside of roof deck.
2. 5/8″ type X fire rated gypsum wallboard both sides of wall.

If these two notes, plus others, appear five times on five different wall sections, one can understand how it could be simpler to write the notes one time in one place. Then, on the wall sections, only the numbers 1 and 2 need appear, with arrows pointing to the materials.

Architects and engineers can present information to the contractor in many, many ways. An architect is more likely to provide a sketch where an engineer will write a paragraph. Drawing a picture or writing an explanation and the range in between describes a difference in plan presentation.

Section 3 Plan iterations from schematic to permit sets

Introduction

The following are terms used to describe specific stages that a "set" of plans goes through. The word set is meant to include all of the plans from architectural and civil to structural and MEP, and all of these move forward in the following progressions.

This following definitions concern the *vocabulary that defines the stages of plan completion.*

Schematic plans

Schematic drawings are those that show the initial design scheme (the word "scheme" does not imply a trick; it is an actual architectural term that can be related to a "theme"). For a house, schematic drawings would be a general floor plan and maybe one outline elevation. Much would be left to draw, but enough would be there for an owner to tell if the design and floor plan is a good start. For a building, schematic drawings show a design scheme that defines the general scope and conceptual design of the project. It includes studying the relationship between building components and materials. At the end of the schematic design phase, the architect will present some rough sketches to the owner for approval.

Presentation drawings

Presentation drawings are not a part of plan sets. They are promotional drawings and are only occasionally done. Their purpose is to explain a design scheme and to sell a project. They may include a perspective drawing, which is a three-dimensional illustration by a good artist that shows off the structure. These drawings are called renderings, and an artist that draws them professionally may be hired. Renderings include textures and shading, trees, and people, none of which ordinarily shows up on architectural drawings. These drawings are often drawn early in the design process, and if so, the rendering may be off-scale a bit, the windows may later change, etc.

Percentage of plan completion

Plans half completed weren't used very much in the past on public work. Having a budget made on plans 50% complete is mostly a new phenomenon, at least on the scale that it now happens. With the advent of construction management, more and more budgeting is done for owners, and the 50% stage of completed plans is now commonplace. Of course, defining what 50% complete means is an arbitrary marker, but the point is that contractors are now asked to provide budgets very early in the drawing process. Providing a budget at 50% completion of the plans means, for the contractor, more budgets will be done later. I'm tired already!

The federal government has started using a design/build process with partial plan completion. Perhaps more of these will be seen in the future. The author bid a project that was for a new USDA building (United States Department of Agriculture). The plans were said to be "70% complete", and upon review, they were very specific, representing the needs of the local USDA office almost completely and seemed even more than 70% complete. However, the architect and engineer that had prepared the plans were 500 miles away, near where the USDA contracting and design people were. The local USDA folks were scientists and needed some new construction, but in this age of minimum government the people handling the construction logistics were far away. They had hired a local architect at their location to prepare some plans to 70% completion, and left it up to the local bidders to provide the last 30% of design work. The contractors bidding the job included an architectural fee in their bids to "finish" the plans.

Design development (DD)

After schematic design is approved, the design development stage is entered. Another elevation is drawn for the house design, the initial one is more detailed, and the floor plan gets some additions like the layout of the kitchen. For a building, the schematic drawings move towards completion, and more plans, sections, details, and elevations are developed. This is where plans get more specific.

Mechanical and electrical equipment and their approximate sizing are determined. The required space for equipment and wiring, for example chases, are shown. The structural engineer provides preliminary sizing of major structural components. The electrical engineer provides a preliminary layout for lighting and equipment.

Working, or construction, drawings

After more owner approvals, the plans are completed, and these final drawings issued by an architect are called "working", or "construction", drawings. These terms stem from the inclusion of wall sections and details that can be used to build from, hence the term construction drawings. These drawings are finished, done. Well, mostly. See Addendums and Plan Changes in Section 8.

Bid set

Sometimes the drawings that the contractor receives are labeled the "bid set". Drawing sets go through a lot of iterations, and at the time of bidding the architect may put a stamp on the plans and label them a bid set. The content of the plans should be the same as working drawings. However, the use of the term bid set, together with a date in the revision notes of the title block, defines a set of plans at a certain moment in time.

Permit set

This is the same set of plans as the completed working drawings, and is the same as the bid set.

What makes this set different are requirements by the building department. They often require additional information besides the plans provided by the architect and engineers. For example, if the structure has wood trusses, the building department may require that the truss plans accompany the owner's set of plans. After all, they are structural, and therefore have to be signed by a structural engineer. It is common practice for the architect/engineer to rely on the truss manufacturer to provide the shop drawings with an engineer's stamp of approval on them. This stamp would be from the truss plant's engineer, not the engineer "of record" (the engineer for the project).

In Florida, because of hurricanes and the high scrutiny of doors and windows, a "product approval code" must be supplied to the building department with the permit application. The manufacturers of doors and windows (and other products) furnish to the State of Florida documentation about the results of wind testing, fastening requirements, etc. The products, once approved by the state, are given a code number that contractors must furnish with the permit application.

A full set of the working drawings is accompanied by separate information, including truss drawings and product approval codes, and becomes a permit set.

Section 4 Civil engineering plan sets

The various civil drawings

The civil plans are drawn to an engineering scale and are numbered C-1, C-2, C-3, etc. The first sheet is a cover page, which contains a table of contents, site location, legal description, and other data pertinent to the site, such as a trip generation study (to show the City Planning Department how much traffic there is, in order to properly evaluate entrances, turning lanes, etc.).

The drawings completed by the civil engineer are called a "civil set" of drawings. The word "set" is also used to describe the entire collection of plans, which is the civil set plus the architectural plan set, the structural plans, etc.

The description of sitework, which is what civil plans are about, can require ten to twenty or more sheets of plans. Only a few are covered here, but a civil set of plans can include a cover page, a survey, a site plan, grading plan with contours, parking lot layouts with striping, handicap ramps, wheel stops, landscaping plans, and more.

The Trip Generation Study shown below illustrates the range of information that civil plans contain. A typical site plan is one that shows topographic lines of existing grade, alongside new lines that represent the contour of the new grade. From these lines the "cut and fill" can be determined, with cutting describing excavation and fill being the need for fill dirt to be brought to the site, or "imported". Cut and fill can be figured by making a grid overlay on a topographic drawing; there are software programs that handle this data by figuring how much cut or fill is required, say, every five feet. All of the pluses (import dirt) and minuses (remove dirt) are determined and a net result determined. The cut and fill may change the grade all around the site but still have a neutral quantity of zero need for fill because the excavation equals the fill needed. This result is said to be "balanced".

See the online resources for diagrams 124.1 & 124.2

Site surveys

The site survey is an important document. Tree locations, the contour of the ground (topographic surveys), existing utilities and buildings, parking lots, and retention areas can all show up on surveys. The architect uses surveys to locate where new construction occurs and where existing conditions are to be changed.

Surveys are drawn by land surveyors. They are usually drawn to one of the engineering scales, but large sites may be drawn at an even smaller scale. The tools of the trade are transits, theodolites, and steel tapes, which now use GPS. The most common survey is a *boundary survey*, which simply establishes the corners and measures the sides of a property. Only horizontal measurements are used, and the surveyor starts out by using a known "benchmark", which is that spike in the ground nearby that has shown up on local surveys in the past. There is a strict protocol for using and reusing benchmarks, which doesn't concern us here.

In addition to a spike in the ground at the property corners, the purchaser of a boundary survey gets a drawing with the corners defined in legal language. Each corner has its own legal description, and each side is exactly so many feet long. A boundary survey will also locate any easements on the property, which is where a second party is given a "right-of-way", such as easements used by local utilities.

A bank requires that a boundary survey be completed when property and buildings are purchased, to ensure they are loaning money for an identifiable specific location. If a new building is constructed, the bank will require a beginning *and* ending survey (we owe this double duty innovation to the 2008 era of shenanigans), to make sure the builder put the building where instructed and not on his own property next door.

For a new building site, the elevations of the ground are needed to study the grades of parking lots, sidewalks, and retention areas and to determine the all-important elevation of the first floor of the building. A type of survey that provides the ground elevations, usually at intervals of one foot, is called a "topographic survey" or "contour plan". This survey adds vertical measurement to the horizontal measurements of the boundary survey. The word "topographic" strictly means the terrain and its contours. These new grades, representing changes in the contour of the ground, can require major earthwork of great amounts of "imported fill", or a "cut" job, which means a net reduction in earth.

Land surveying is a separate trade and profession. Sometimes a civil engineering firm that does a lot of site planning work employs surveyors in-house. A survey drawing will accompany the civil engineering portion of the plans. The survey is part of the civil plan set, and civil engineers and surveyors speak the same language.

An architect finds a survey of existing conditions helpful when planning a new building or addition.

Builders often hire surveyors at the beginning of construction for the specific purpose of locating the corners of a building. These locations, of course, must be exact; no one wants to find out a few months later that the building is out of square or in the wrong location. Hiring a surveyor for layouts is good insurance for a contractor. Saving money by having the superintendent or well-meaning Larry the lead carpenter do it may save money in the short run but may be a costly mistake.

The contractor often tells surveyors to provide "offsets". Since the excavation of footings would remove any stakes placed at the corners of a building, the surveyor places a stake five feet away, or ten feet away, or whatever the contractor wants, from the building corner. Using the nearby offsets, Larry can take it from there and build batter boards and pull strings.

Site plans

Depending on the size of the job, some of the following may be shown on separate sheets. The work of site plans is to describe parking lots and their slope (drainage to catch basins), lighting, curbing, and retention basins with contours. Site plans provide the dimensions of sidewalks. The runs of underground water, sewer, and electrical service are all shown on site plans. So are their details, such as sections (such as through catch basins), and the piping of sewer lines and electrical conduit. In the language of civil engineering, new work shown on a property is known as an "improvement".

Site plans are reviewed by the planning staff at cities and counties, not the building department. They review everything touching the ground, the footprint. Of great importance are drainage and access, parking lot size, signage, and handicap parking.

Landscaping plan

A landscape architect is usually hired for this plan, not a civil engineer. Plant symbols are used and shown in a plant legend. Many localities have strict protocols on new planting, which is usually accompanied by an irrigation plan showing water lines, a meter, and control valves.

Section 5 Architectural plan sets

Architectural cover sheet

A cover sheet for a small project is shown below. The information on it would be on several sheets for a larger project. For example, the life safety plan would have its own sheet, and the structural information and codes might run on for several sheets. A map, showing where the project is, is almost always on the cover sheet, and it locates the project in a couple of ways. One map will depict the project and perhaps just a few surrounding blocks or buildings, and another will show the entire region.

Four items on this cover page are enlarged in Figures 1, 2, 3, and 4, following the overall cover sheet.

See the online resources for diagram 125.1

The cover sheet, by convention and by law (code), serves some very specific purposes. Requirements vary by county and the items shown here are not intended to be a complete representation.

One purpose of a cover sheet, whether it is for architectural drawings or for civil, structural, or mechanical engineering, is to contain a table of contents naming all of the drawing sheets.

See the online resources for diagram 125.2

The cover sheet often contains summary structural code information, and the applicable building codes are often listed in one, two, three fashion. This cover sheet covers specific wind load design (see below).

See the online resources for diagram 125.3

An architectural cover sheet contains important building code classifications. It is common practice to "reference" a model code such as the International Building Code (IBC).

The IBC separates buildings into "occupancies", such as a school, church, or factory. Occupancy is a big subject for building departments, because "like kind" occupancies share the same fire code and other requirements.

The International Building Code uses the following classifications:

1 Assembly. Groups A-1 through A-5.
2 Business. Group B.
3 Educational. Group E.
4 Factory and Industrial. Groups F-1, F-2.
5 High Hazard. Groups H-1 through H-5.
6 Institutional. Groups I-1 through I-4.
7 Mercantile. Group M.
8 Residential. Groups R-1 through R-4.
9 Storage. Groups S-1, S-2.
10 Utility and Miscellaneous. Group U.

See the following occupancy classification from the architectural cover sheet:

See the online resources for diagram 125.4

The title block, usually located on the right side of the cover sheet, contains some important information. The name of the architect and project name appear here, but this is also where the date of the plans is. The issue date of the plans is important because of the many iterations of plan sets. The estimator must ensure that he or she is working with the latest set of plans, or revision. There is a special place for revision dates, and it is often in the lower-right-hand corner of a cover sheet; see below.

See the online resources for diagram 125.5

The following are typical parts of cover sheets from various projects:

See the online resources for diagrams 125.6, 125.7, 125.8, & 125.9

Floor plans

A floor plan is a drawing to scale with a view from above (in "plan view") showing the relationship between rooms, spaces, and other features at one level. Dimensions are usually wall-to-wall lengths. Cabinets and other features below four feet high are shown with solid lines, and symbols are used to denote sinks and toilets. Dashed lines are used to locate upper cabinets, duct chases, and perhaps changes in ceiling heights. Floor plans also contain notes about finishes or construction methods. Doors and windows are numbered or lettered.

Reflected ceiling plans

These are plans, seemingly from above looking down, at the products in a ceiling. However, the term "reflected ceiling" is derived from a view of the ceiling as if it were *reflected from a mirror on the floor*. These show various materials, such as "hard" ceilings, which are often gypsum board, and acoustical ceiling layouts. They also often provide the ceiling height of the individual rooms.

Lighting supported by the acoustical grid is typically shown on reflected ceiling plans. It will not have electrical switches and other electrical work, but the fixtures will be located, as are supply and return registers for air conditioning. These plans are part acoustical ceiling grid and tiles, part electrical, part air conditioning, and part anything else that might be at the ceiling line, such as smoke detectors, fire sprinklers, or security cameras.

Cubicle curtain track, in doctors' offices and hospitals, is located on reflected ceiling plans. So is the location of operable walls, or folding partitions of the type that are in the large meeting rooms of hotels and churches.

Life safety plans

The main purpose of a life safety plan is to provide the distance from the farthest point inside a building to the exterior door. This is called the path of egress, and it is usually shown with a long arrow traveling down corridors.

Another use of the life safety plan is to show the locations of fire extinguishers and cabinets, because they are often placed in corridors and near exits.

Because of the corridors being shown in plan, this is a good place to illustrate fire rated corridor wall construction, often shown shaded and with notes, "Extend 1-hour wall to roof deck" or other instruction.

Sometimes multiple paths of egress are shown on life safety plans, especially in large buildings, with distances (in feet) given from various locations to exit doors. In multi-story construction, a life safety plan is provided for each floor, and if the floor area is large, more than one life safety plan may be shown.

Life safety plans are a requirement of building departments. They have to be included in a set of commercial plans (not residences), and are often near the front of a set of plans. A simple life safety plan can sometimes fit on the cover sheet as shown below.

See the online resources for diagram 125.10

Elevations

Elevations are two-dimensional drawings of the vertical structure. The plans typically contain four drawings with an exterior view – north, south, east, and west. There may also be any number of interior wall elevations that view the walls of bathrooms, kitchens, or the boardroom.

The drawing below has typical notes found on elevations.

See the online resources for diagram 125.11

Sections

A section is a larger view of a part of a smaller scaled drawing. A section cuts through an object and what is viewed is actually a "cross-section". Review the concrete footing pad with a concrete square pier on top in the foundation sheet S-l. The perspective drawing and the concrete pier plan show where the sections are cut; the three sections – A, B, and C, are shown at the bottom of the drawing.

See the online resources for diagram 125.12

The concrete pier plan is a view from above the footing and pier. Imagine a knife cutting in the direction of the dotted lines shown in the perspective drawing. After these slices are cut, look in the direction of the arrows shown on the plan. The results are the edges shown in the three sections A, B, and C.

The dashed and solid lines are properly shown on Sections A, B, and C. Note the upper column pier is shown as a dotted line in Section A because it is "beyond". In Section B, the upper pier is a solid line because the sectional view cuts directly through the object. The pier is not shown in Section C because it is "behind" the viewer.

The most difficult thing about sections is interpreting how far along a wall the section is relevant. How long or how far is that section good for? With wall heights and bearing conditions changing all over the place, it can be hard to tell where one construction starts and the other stops.

Staggered sections

Usually, when that section knife makes a cut through an object, it extends in one plane across the entire object. The following is an example of when the view changes partway across a section, to a different plane. It is actually two different sections, two views, but through this drawing device, the two planes are shown in one elevation drawing.

See the online resources for diagram 125.13

Section A cuts through wall A, and wall A is going to be prominently displayed on the left side of Section A. However, instead of continuing horizontally straight to the right, the drafter wants to also show wall C, so the section line moves north in the middle of the drawing before continuing horizontally east to cut through wall C. Section A will now be able to define walls A and C. Wall B can appear "beyond" on the left side of Section A, but not the right. This staggered Section A-A drawing will be of two planes, the section knife cut in two places. Staggered section lines are usually "connected", as shown in the above version 1 sketch, not like the below in version 2.

See the online resources for diagram 125.14

The floor plan has been shown in two versions, 1 and 2, the only difference being in how the section lines are drawn. They mean the same thing. Section A-A below looks the same for versions 1 and 2.

See the online resources for diagram 125.15

Details

A detail is a larger view of part of a smaller drawing. The detail is an enlargement of any area or portion of the other plans – architectural, structural, mechanical, etc. The scales of 1″ to 3″ are often used for these sketches. Details A and B below depict the pattern of floor tile, a horizontal surface. Detail A arranges the floor tile in the room on the left to be in a checkered pattern with alternating light and dark tiles. Detail B indicates the floor tile in the room on the right to be placed diagonally.

See the online resources for diagrams 125.16 & 125.17

Detail C is used to define a corner guard. Wherever "C" is located in plan (there are four total), a 4′ high stainless-steel corner guard is to be placed. In the lower right of the drawing is a small "elevation" showing the top of the corner guard to be at 4′-4″ AFF (above finished floor).

Schedules

A schedule is made of rows and columns and collects a lot of information in one place instead of having notes spread out all over the plans. Schedules can be made of components from footings to floor and wall finishes. They are used for products that might appear again and again in various locations. For example, a door schedule takes doors found on multiple floor plans and makes a list of them in one place. All kinds of accessory information can then be given about scores of doors on a schedule, which saves the designer and estimator a lot of time.

A finish schedule can contain specifications and lists of manufacturers, as well as selections of colors or carpet. A finish schedule can be part plans, part specifications, and part selections of finishes. Some trades can use schedules to group room numbers with similar floor or ceiling material. This includes flooring contractors and acoustical ceiling subcontractors. Another trade, gypsum board, is often given instructions on a finish schedule, such as the "level 4" comment on the finish notes at the bottom left of the schedule. Finish schedules are "read and interpreted" just like the plans are.

The schedule is a collection sheet for the architect to document selections that have already been made, like the black base that is the only option for baseboard on this finish schedule. No manufacturer is listed, so that must be in the specifications. The carpet for this project has already been selected; see CPT-1 and CPT-2, with a manufacturer and style selected. Done. There are even notes about thresholds and transition strips; see the finish notes. However, a submittal would still be made, as always!

No paint colors are named on the following room finish schedule; they have not been selected yet. However, in the finish specifications at the bottom right of the schedule, the number of coats of paint is given for the walls and ceiling. In addition, the instruction is given to use a flat sheen on the ceilings and epoxy paint in the restrooms. After the bid award, the painting subcontractor using this finish schedule would need to supply a color chart from one of the manufacturers in the specifications, along with product data about the paint.

A useful part of a finish schedule is providing ceiling heights, as well as ceiling materials. Sometimes, but not always, these heights are repeated on the reflected ceiling plan. There are many times that an estimator can move from viewing wall sections that lack heights and confirm "where the ceilings are" by looking at the finish schedule. Interior nonbearing walls are sometimes noted to be built "one foot higher that the ceiling". The estimator ends up determining the wall height from the finish schedule or reflected ceiling plan.

The ceiling types found in this project are shown in the "ceiling" column of the finish schedule and are either painted gypsum board or suspended acoustical ceilings. In the finish notes at the bottom left of the schedule, it is found that WR board, which is water resistant gypsum board, is to be used on the walls and ceilings of the restrooms.

Some examples of structural components that appear on schedules are footings and beams. Imagine a sea of footing pads dotting a large site, say twenty-four different sizes labeled A to Z. Each one occurs multiple times and they have various configurations and sizes of rebar. On a footing schedule, their length, width, depth, and rebar notes can be shown one time in one place. The plan doesn't have to have a lot of repeating information cluttering up the drawing; the plans are more organized, and the designer saves time. The estimator can easily count the quantity of A footings, B footings, etc., from the plan and then use the schedule to get other information.

Concrete beams can be organized by the structural engineer, just like footing pads are. Beams can occur above doors and windows and other openings all over a building, but many of them might be the same size and have the same reinforcement. Call it beam 1, describe it in a schedule, and beam 1 might occur in multiple locations.

See the online resources for diagram 125.18

For an example of a door schedule, see the chapter about doors.

Section 6 Structural plan sets

Structural foundation plan

Structural plans deal with load-bearing components of the structure. Structural plans and details are more dimensionally exact than architectural plans but still require extensive shop drawings, primarily because of *connections*. The connections of piece to piece are "general", and steel detailers provide the exactness of steel connections. More of this is discussed in the chapter on structural steel.

Materials primarily used for load bearing, and thus considered structural, are concrete, masonry (blocks, bricks, and stone), steel, and lumber.

The bottommost part of a structure is usually concrete. Continuous footings, footing pads, elevator slabs, and piles are all found at the bottom of a building. Large buildings are often supported by concrete, wood, or steel, "piles" that are driven into the ground underneath footings or pads. A foundation plan looks down from above and shows the outlines of these components. The structural engineer is given the floor plan(s) and elevations from the architect to work with.

A foundation plan is often accompanied by a footing legend, which lists in one place many different sizes of footings and pads. In addition to sizing the concrete, it will contain extensive reinforcing, and the work of structural engineers is to size and locate the rebar.

The foundation plan is usually the first sheet of the structural plans, unless the first one begins with a page of specifications, as they often do. The first sheet is named S-1, and subsequent sheets S-2, S-3, etc. A typical footing plan, or foundation plan, is usually drawn at 1/8″, 3/16″, or 1/4″ scale.

Structural slab plans, slab and footing plans

Slabs with thickened edges deep enough to support loads, without the use of separate footings, are called "monolithic slabs". These plans are drawn to the same scale as floor plans and take an entire sheet or more to define them. Mono slab plans often refer to many details and sections because of multiple loading requirements. For example, columns along the perimeter spaced 20′ apart may support heavy roof loads requiring a large deep footing while the slab edge between them may only support a non-bearing wall and be supported by a shallow slab edge.

A combination slab and footing plan defines both footings and slabs. These drawings are used on uncomplicated structures where the plan will not be busy with too much information.

Structural wood or steel floor framing, ceiling framing, roof framing

Floor and ceiling framing plans are the layouts of (usually) horizontal pieces placed in ladder-like fashion side by side in repeating rows. Roof framing plans can depict horizontal or sloping structural "elements". These elements can be wood joists or rafters, wood trusses, steel bar joists, steel trusses, and wood or steel beams and girders.

Plywood (used as a floor surface) is often placed on top of floor joists, while various kinds of steel decks are placed on top of bar joists. Floor framing plans often provide the location of vertical columns underneath and define the horizontal membrane that fastens on top of the structural members, often plywood or steel decking.

Framing plans are often drawn to the same scale as floor plans, either 1/8″ or 1/4″.

Section 7 Mechanical plan sets

Plumbing plans

The two major parts of plumbing plans are water supply and drainage. Water is provided under pressure from a municipality, and typically enters a property at a meter that counts the gallons of use. Drainage from plumbing fixtures flows by gravity away from the property. Water drainage from sinks is mixed with waste from toilets. The plumbing plan provides the sizing of water and sewer piping and fixture location. The plumbing plan typically shows the route that pipes will be run for water supply, drainage, and vents.

A "plumbing riser diagram" is an perspective layout of the path of piping with tree-limb like offshoots running over to fixtures. A trap is the U-shaped pipe beneath a fixture. The purpose of the U is to hold a small amount of water that provides a seal. Sometimes traps are a part of the fixture and can't be seen. Drainage lines require vertical "vents", which vent air up and out. Riser diagrams can be drawn for hot and cold water coming in, or for wastewater or sewage going out.

HVAC plans

Air conditioning is shown on mechanical plans, which are typically numbered M-1, M-2, M-3, etc. These drawings are made by mechanical engineers. A ductwork plan shows (viewed from above) the mostly horizontal layout of ducts. The size of the duct is often labeled directly on the duct run. Often, a continuous sheet metal duct (above ceiling) transitions to a flexible material where it bends down to the ceiling level. These flexible ducts are often non-metal and are round. They connect to ceiling air conditioning diffusers, which direct air into a room (or return air in the other direction). Supplies and returns are sometimes called grilles or registers.

If a duct goes through (penetrates) a fire rated wall, it may have a "damper" located in the duct and noted on the plan. A fire rated damper can "close off" the opening and keep fire from traveling inside a duct room to room.

Return air ducting is shown in the same way that supply ducts are. The size of return and supply grilles are sometimes shown in legends.

The kinds of HVAC units, quantities, and location are shown on the mechanical plans, as are the thermostats. The exhaust fans used in restrooms are part of the mechanical plans, although the electrical trade often is given the responsibility of installing them. That is because most of the work involved is in wiring it. Other than that, the labor is to position it where it goes. This is an example of contractors of different trades dividing the scope of work up within a given specification of one trade. Division 15 contains air conditioning, which is where exhaust fans are found. The exhaust fan product cost is picked up by the vendor for the A/C trade (material cost paid for by the A/C sub), then supplied later to the electrician, who installs it (and places the labor cost in the electrical estimate).

In multi-story construction, ductwork sometimes runs vertically, floor to floor, located in rooms that house the "duct chase".

Electrical plans

These drawings are made by electrical engineers. Power plans are used to show receptacles. A lighting plan shows fixtures and the wall switches that turn them on. These plans are usually drawn to the same scale as floor plans.

Communication wiring is shown on electrical plans. This is telephone (voice), computer (data), and fire alarms.

Power is brought into a building in a single large wire, called a "service". Sometimes the service is shown on an electrical site plan. When the service enters the building, it is connected to an electrical "panel", where power is divided into circuits and the wiring (within conduit) is run to receptacles and equipment. Each circuit may have several receptacles on it. "Dedicated" circuits are noted, which means that power is distributed to a single receptacle. The circuits on panels are defined on the electrical plans.

Section 8 Plan revisions

Addendums

An addendum contains clarifications, modifications, additions, or deletions to the plans or specifications during the bidding process. Or, the addendum may extend the bid date, a common occurrence. Often, addendums are prompted by RFIs (requests for information), written by bidders and sent to the architect. There can be multiple addendums, collectively known as addenda (plural), each one being a separate list with a specific date. The plans may remain unchanged even with a long list of addenda.

When a bidder turns in a bid, each addendum must be "acknowledged" on the bid form. A bidder must affirm that each addendum has been read and the bidder has full knowledge of it. If addendums are not acknowledged, the bid is subject to being rejected.

Occasionally, a minor conflict that was thought "cost neutral" at bid time twists and turns into a major problem, and out of nowhere a $25,000 bust is looming. The point is not to get into a "conflict" over conflicts in the first place. The time to avoid this is during the bid process, during which the plans and specifications are often amended several times due to questions by the bidders. Contract language in the industry, such as in AIA 201, requires that contractors report discrepancies to the architect.

An RFI should be made to the architect concerning mistakes or errors in the plans. The architectural profession is charged with the large task of describing, with scaled plans and written notes, the entire components of a building or renovation. Because estimators have to thoroughly study the plans, it is they who find plan mistakes and write questions that are answered by addendums. Addendums must be studiously considered by the estimator, as costs can shift across a range of trades. It is to everyone's benefit that proper revisions are made.

Plan changes

When an addendum item directly effects a plan sheet, for example a note is added to a detail or an existing note is corrected, the note may be circled on the plan (bubbled) to highlight it. A lengthy addendum may cause the plans to be bubbled in many places. This is a change to the plans, called a revision, and it has a specific date.

When revisions are made to the plans, the "revision date" is noted in the title block. A revision is not a new version or iteration of the plans (unless there are so many changes that new sheets are issued), but the plans have been altered or clarified. The architect revises the date in the title block to differentiate that set (all of the sheets within a set get the same revision date, whether or not a change is made on that particular sheet) from the one before. Estimators can also use a "revision date" to reference their bid, tying the price to a definable set of plans with a specific set of revisions.

There may be several addendums without the plans being revised with a new date. However, once the plans change by even one item, they become a different document and when "reissued" become revision # 1. If a plan sheet is affected by numerous revisions, the entire sheet might be reissued. Sheet A-4 may be "reissued" in addendum # 2.

Record drawings

Record drawings are the same as "as-builts". As-built drawings indicate the actual routes of pipes, wiring, or other item, usually in the ground or in a concealed space, which are not placed exactly where shown on the plans due to "as found" conditions. The "record" of where the pipe was actually routed is often first noted on superintendent plans and perhaps later transferred to an office set that is submitted to the architect as part of the closeout documents.

3

THE SPECIFICATIONS

Section 1 Introduction

The specifications are in a specific format, arranged by Division. They are contained in a book called a project manual, which is often a lengthy document. While the specifications are the main part of a project manual, the information to bidders (ITB), which is from the *owner*, is at the beginning of a project manual.

The technical specifications begin with Division 2. In the old CSI format (Construction Specifications Institute), still in practice with many architects and builders in 2018, all construction work is contained within 16 divisions, with the first being Division 2 Sitework and the last one being Division 16 Electrical. With the updated CSI format, the divisions range from 2 through 50. Of course, new trades have not been invented – the new format is just a more specific breakdown of the work.

The technical specifications, *prepared by the architect and engineers*, are the instructions concerning all of the construction trades. They include concrete, masonry, doors, and the MEP trades. The elevator is also included, if there is one. *The technical specifications, however, do not include Division 00 General Conditions or Division 01 General Requirements. These two divisions, called nontechnical specifications, are not about construction parts and pieces and are discussed separately at the end of this book.*

Note the word *prepared* in the first sentence in the preceding paragraph. Architects get specification language from others; they do not originate it. Manufacturers, trade organizations, and building codes are the primary sources for specifications.

Section 2 The old 16 divisions and the new CSI master format

In the chart below, the old format is on the left and the new one on the right. This is an abbreviated list of the CSI format, as only the division "names" are given; for example, look at Divisions 2 through 5 (Sitework through Metals). Note how the CSI has kept the same numbering system for the early divisions of work, including their breakdown into parts, as shown for Division 6, which is divided here into seven parts. *Note that a full breakdown of the CSI format would include dividing up these seven parts.*

Divisions 7 through 14 remain the same, and then the new format on the right expands into many more divisions.

See the online resources for diagram 132.1

Section 3 Three parts of every specification

General, products, execution

Consider that a project manual is open to Division 6400 and it is named Cabinetry instead of Architectural Woodwork. That is the prerogative of the architect creating the specifications for a specific project.

However, all divisions of work use the same format to describe the products within them. This is an all-important feature of specification writing. From Division 01 General Requirements, but excluding Division 00, then continuing through all of the technical specifications, whether there are 16 divisions or 50, their format has three parts:

Part 1 – General
Part 2 – Products
Part 3 – Execution

Part 1 General first gives a summary of what products are being covered in this section of the specifications.

Part 2 Products is often the quickest way to find information about a product, such as who the approved paint manufacturers are, or to find a model number for an access panel or roof hatch, etc.

Part 3 Execution is not about what happens to estimators who make mistakes. It concerns installation, labor, and tolerances of work in place. This section defines the levelness of concrete slabs and the vertical tolerance of walls from blockwork to carpentry. Part 3 concerns quality of jobsite construction.

A more in-depth look at the three subparts of specifications includes these often used categories:

Part 1 – General

1.1 Summary
1.2 Related sections
1.3 Quality assurance

1.4 Referenced standards
1.5 Definitions
1.6 Submittals
1.7 Warranty
1.8 Delivery, storage, handling
1.9 Extra materials

Part 2 – Products

2.1 Manufacturers
2.2 Substitutions
2.3 Materials
2.4 Components
2.5 Accessories

Part 3 – Execution

3.1 Installation
3.2 Field tests
3.3 Preparation
3.4 Tolerances
3.3 Cleaning
3.4 Protection

The items within the three parts above are almost always a part of the specifications. There are other topics, but these are the major subjects that repeat over and over from project manual to project manual, job to job.

Part 1.1 Summary

The Summary describes a general or overall assembly, or group, of products, while Part 2.3 Materials lists specific products. For example, Part 1.1 Summary of Division 9 Exterior Gypsum Sheathing and Decking may state, "This section includes gypsum sheathing and deck products designed specifically as a substrate for exterior wall and roof assemblies." This is a general list. Part 2.3 is where specific materials are listed.

Or, within Division 6 Carpentry, Part 1.1 Summary may read, "This section includes cabinets, the tops, and the cabinet hardware."

The Summary language for Division 3 Concrete may state, "This section includes cast-in-place concrete, concrete formwork, reinforcement, concrete materials, mixture design requirements, placement procedures, and finishes for footings, slabs, etc."

Part 1.2 Related sections

Related Sections are helpful because they tell what is *not* being covered. Perhaps the estimator is searching for access doors, and looks in Division 8 Doors and Hardware, which would be logical, but they are not named there in Part 1.1 Summary. A quick look at Related Sections of Part 1.2 states that access doors are a "related" item and can be found in the mechanical portion of the specifications.

Consider Division 6400 Architectural Woodwork, which might define cabinets or high-end woodworking. Related sections to this would include both rough carpentry and finish carpentry. If Division 3300 Cast in Place Concrete is being reviewed, its related sections would include masonry grout and mortar mix.

Part 1.3 Quality assurance

This part of the specifications often contains requirements for installers of the product. Many construction products (for example roofing) and equipment require training by the manufacturer so that compliance with installation procedures are met. The company installing the roof may be required to have five years of experience in work of this type and a letter verifying their training from the manufacturer.

Another type of quality assurance is a mockup, the construction onsite of a small section of the work. This could include examples of stucco or plastering work, showing the desired texture. Or, the mockup could be of a full size piece of cabinetry, including the countertop and hardware. Another common mockup is the construction of a brick wall showing the desired mortar color and type of mortar joints (concave, struck, etc.; see Part 4 Masonry).

Part 2.1 Manufacturers

The architect will name one or more manufacturers that are acceptable to use. Often, one product, piece of material, or kind of equipment is selected, and the data from its manufacturer is used as a standard. The architect uses manufacturer's literature to write a specification, often copying it verbatim. Then, other manufacturers are named in this section that are known to produce similar goods. Products of these other manufacturers can be used for bidding (the named manufacturer may not be competitively priced) as long as they have a product considered "equal".

The language "or equal" can be used in the specifications whether one or more manufacturers have been named in the specifications.

Part 2.2 Substitutions

Language in this part of the specifications often states that the risk of providing "or equal" products lies with the contractor. If a supplier quotes a contractor items that technically are not equal, and they are used in a contractor's bid, the architect may later disapprove the submittal of these products. Just because substitutions are allowed, the substitution will have to be deemed equivalent to the product named in the specifications.

Substitutions are sometimes not allowed. A specific piece of equipment may be needed to accompany existing equipment by a certain manufacturer, or a product may be needed that an owner's personnel already knows how to service and maintain.

Part 2.3 Materials

For the gypsum sheathing and deck products mentioned above in Part 1.1 Summary, Part 2.3 Materials might state that the wall sheathing be 5/8″ DensGlass Fireguard by Georgia-Pacific and the roof sheathing be 1/2″ fiberglass gypsum roof board, and then define the fasteners. The Part 1.1 Summary is "in general" and Part 2.3 Materials is "specific".

Another example is taken from the specification within Division 9 for Resilient Base and Accessories. While the summary in Part 1 may state, "molded rubber products", the products and materials in Part 2 will be specific and list rubber base, corners, adhesive, and stair nosings.

Part 3 Execution

This part of the specifications concerns the installation of the material, product, or equipment. If it is a manufactured item, the installation instructions will come from the manufacturer or a trade organization. It is rare for an architect to originate specifications, but occasionally special owner instructions or needs may be conveyed here that is beyond that of a manufacturer or the usual best practices covered in publications by trade organizations. Here are some typical instructions for installing wood casework:

1 Verify backing and support framing (*in other words, have something to nail to*).
2 Set and secure casework rigid, plumb, and level.
3 Carefully scribe casework abutting other components, with maximum gap of 1/32 inch.
4 Conform to AWI (Architectural Woodwork Institute), AWS Section 10 (Architectural Woodwork Standard) requirements for the following:

   a Smoothness.
   b Gaps.
   c Flushness.
   d Flatness.
   e Alignment.

PART 2
Estimating

1

QUANTITIES

Section 1 Quantity surveys or takeoffs?

Quantity surveying and takeoffs defined

In the United States, the collection of construction quantities required for pricing an estimate is the unglamorous term "takeoff". A more apt term is a "quantity survey", which more appropriately defines studying plans in search of quantities (hence the word survey, which can mean "to find"). However, this term was long ago preempted by the British and it now means, at least in England and many other countries, the whole field of estimating, from feasibility studies to pricing, the whole nine yards (an American term). The quantity surveyor abroad will have a degree in Quantity Surveying, may have an independent business, and might work for owners, architects, engineers, and contractors.

The price of a project, and a breakdown for it, depends on where in the food chain that the estimator is. The owner wants a total price, the architect wants the project broken down into twenty to forty or so schedule of value items, but the takeoffs in this book are of labor and material that a construction company is going to complete in-house. In other words, they need the information! There is no need to count the concrete quantity if a subcontractor is including its cost in their bid (unless a construction manager (CM) is preparing a budget, in which case a breakdown for formwork and rebar will probably not be completed, just total yardage of concrete so that a ballpark cost per yard can be used).

Whoever is doing the work will need the quantity of edgeforms, grading, mesh, concrete, expansion joint, etc. because a company project manager is going to (if the job is awarded to them) buy it and company employees are going to do the installation. How to count these individual items is the subject of the detailed takeoffs presented in this text. The estimator must list the individual tasks of performing concrete sidewalk work, the forms, grading, pouring concrete, etc., then figure labor and material costs for each item on a pricing sheet. This starts with counting the materials in their various "units of measure". There is a lot of arithmetic involved, even with a simple operation of pouring a concrete sidewalk – if the individual tasks are counted.

The quantity for each item, calculated on a takeoff, is assigned labor and material separately on the estimate. *A quantity survey (QS) doesn't have anything to do with pricing! Dollars appear on the estimate.* The work of the QS involves units of measures such as L, W, Ht, and cubic yards. It is the estimate that includes *pricing units of measure* such as material unit costs, extended cost of material, work-hours, labor rates, and the resulting dollar cost of labor.

Estimators look askance at a "square foot" price, which to them means simply a "guess" or "ballpark" number, unless it is derived from an itemization, or breakdown, of individual costs adding to a total. And itemized costs require itemized quantities, which are the purpose of a takeoff.

Major subcontract items like mechanical and electrical work are not *a part of the quantity survey*. The prime bidder is going to receive lump sum bids for these, and major subcontract prices are routed straight to the estimate, where the dollars are.

A type of bid that owners sometimes use, called a "unit price bid", helps to define takeoffs and quantity surveys (they define what takeoffs are not). An engineer might prepare, for the bidders to use, a collection of work items that might begin with 100 cubic yards of concrete footings and end thirty items later with 100 squares of roofing. The contractor is presented with a format of several, or scores of, individual prices. This is typical for highway and road construction bids – say 10,000 sy of asphalt paving or 10,000 sf of 4″ concrete sidewalks. The contractor fills in the unit prices, one by one. Occasionally this form of bid breakdown is used for general construction. The bid sheet may have 30 or 100 unit prices which, when extended to subtotals, with all of these totaling a base bid. *What this means for the contractor is that 30 or 100 individual estimates have to be created, all of them having a breakdown of individual work items.*

The term "quantity survey" is not used much in the U.S., and an academic degree is not offered for it. At most American universities, there are two courses in estimating (one for quantities and one about pricing), and the term takeoff is used. However, the course names often ignore the issue by being named Estimating 1 and 2, and when the term "quantity survey" is used in the United States, it usually means a takeoff, which is the determination of construction quantities from a set of building plans and specifications.

In this book, the term "quantity surveying" is used to describe the collection of all spreadsheet quantities required for the estimate. The term quantity surveying is such a good term it shouldn't go to waste! The QS is the group of quantities for everything – concrete, masonry, carpentry, etc. The individual spreadsheets for concrete, masonry, and carpentry are called takeoffs. The concrete quantities are calculated on the "concrete takeoff", the blocks on the "block takeoff", etc. When describing the plan reading and arithmetic involved to calculate these quantities, the words *count*, *quantify*, and *takeoff* are all used. To "complete a quantity survey" or to "do a takeoff" means reading the plans and doing the arithmetic to quantify the "work descriptions". The people employed to do this work are the estimators of construction companies.

The takeoff spreadsheet format

In this text, standardized "spreadsheet forms" are used for counting concrete, masonry, etc. The "units of measure" listed in the column headings give the student a format to follow as the rows of descriptive work items are completed left to right. The same form is used to count carpentry work for any project plans. Concrete is always counted using the same form, and so on.

A full quantity survey of a building or renovation can require an enormous amount of data collecting. Accuracy is aided by standardized uses of arithmetic. There are takeoff rules of measuring, like "rectangles are easier to count than triangles", and shortcuts that get the right answer in the least number of steps. Whether measuring with a "mouse" or a "scale", using the same methodology over and over limits confusion and mistakes and allows quick organization of "like kind" groups.

The purpose of a takeoff is to determine quantities for *some* of the items on plans. The determination of quantities is not just a list of or "bill of materials". It is not a lumber list or a list of pieces of steel. The takeoff and estimate includes descriptions of labor that are going to be missed in simply an accounting of material. Huge amounts of labor may not even appear on the plans, such as excavation for concrete, formwork, shoring and wall bracing, material handling and more. The purpose for most of this counting, whether it is called a takeoff or a QS, is to *define* "in-house work activities", giving them a name and quantity, whether labor, material, or equipment.

Some counting is also completed for material that the contractor is going to purchase but hire a subcontractor to do the installation, for example brick or block labor. Subcontractors can be hired for labor only, figuring their work with a unit price (such as per block), while the prime contractor furnishes the material.

The rows of a takeoff begin with naming a "description" (a work activity of labor, material, or equipment), which sounds easy enough. But, consider the numerous descriptions required when there are a hundred concrete pours at various elevations. When the counting is done, what should the breakdown look like? Should all the footings be counted together, or should they be separated by elevation, which is how field work would proceed? Is that line item 43 labeled just a footing? Would a better description be "footing C lower level"? This is the formatting of "descriptions of work" that occurs on horizontal rows of a takeoff.

Rows of a takeoff do not need to be cost coded. Work items are sequentially numbered in the order taken off, with like kind items grouped together. Later, when these are transferred to the estimate, they are arranged in cost code order.

Many of the column headings on the quantity survey, the lengths, widths, and heights, are not forwarded to the estimate because they are not needed there. Only the end results of the quantity survey – the CY (cubic yards) of footings and slabs, the LF (linear feet) of 2 × 8 fascia, etc. – are shuffled forward to the estimate. These ending quantities are shown in bold type on the takeoffs in this textbook, and since they are usually the result of left to right multiplication, they are found in the right-hand column of takeoffs.

Using the takeoff spreadsheet

Estimators are creatures of habit when it comes to gathering information from the plans and formatting the results on a takeoff. There are suggestions in this book such as "start at this corner of the plan and count clockwise". Doing tasks the same way every time minimizes error. Exceptions are not good; just because the plans are weird (not the design but oddly presented drawings) doesn't mean the organization of the takeoff should be altered. Make the plans conform to the company format.

One must abide by a good set of data collecting rules to properly describe the work and get the quantities counted correctly. The quantity survey can generate a large volume of information. The estimator proceeds in an orderly and organized manner so that the figures make sense when completed. Getting the answers right, with means for backtracking, is far more important than speed.

Quantities for some of the more minor subcontracts are *a part of the QS, such as concrete finishing or block labor – when the material is purchased in-house but a sub performs the labor.* The quantities and units of measure needed depend on how the subcontractor prices the work. If the concrete finisher charges per square foot of slab area, then the QS will contain the total square feet of the slab. If the mason raises the price for every 8′ in height, then the QS needs to have the block count divided up into 8′ lifts because that is what is needed on the estimate. Takeoffs are made to serve the estimate; the takeoff must deliver the correct quantity to the estimate and express it in the unit of measure needed for pricing.

The whole point of doing a quantity survey is to describe and count the units of measure needed to price the job on the estimate. The quantities are expressed in appropriate *pricing* units of measure, such as each, square feet, and cubic yards. These measures are forwarded to the estimate, the step after the QS – some quantities to calculate labor costs, some quantities to price the material, some to do both.

Section 2 Takeoff rules and standard procedures

Some rules about takeoffs include:

1 Combine like kinds.

This first rule about takeoffs is to group items that cost the same. This provides the most information in the least number of rows. There will be many examples of this in the following chapter sections. Since the purpose of the quantity survey is to feed information (descriptions of work and their quantities) to the estimate, combining items of work that costs the same makes sense. If there are twenty identical concrete pads (say 2′ square × 6′ deep) to pour for twenty light poles, the concrete quantity should be on one row.

2 If it is different, keep it separate.

Isolate differences that have unlike unit costs. Consider the light pole footings above. If five of the twenty footings are located on a hill in rocky ground, and fifteen are located on flat ground with good access, the estimator realizes that the labor to prepare and pour the footings on the hill will take longer than those on the flat. The work should be separated into two descriptions and two quantities.

Fire rated plywood, for example, should be kept separate from other kinds of plywood. If it's different, keep it separate, and that is formatting by row. While the labor might be the same, the material cost is different, and only one unit price for material can appear on a row, so two rows are used.

Often quantities of the same material are kept separate simply because of location. If different rooms or floors are separated, it may help the project manager or superintendent later because separate lists appear for each one. The estimator can better assign material handling hours if the paneling on the first floor is separated from the paneling on the eighth floor.

Quantities of the same material, say 1/2″ plywood, might be kept separate because of purpose. Detail "A" might use 1/2″ plywood for exterior wall sheathing while detail "B" shows the same plywood being used to build interior shelving. Most of the plywood for the wall sheathing would be used in whole pieces and be installed quickly; the plywood for the shelves will be cut up into smaller pieces and will take longer to install per square foot. The quantities are kept separate so that they can be sent separately to the estimate where they will be assigned different labor factors. This will result in more hours per unit for the shelving and a higher dollar costs per square foot.

3 Narrowly define the descriptions of work.

Taking the time to use three or four select words to describe the work is essential on a long takeoff. This helps to keep track of what's been counted as well as insures that the work item is assigned its proper cost code when it is transferred to the estimate. *Location* is an important factor in naming descriptions and keeping things separate. If the same paneling is being installed on the walls in various locations of a building, or on different floors, keep it separate. When takeoff items are separated, the following apply:

a It is easier to check the quantity later for accuracy.
b It is easier to isolate if change order pricing is requested.
c There might be a difference in labor at some locations (upper floors, for example), and consequently it needs to be shown separately on the estimate.
d Purchasing by area is enabled.
e Scheduling descriptions and labor hours (used for figuring days of work) are already separated for concise parts of the work.

4 Check the sum of the parts against the whole.

Examples of this occur often in the world of takeoffs. Consider five footing types A–E. The estimator counts the As, then the Bs and so on. This is fine, but the count is not done when there is a total for each type. Say there are sixty-seven total footings after all of the individual footings are counted. Then, the estimator counts all of the footings, *regardless of type*, and counts sixty-eight. One footing has been missed. *Always take the time to check the sum of the parts against the whole.*

5 Don't use the takeoff for an arithmetic pad.

The is *not* the place to count endless geometrical shapes nor is it an adding machine to show arithmetic. If a concrete slab is shaped with odd rectangles and triangles, the takeoff is not the place to add them all up! The plans themselves are the best place to do this – whether by hand or mouse.

6 Take off structural components first and start at the lowest elevations.
7 Determine quantities expressed in *units of measure* needed on the estimate.
8 When counting large quantities, if a small number is multiplied by a large number, *carry them both to the same decimal place!* Yes, a 4″ slab thickness is 0.33 feet. However, if the slab is 20,000 square feet, the 4″ slab is 0.33333″ thick (6,666 cf, or 246.9 cy), not 0.33″ (6,600 cf, or 244.4 cy). Be careful when converting inches to decimals!

Section 3 Summary

If the student does not know how to measure the length of a winding block wall by hand, with a scale and a set of hard copy blueprints, there will be mistakes made when using computer devices. The best teaching aid is a set of plans, a measuring scale, and an Excel spreadsheet. A device cannot be relied on to get the right answer if the hand and eye lack judgment.

The following abilities are used by estimators to complete a quantity survey. These key elements are (1) plan reading, (2) construction techniques, (3) geometry, and (4) organizational skills.

1 The proper interpretation of plans is paramount to an accurate estimate. It requires experience obtained by the review of hundreds of sets of plans. A building is a geometric shape with floors at various elevations and sizes, from basements to penthouses. The estimator must have the ability to visualize the structural components and spaces of a building, recalling them at will and remembering the path of already counted components.
2 Experience of a construction trade, of how to build utilizing trade components, is essential. The estimator has to know all of the steps involved in separate construction operations before they can be listed and assigned labor. Knowledge of construction sequencing allows for the estimator to account for a full "scope" of the work and not leave anything out of the estimate. Knowledge of the industry must be known in order to understand product purchasing and the submittal process. Whether a piece of lumber is counted by the linear foot or by each depends on how it is priced at the lumber yard, a pricing issue. The takeoff must count things in the way they are priced on the estimate.
3 Many judgments are made, such as whether it is best to count concrete by centerline or rectangle, or lumber by linear feet or each. The centerline or rectangle method of counting quantities of concrete footings depends on the configuration of the foundation and the mental agility of the estimator. The estimator must have keen three-dimensional spatial recognition and be able to envision where products are positioned from viewing two-dimensional plans.
4 The estimator must have organizational and formatting skills. He or she must measure a small component of the building, using specific units of measure, and place the result in a specific location on the takeoff. It includes the formatting of columns and rows, an all-important function, seemingly a simple process. The results of this arithmetic, the totals, are forwarded to the estimate where they are organized by the company cost code. The estimator counts quantities and calls this a takeoff, and then prices them on an estimate.

The estimate is formatted to receive costs in very specific locations *depending on the type of cost* and numbered by the company cost code. An estimator knows *exactly* where something should be even if looking at a long ago forgotten job. Recall the story of the Greek palace, a large building with many rooms that the ancients used as a memory recall device. If the Greeks wanted to remember ten things, or ten parts of a speech, they would mentally associate item number one inside one of the palace rooms (ornately furnished for easy recall), form a mental picture, and memorize it.

How to go about filling in the rows (descriptions) and columns (measure and count) on a spreadsheet is what quantity surveying and estimating is all about. When a good takeoff or estimate is completed, you should be able to read it like a book.

2

PRICING

Section 1 Introduction

The usual two- or three-week process of bidding a job involves counting quantities while getting quotes on products and discussing the job with subcontractors at the same time. The sequencing of these tasks marches to a repeating pattern. However, the time spent on each bid is tailored to the kind of job it is for each contractor. Note the three general tasks in the first sentence of this paragraph – counting, product quotes, and subcontractor pricing. The estimator must judge the proportion of these to properly manage the available time.

First, if the project involves much in-house work, the estimator must get busy right away counting the quantities. Second, important product quotes require the timely distribution of plans and specifications to suppliers, giving them enough time to contact their manufacturers if needed. Third, many of the subcontractors will need a week or two to bid the job, and the estimator may spend hours on the phone discussing the work with them. The time and study spent on these three general categories depend on the scope of each one.

It is often the case that when the estimator first obtains the plans, takeoffs are started. But very soon, there is often a mix of counting and pricing. Assembling the estimate begins, which is composed of several separate sheets. Quantities from takeoffs are forwarded to one part of the estimate. The estimator does his/her part and figures hours adjacent to quantities, and then multiplies hours by a labor rate to equal dollars. Product quotes and subcontract prices are placed in other parts of the estimate. This chapter covers how the estimate is formatted to receive these costs.

Takeoffs require the collection from the plans of many units of measure – L/W/Ht, etc. They also require careful grouping of work into "like kind" descriptions and the separation of labor and material that is priced differently. *These distinctions and differences and multiple units of measure are not used in the counting of some divisions of work.* The heavy lifting for the estimator is done with takeoffs for concrete, masonry, and carpentry; *the method changes for the counting of doors, windows, marker boards, and bath accessories.* The unit of measure for them is simply *each*, and they are easy to count compared to the process of counting cubic yards or the quantity of block.

Furthermore, there is an important difference in the physical subject of what is being counted. That is, the early divisions of work concern many *materials*. Divisions 7 and higher contains mostly *products*. These are important distinctions for the estimator. Materials are often minor expenses that the estimator may not get quotes on, like form lumber. There are dozens of estimate line items that an estimator "plugs in" a material cost for, such as common lumber or concrete (if there is a large amount of lumber or concrete, the estimator would request specific quotes).

There will also be quantities for such things as wall bracing or temporary barricading of window openings that have a summary unit of measure of "each" or "item". The estimator recognizes that there will be time (labor expense) expended for an activity, but there is no corresponding detail on the plans. In order to figure out the amount of material required, the estimator would have to stop and prepare a shop drawing or sketch. In the example of temporary window barricading on a renovation job, instead of trying to figure the exact count of 2 × 4s and plywood for fifty window openings, the estimator might take a shortcut and "plug" in a material price of a hundred or two dollars for "each" window. The determination of the quantity of 2 × 4s and plywood is kicked down the road to the project manager (PM).

"Products", on the other hand, can be expensive, and the estimator typically gets quotes for them. Often, they are not counted on a takeoff. There is no need to list, for instance, eight each 36″ grab bars on both a takeoff *and* estimate. The quantity of eight grab bars can *be placed directly on the estimate*, as well as sixteen door frames and sixteen doors, or a dozen marker boards. The products must still be quantified, but their totals are not always first determined on a takeoff sheet. Just add them up and write them down on the estimate.

The quantities of several divisions of work are sent straight to the estimate.

Contractors arrange their costs into three categories on the *right* side of estimate sheets. They are Labor, Material, and Sub (a fourth column for equipment costs is sometimes used but not in these case studies). Together with a column in the middle to receive the quantities from takeoffs, *an entire estimate can simply be one long list* of costs and prices starting with Division 1, then Division 2 Sitework, Division 3 Concrete, and so on down the page. Add it up at the bottom, with adjustment made for sales tax and labor burden.

In this text, *instead of one long estimate sheet* added up at the bottom, the estimate is formatted with three different sheets for adaption to three different kinds of costs. They are the *unit price sheet*, the *P/S sheet* (for products and subs), and the *job overhead sheet*. A fourth *summary sheet* adds them all up.

This text is more concerned with determining quantities than pricing. There are dozens of takeoffs shown but only a few full estimates with pricing sheets.

Section 2 Unit price sheet

Takeoff quantities are placed here

After the estimator has completed a quantity survey, has multiplied L/W/Ht and produced units of measure such as cubic yards, quantity of block, and linear feet of lumber, the takeoff quantities are sent to a "unit price" sheet for pricing. The ending units of measure from takeoffs are sent to a specific part of the estimate (see columns 6 and 7 of the following unit price sheet). The work of the unit price sheet is to price takeoff quantities, turning them into dollars of labor, material, and subcontractor unit pricing (not lump sum).

The unit price sheet should be the first part of the estimate completed. These costs are the result of time-consuming takeoffs, which cannot be put off until bid day. This part of the estimate, the evaluation of in-house work, the smaller material packages, and smaller sub pricing, should be finished before the big quotes and bids are received.

Refer to following unit price sheet for an office building project to see how it is formatted. A feature of this part of the estimate is the three-unit price columns to the left of the quantity column. There is one for labor (a production unit), one for material, and one for subcontractor pricing.

Use the unit price sheet for multiple rows of labor.

Work-hour units

Column 3 is a work-hour production unit used to determine hours in column 8. These hours are multiplied by a rate per hour in column 9 and equals labor cost in column 10. Productivity is a separate subject apart from this text, but a short explanation of it follows. One method of estimating hours of labor is called the "work-hour unit" method. One work-hour unit means that it takes an hour to install one unit of measure. The work-hour unit is multiplied by the quantity and the result is the number of hours. An advantage to using work-hour units to price labor is that they do not change. However, when labor is priced by a "dollars per unit" method, it has to be updated when the hourly cost of labor increases.

There are many estimating methods used to calculate labor. Formwork can be estimated by "dollars per square foot of surface area". The labor to erect structural steel is often estimated in "dollars per ton". There are others.

Labor production varies due to multiple factors. Every job and every location is different, and an estimator knows that the work will proceed within a given "range" of work-hour units. Nothing is exact, and there is no way that a memorization of productivity rates can be used across the board. In fact, the range of work-hour units is very broad within each category of work.

A simple example of work-hour units can be shown by the task of installing paneling inside a conference room. Assume the ceiling is 8′ high and there are twenty sheets of 4′ × 8′ paneling required. A "rule of thumb" number that estimators share is that paneling will take about an hour per sheet to install, which includes some handling and setup time inside a building. Ceiling cove and base is figured separate. Note that dividing 1 hr by 32 s.f. (one sheet of paneling) is 0.03125, about 0.03. (A room with 10′ high walls would take longer per sheet because of the extra cuts (say 0.04), and walls with many turns in them would take longer, and so on.)

There is a total quantity of 640 square feet (20 sheets × 4 × 8). When this quantity is multiplied by 0.03, the result is 19.2 hours (about an hour per sheet). *The number ".03" is a work-hour unit.* If an estimator is familiar with and confident of a work-hour unit, it can be multiplied by 6,400 sf (192 hours) or 64,000 sf (1,920 hours) given the same conditions. Work-hour units stay the same, requiring no updates. The same carpenter will install the same 20 sheets of paneling in the same conference room year after year in approximately the same 19 hours, even though his/her wage rate may go up.

Contrast the paneling example with 50 sheets of exterior plywood installed on straight exterior walls 8′ high, again using a product that is 4′ × 8′. Production with the plywood is faster, it is outside and not inside a conference room, and there's a large stack of plywood nearby (less handling and a larger quantity), whereas the paneling must be carried inside. The sawdust is blown away outside by the wind; while inside it has to be cleaned up. Although the paneling is thinner and lighter than the plywood, the estimator realizes the paneling requires more handling and precision cutting.

A range of work-hour units for this kind of plywood sheathing work is between 0.01 and 0.025 with a typical production rate of about 0.015. Note the difference between 0.01 and 0.025 is a "two and a half to one" factor, which can account for different job conditions such as multiple stories, staggered levels, lots of corners, etc. Selecting 0.015 (midway in the range) for the given example of an easy exterior plywood installation, it is half the paneling work-hour unit of 0.03, or one half hour per sheet (0.015 × 32 sf = 0.48 hr).

This is an example of the range in labor production for 4′ × 8′ products. Installing gypsum board, another sheet product, is faster than the exterior plywood example. An average production rate for it is around 0.01 or even faster. Crews that install gypsum board often do this one task all day every day and become incredibly fast.

The lesson is that installing products that come in sheets can have quite different production rates. There is a four to one difference in productivity in the examples above (0.01 to 0.04), between gypsum board and paneling. It is the estimator's judgment within these ranges that accounts for thousands of work-hours of work on an estimate.

Material unit prices

Column 4 contains the material unit price. When multiplied by the quantity, the material cost for the row is shown in column 11. A unit price for concrete or block is very different from a quote for marker boards or projection screens (product costs), which are for lump sum amounts. The estimator places the cost of the projection screens on the P/S sheet (this pricing sheet is introduced next) in one lump sum number. However, various unit costs for concrete might appear many times for multiple pours on the unit price sheet, and even if quoted by the plant as a guaranteed unit price per cubic yard, the contractor is at risk for the quantity.

Subcontractor unit prices

Column 5 contains the subcontractor unit price. When multiplied by the quantity, the subcontractor cost is shown in column 12. The unit price by a sub to lay shingles, $100 per square, and to install brick for $2 per brick, is for labor only. While appearing in the "Sub" column, these prices are not like the lump sum bids for major sub quotes (placed on the P/S sheet) that include labor *and* material, although these unit prices may include some tools and equipment. The unit prices in column 5 are for subcontract labor only and are not a guaranteed number. The quantity can be wrong or the bidders may realize later that they figure quantities a bit differently.

A partial (only a few items are shown) unit price sheet for an example office building project is below. *This sheet of the estimate has more arithmetic on it than the others, and that is its purpose, to do the work of pricing rows and rows of labor and material from the takeoffs.*

			UNIT PRICE SHEET									222.1
name	OFFICE BUILDING				JOB #:				date:			
code	DESCRIPTION	Work hr unit	Mat'l unit	Sub unit	QTY	unit	Hrs	$ rate	Labor	Mat'l	Sub	Total
3110	Place 1/2" x 4" exp jt	0.025	1	0	500	lf	13	25	313	500	0	
3210	All # 5 footing rebar	0.02	1	0	2,000	lf	40	25	1,000	2,000	0	
3210	Tie wire	0	10	0	2	roll	0	25	0	20	0	
3220	All 6 x 6 #10 slab mesh	0.004	0.5	0	10,000	sf	40	25	1,000	5,000	0	
3320	6 mill slab visqueen	0.0015	0.25	0	10,000	sf	15	25	375	2,500	0	
3320	Curing compound, spray	0.75	15	0	50	gal	38	25	938	750	0	
4100	Lay brick	0	2	2	25,000	ea	0	25	0	50,000	50,000	
7200	Lay shingles	0	200	100	110	sq	0	25	0	22,000	11,000	
1	**2**	**3**	**4**	**5**	**6**	**7**	**8**	**9**	**10**	**11**	**12**	
							3 x 6		8 x 9	4 x 6	5 x 6	

Estimating judgment Supplier Takeoffs
Subcontractor unit price

Section 3 P/S sheet

Product quotes and subcontractor bids

Both major lump sum *product* quotes and major lump sum *subcontract* prices are placed on the P/S sheet of the estimate. Quotes are placed in the material column 7 (see the following office building project) and sub prices in the subcontractor column 8. *Make a list and add them up.* Call it the "P/S sheet", short for major Products and Subcontracts that are lump sum prices, a category of costs that usually don't involve takeoffs.

By definition, lump sum subcontract prices include *labor and materials*, and these bids are sent straight to the P/S sheet. They are not a part of the QS. These site contractors, structural steel companies, painting and plumbing businesses, and air conditioning and electrical subcontractors provide bids that are one big lump sum number. When added to the storefront, flooring, and painting costs by even more subs, and the acoustical ceilings, metal stud, and gypsum companies, this is the largest part of the bid, often fifty, sixty, or seventy percent of the project cost for the prime bidder.

To finally receive lump sum numbers might be the result of weeks of discussion between estimator and subcontractor or supplier, but once the estimator decides to use a price the estimate work is uncomplicated. The estimator receives major quotes and subcontract prices at about the same time. Often, they are received late in the bidding process or on bid day. This timing factor, in addition to their lump sum nature, figures into how they are handled on the estimate. Since they may change on bid day, it is best to have them all in one place and not scattered across a multipage estimate! They are similar "like kind" costs in estimating language.

Suppliers and distributors

The material cost for many products comes to the prime bidder in quotes from suppliers, distributors, and fabricators. Quotes for major materials can be of significant dollar amounts for such things as doors and hardware, wood or steel trusses, bathroom accessories, toilet partitions, projection screens, signage, and marker boards. Some of these can be big ticket items and are sometimes counted by the suppliers themselves, who base their quote on specific quantities. The quote for each group is one lump sum number, and even if the quotation shows the unit cost of various marker board sizes, the estimator doesn't enter the material cost of each sized product on the estimate unless there is a reason to. It is easier and there is less chance for error if one total amount is used.

Major products for a building or renovation can be in the range of ten, twenty, or thirty percent of the total costs of a project for the prime bidder. Before bid day, the estimator should enter a "plug number" (a best guess) for all the subcontractor prices and product quotes. *It is important for the estimate to be completely filled out and the total bid cost approximated the day before the bid.* With plug numbers in place, the estimator and management can review the estimate and see approximately what amount the bid is going to be. With this information, they can discuss job duration, markup, and bonding.

Sometimes the supplier will figure a quote without contractor involvement. Door and hardware suppliers often get the plans and specifications on their own and quote the contractors bidding the job. The quantities here aren't usually in question; the door numbers are listed on a door schedule containing frame, door, and hardware requirements.

Suppliers do not state in their quote that they are providing *all* the lockers for a project, or *all* the lockers on the plans. They will review the plans and come up with a quantity, or rely on the contractor's quantity, and provide a quote for a certain number of products, which they buy from manufacturers that they are distributors for. If conflicts about the quantity arise, their quotation is only good for the quantity on their quotation. It is the prime bidder that is at risk for quantities.

Nevertheless, the estimator views quotations as "lump sum" numbers on the estimate as opposed to unit prices.

Quotations for major materials are received in writing with sales tax and delivery excluded or shown separately. Sometimes third parties such as trucking companies are hired for deliveries, and these costs are called freight and can be a big expense. The estimator can't often rely on suppliers to give advice about freight costs; the suppliers' concern is the quote for their own material. The location of the job and how much it costs to get the product delivered is up to the estimator.

There may be other considerations with the quotation total besides sales tax and freight. An hour before the bid, the estimator may see "Thirty-two type A windows" listed on a quotation when his/her count is thirty-four. Or, the low quote for cubicle curtain track is from a supplier using a manufacturer not listed in the specifications and unfamiliar to the estimator. These are issues for the estimator's judgment.

Usually a lump sum material quote carries the estimator's hope that it is a firm number without risk. There should not be a problem unless the quantity is in doubt. For example, if there are two roof hatches prominently displayed on a roofing plan and the estimator has a quote for them from the specified manufacturer, spelling out the size and model number, it is safe territory for the estimator.

The estimator takes the quote, adds the tax and freight, and comes up with a total cost that is written directly on the quotation. Judgment will determine how much is added to a quote for freight, or for incorrect quantities, or whether to accept or reject quotes that may not meet the specifications. It is best to place the tax right on the quote before the total number is entered on the estimate. Sales tax varies by county, which prevents a single sales tax rate being multiplied by all quotes after they are placed on the estimate.

Fabricators

The situation of a fabricator providing a quote is quite different from a product quote or subcontractor price. A truss plant will be required to provide engineered (signed and sealed by an engineer) shop drawings if awarded the job. The bidding plans for the roof may be sketchy (diagrammatic) with no bracing and blocking or detail shown. But if the roof spans and slopes are properly shown and dimensioned on the plans, the fabricator is expected to bid the job whole, and not charge for extras later. A truss quote is provided "with engineering", meaning that a fully designed and safe truss system will be provided for a firm price. *It is a "design and build" project for the fabricator. The contractor is not at risk for the quantity of material or product.*

Office building P/S sheet

In-house labor is usually figured on the unit price sheet. However, the exception is when the labor is simple to figure and can easily appear on one row of the P/S sheet, the same row as the product cost. Review the examples of windows and skylight for the office building.

First, see the "thirty-three type A" windows. The 99 labor hours and labor cost of $2,970 can easily be shown there on one row, the same row that contains the material cost. *This is an example of labor appearing on the P/S page*, when it is an easy one row calculation of hours and dollars of labor. The hours are placed in column 4. These windows are all the same size so they take the same amount of labor to install.

What if the windows for the office building were all different sizes, large and small? This would require separate descriptions on separate rows so that different hours could be assigned to each size. It would be better to use the unit price sheet in this case. It's not good for the P/S sheet to run on and on and become lengthy, because it can be burdensome on bid day when a tight focus is needed on the big numbers that keep changing.

In the above window example, the cost for labor and material can be separated and tallied on two separate sheets. A multi-row breakdown of labor can account for various sized windows on the unit price sheet. However, the total material cost of the windows, plus tax and freight, could be placed on the P/S sheet. The company cost code for windows would appear on both sheets.

One skylight costing $3,200 in a lump sum material quote can appear on the P/S sheet with 32 estimated labor hours shown on the same row as the material. There is no need to list a row on the unit price sheet when the estimator just wants to place 32 hours beside the material cost.

The P/S sheet is for the big lump sum numbers. *It is better to keep it that way* so that on bid day an entire page can be quickly scanned, the estimator viewing hundreds of thousands of dollars, or millions, at a glance. These are the numbers that may change all morning because of competitive bidding. It can be confusing if the computer screen is cluttered with rows and rows of items that could be on the unit price sheet.

P/S SHEET (major products and sub prices) *223.1*

job	*OFFICE BUILDING*		*number*				*date*	
Code	*DESCRIPTION*	*QTY*	*HRS*	*RATE*	*LABOR*	*MAT'L incl tax*	*Sub*	*Row*
8400	Type A windows	33	99	30	2,970	10,494	0	1
8500	Skylight 4′ × 8′	1	32	30	960	3,200	0	2
9400	Painting	LS	0	0	0	0	150,000	3
15200	Air Conditioning	LS	0	0	0	0	500,000	4
16000	Electrical	LS	0	0	0	0	750,000	5
			131		3,930	13,694	1,400,000	
1	**2**	**3**	**4**	**5**	**6** 4 × 5	**7**	**8**	**9**

Only two abbreviated sheets (of four) of the office building estimate have been shown. A job overhead sheet and a summary page are introduced in the church estimate of Section 4.

Section 4 Church estimate

Church scope of work

The shortened (not all trades are listed) scope from a church job is used to illustrate how the estimate pages are used. It is a 30,000 sf building, 100′ × 300′.

Scope of the work:

Include:

30,000 sf mono slab with thickened edges.
Concrete material by GC, concrete placement by unit price subcontractor.
1,500 LF of interior frame walls 10′ high.
1/2″ roof sheathing and fascia, roof slope 5/12.
Installed lump sum by framing subcontractor.
Lumber purchased by GC from a lumber yard.
Wood trusses installed by framing subcontractor, purchased by GC from a truss fabricator.
100 doors installed in-house by GC, purchased by GC from a supplier.
Toilet partitions installed by GC in-house, purchased by GC from a supplier.
Fire extinguishers installed by GC in-house, purchased by GC from a supplier.
Plumbing, subcontract lump sum.
HVAC, subcontract lump sum.
Electrical, subcontract lump sum.

Church unit price sheet

The unit price sheet for the church is shown first, the P/S sheet second. The quantities used in this estimate would come from takeoffs not shown here.

The 650 cubic yards of concrete for the church job is shown in row 1 of the unit price sheet. The price of $125 per yard, in column 4, is a quote from a concrete plant specifically for this job. The quote is "per cubic yard", and the contractor is at risk for the quantity. However, the supplier has provided a concrete quote, lower than the everyday price for concrete, and it is good for the duration of the job.

Row 2 of the unit price sheet shows the arithmetic to arrive at the slab labor of $22,500 in column 12, which is being provided by a subcontractor for $0.75 psf, shown in column 5, who has the trowel machines and tools for the work. The concrete placement for a slab is a common occurrence and the estimator, knowing the range of prices that this work usually goes for, may or may not discuss the job and seek a quote from a subcontractor while preparing the estimate. The cost for the work is usually quoted "per square foot". Note that these kinds of quotes, such as a mason laying block per "each", are quite different than the lump sum price for the plumbing or electrical work. A sub working by the unit may get a different count than the estimator; the contractor is at risk, maybe not a large risk, but this kind of sub price is not a guaranteed lump sum.

The unit price of seventy-five cents per square foot is placed in column 5 where it can be multiplied by the area of thirty thousand square feet in column 6.

The second construction trade on the church project is Division 6 Carpentry. The unit costs quoted by a local lumber yard are shown in column 4, and the cost of the material is extended to column 11. The labor column for carpentry is zero; see rows 3–7, because there is no hourly work (this work is done by subcontract; for this cost see the P/S sheet).

The next several items are frames, doors, hardware, and bath accessories. The quantities for these items are taken from the plans and placed directly on the estimate without the use of a takeoff. However, if the estimator wants to keep track of quantities by floor or by building, this separation can be organized on a takeoff as needed. Or, if the number of doors and hardware information is greater than this example, a takeoff may be needed to stay organized and maintain accuracy. Often the hard copy of the plans themselves (on or beside the door schedule) can be used to write and collect quantities on.

The material lump sum costs for these, including tax and freight, appear on the P/S sheet. The labor for them is shown on the unit price sheet, because these major materials have components that are "unlike" in labor production. A mirror is different from a tissue holder. Work-hours are assigned to them separately. The only material cost shown on the unit price sheet for doors and bath accessories is $500 on row 8 that the estimator has included for handling and protection of doors.

	No in-house labor for carpentry	*A purpose of the Unit Price sheet - mat'l unit price × qty = material cost*	*Unit price concrete sub cost here*

UNIT PRICE SHEET

224.1

name CHURCH						JOB #:			date:			
code	DESCRIPTION	Work hr unit	Mat'l unit	Sub unit	QTY	unit	Hrs	$ rate	Labor	Mat'l	Sub	Row
3320	Concrete pour slab material	0	**125**	0	**650**	cy	0	25	0	**81,250**	0	**1**
3320	Concrete pour slab labor	0	**0**	0.75	**30,000**	sf	0	25	0	**0**	22,500	**2**
6111	2 x 4 pt base plate	0	**1**	0	**2,200**	lf	0	25	0	**2,200**	0	**3**
6111	2 x 4 top plate	0	**1**	0	**2,200**	lf	0	25	0	**2,200**	0	**4**
6111	2 x 4 x 10 studs	0	**1**	0	**20,000**	lf	0	25	0	**20,000**	0	**5**
6112	2 x 6 fascia	0	**2**	0	**1,000**	lf	0	25	0	**2,000**	0	**6**
6130	1/2" roof sheathing	0	**1.25**	0	**35,000**	sf	0	25	0	**43,750**	0	**7**
8100	Handling for all drs	**0.33**	**5**	0	100	ea	**33**	25	825	**500**	0	**8**
8100	Install HM frames blk wall	**1.5**	0	0	50	ea	**75**	25	1,875	0	0	**9**
8100	Install HM frames std wall	**1**	0	0	50	ea		25	0	0	0	**10**
8100	Install HM door slabs	**1.5**	0	0	50	ea	**75**	25	1,875	0	0	**11**
8200	Install wood door slabs	**2**	0	0	50	ea	**100**	25	2,500	0	0	**12**
8300	Install group 1 hardware	**3**	0	0	25	ea	**75**	25	1,875	0	0	**13**
8300	Install group 2 hdw	**2**	0	0	25	ea	**50**	25	1,250	0	0	**14**
8300	Install group 3 hdw	**5**	0	0	25	ea	**125**	25	3,125	0	0	**15**
8300	Install group 4 hdw	**7**	0	0	25	ea	**175**	25	4,375	0	0	**16**
10100	Install tissue holders	**0.7**	0	0	**40**	ea	**28**	25	700	0	0	**17**
10100	Install soap dishes	**0.7**	0	0	**16**	ea	**11**	25	280	0	0	**18**
10100	Install 18 x 24 mirrors	**1.5**	0	0	**24**	ea	**36**	25	900	0	0	**19**
10100	Install handicap mirrors	**2.5**	0	0	**8**	ea	**20**	25	500	0	0	**20**
10100	Install waste receptacles	**1.5**	0	0	**8**	ea	**12**	25	300	0	0	**21**
10100	Install 36" grab bars	**1**	0	0	**40**	ea	**40**	25	1,000	0	0	**22**
10100	Install 42" grab bars	**1.1**	0	0	**40**	ea	**44**	25	1,100	0	0	**23**
10100	Install paper towel holders	**0.7**	0	0	**8**	ea	**6**	25	140	0	0	**24**
							905		22,620	151,900	22,500	
1	**2**	**3**	**4**	**5**	**6**	**7**	**8**	**9**	**10**	**11**	**12**	**13**
							3 × 6		8 × 9	4 × 6	5 × 6	

A purpose of the Unit Price sheet - using production units to obtain hours. Clmn 3 × 6 = 8.	*Lump sum material quotes for doors and bath accessories are shown on the P/S page, not here.*

Church P/S sheet

A framing subcontractor has bid the carpentry work on the school project. It is a lump sum price to install interior walls, trusses, sheathing, and fascia. The subcontractor arrived at the price of $40,000 by a square foot method, and it is entered on row 1 within the Sub column of the P/S sheet.

The material quotes for trusses, doors, bath accessories, toilet partitions, and fire extinguishers are all quoted lump sum. These five quotes are placed in column 7 on rows 2–6 and include tax and delivery costs.

Consider the labor cost for the 40 each toilet partitions and the 12 each fire extinguishers (F/E). The estimator and supplier arrive at the same quantities, and they are placed on the P/S sheet. The unit of measure for them is "each"; they are for like kind items, and the labor is easy to figure. Both labor and material are shown on rows 5 and 6. The estimator uses 8 work-hours per stall to handle and install toilet partitions, and 1.25 work-hours per F/E. These work-hour units do not need to be shown; it is an obvious and simple calculation appearing on one row.

In addition to the framing subcontractor, there are three other subcontract prices shown on rows 7–9 for the plumbing, HVAC, and electrical trades. The low plumbing bid is $300,000. The low HVAC and electrical bids are $400,000 and $500,000. The estimator will have discussed the job with all of them multiple times in the preceding weeks. Scopes and durations, plan issues and addendums all have been considered. These bids are received by the estimator late in the morning on bid day. Sometimes subcontract bids are received after 12 or even 1 pm, and the bid opening with the owner is usually scheduled for 2 pm.

The costs appearing on the P/S sheet are the ones that are subject to change on bid day. It is helpful to the estimator, because a lot of information is received in a short amount of time, to have these numbers all in one place. It helps to avoid confusion and mistakes if the focus can remain on one sheet, one screen of information, instead of scrolling around on a multi-page document.

An example of labor shown on a P/S page is BOLDED below. The labor for doors is **not** placed here because more rows of information is required. See the doors, frames, and hardware labor on the unit price sheet.

224.2

P/S SHEET (major materials and sub prices)

name CHURCH NUMBER date:

code	DESCRIPTION	QTY	HRS	$ Rate	LABOR	Mat'l include tax	SUB	**ROW**
6000	All carpentry labor	LS	0	0	0	0	40,000	**1**
6400	Wood trusses	LS	0	0	0	50,000	0	**2**
8100	Doors, frames, hardware	LS	0	0	0	100,000	0	**3**
10100	Bath accessories	LS	0	0	0	10,000	0	**4**
10400	Toilet partitions	**40 ea**	**320**	**25**	**8,000**	30,000	0	**5**
10600	Fire extinguishers	**12 ea**	**16**	**25**	**400**	2,000	0	**6**
15100	Plumbing	LS	0	0	0	0	300,000	**7**
15200	HVAC	LS	0	0	0	0	400,000	**8**
16000	Electrical	LS	0	0	0	0	500,000	**9**
			336		8,400	192,000	1,240,000	
						↑	↑	
1	**2**	**3**	**4**	**5**	6	7	**8**	
					4 × 5			

These columns used to list lump sum quotes and sub prices.

Church job overhead sheet

This part of the estimate is where supervision, scaffolding rental, cleanup, and testing are shown. The column format is the same as the P/S sheet, but these two sheets are kept separate because they are for different kinds of costs and figured at different times.

Many of these costs depend on job duration. They are also influenced by how many work-hours are estimated and whether the project consists mostly of subcontracts or in-house labor or both. Will an assistant superintendent be needed as well as a superintendent? Is scaffolding being provided for the subs or are they providing their own? Now that the concrete has been figured, what are the testing expenses? Do the specifications require builders risk insurance?

An example list of job overhead is shown below. *It is best for the estimator to figure these costs after getting the rest of the estimate almost completed so that many of the above questions are answered.* This is where to place the anticipated expenses for tools, gas and travel time, job trailers, and dumpsters. The single biggest expense on the job overhead page is often supervision.

JOB OVERHEAD SHEET								*224.3*
job	*CHURCH*	*NUMBER*				*date:*		
code	*Description*	*Qty*	*Hrs*	*$ Hr*	*Labor*	*Mat'l incl tax*	*Sub*	*Total*
1100	Permit	LS	0	0	0	5,000	0	5,000
1100	Supervision	1 yr	2,080	30	62,400	0	0	62,400
								0
1210	Rent temp toilet	12 mo	0	0	0	2,400	0	2,400
								0
1220	Gas/Travel Time	N/A	0	0	0	0	0	0
								0
1230	Truck Allowance	N/A	0	0	0	0	0	0
								0
1310	Test Concrete	6 ea	0	0	0	0	1,500	1,500
1310	Test Dirt	6 ea	0	0	0	0	1,500	1,500
								0
1320	Builders risk insurance	N/A	0	0	0	0	0	0
								0
1330	Tools and Expendables	LS	0	0	0	4,000	0	4,000
								0
1340	Project sign 4′ × 8′	1 ea	4	25	100	750	0	850
								0
1500	Scaffolding erect and remove	3 mo	80	25	2,000	2,500	0	4,500
								0
1600	Trailer/Phone/Mob-Demob	12 mo	40	25	1,000	3,000	0	4,000
								0
1700	Utilities	12 mo	0	0	0	3,600	0	3,600
								0
1800	Punchlist	LS	16	25	400	50	0	450
								0
2260	Dumpsters	8 pulls	0	0	0	3,200	0	3,200
								0
2700	Cleanup interim	12 mo	384	20	7,680	0	0	7,680
2700	Professional clean	1 ea	0	0	0	0	2,000	2,000
			2,604		73,580	24,500	5,000	103,080
1	**2**	**3**	**4**		**5** 4 × 5	**6**	**7**	**8**

Church estimate summary

A row-by-row explanation follows:

Row 1: Four summary numbers are copied from the unit price sheet – total hours, labor, material, and subcontract costs.

Row 2: This row exists to do one important function – to add sales tax in one place to all the untaxed material on the unit price sheet. Unlike the major materials and products that are placed on the P/S sheet, which may be subject to varying sales tax percentages, unit price costs are usually local and subject to the same sales tax, allowing it to be multiplied one time.

Row 3: Four summary numbers are copied from the P/S sheet – total hours, labor, material, and subcontract costs.

Row 4: Subtotals.

Row 5: Four numbers copied from the job overhead – total hours, labor, material, and subcontract costs.

Row 6: Subtotals.

Row 7: This row exists to do one important function – to add labor burden in one place to all the labor costs in the job. The labor within a subcontract price is not subject to the bidding contractor's labor burden; the subcontractor pays it, and it is included in the subcontractors' bid. Labor burden is a percentage (unique to each company but often about 30%) added to labor costs, and consists of at least the following:

Worker's compensation insurance.
Social Security.
Medicare.
Federal Unemployment.
State Unemployment.
Fringe benefits (including holidays and vacation time off).

Bond costs vary but typically range from half a percent to two percent, with one percent used here. Note that the cost of the bond is for one percent of the total bid, not costs, so multiplying one percent times the total cost and markup will be short of the number. The trick here is to multiply the total cost and markup first times the bond cost (.01), then multiply the result times the bond cost plus 1 (1.01). It works.

ESTIMATE SUMMARY						*224.4*
job:	*CHURCH*	*number*			*date*	
Description	*Amount*	*Hours*	*Labor*	*Material*	*Sub*	*Total*
Unit price sheet		905	22,620	151,900	22,500	197,020
Sales tax on unit prices	0.06			9,114		9,114
P/S sheet includes tax		336	8,400	192,000	1,240,000	1,440,400
Subtotals		1,241	31,020	353,014	1,262,500	1,646,534
Job overhead sheet		2,604	73,580	24,500	5,000	103,080
Subtotals		3,845	104,600	377,514	1,267,500	1,749,614
Labor Burden	0.3		31,380			31,380
Total Cost		3,845	135,980	377,514	1,267,500	1,780,994
Markup	12.5%					222,624
Total cost and markup						2,003,618
Bond						20,237
Bid amount						2,023,855

3

A SHORT HISTORY OF BONDING AND LIENS

Section 1 The Heard and Miller acts

Good practices have been forced on the industry, not created by it.

Three subjects here concern the bid price, payment to subcontractors and suppliers, and performance of the construction in a set amount of time. These things are considered of utmost importance to owners, and it is they who have made them key components of the competitive bidding system.

Some history explains the causes from long ago that lead to the use of bonds in contracting. Three bonds are required on public projects – a bid bond, performance bonds, and a payment bond. The government was compelled to tighten the control over bidding jobs, performing them, and paying for them. Throw in architects and engineers that inspect it all and that doesn't leave much that contractors do without watchdogs. Maybe put up the sign.

The Heard Act was enacted a long time ago in 1894. It required contractors to provide some guarantees on federal contracts. Bid bonds were created to force contractors to stand firm on their price. Plus, a performance bond guaranteed that they would actually go out and perform the construction for their bid price, or the bond would be used to hire another contractor. Finally, the government was tired of contractors leaving town without paying subcontractors, so payment bonds became a requirement. All three of these assurances were needed to protect the taxpayer, the public.

These three guarantees are *ethical behavior embedded in the law*. However, on private jobs, bonds are often not used. There is no guarantee on a private job that a contractor will honor a bid. Performance of the contract and payment to subcontractors is left up to the contract and lien laws.

The Heard Act was supplanted by the *Miller Act in 1935, which is still current law*. The big three categories of policing the construction industry – firm and timely bids, performance of the project, and payment, have been around a long time. *They are the bulwarks of the competitive bid system.*

The Miller Act requires sureties (divisions within insurance companies) to provide guarantees to the federal government on jobs above a certain amount. Most states have enacted "Little Miller Acts", as they are called, which have the same requirements as their federal counterpart, to require bonds on all jobs above one or two hundred thousand dollars.

The Miller Act places three bond responsibilities on the surety. All of them support the competitive bid system. They are a firm bid that won't change and payment and performance of the work guaranteed with bonds. Bid bonds, payment bonds, and performance bonds promote a high standard that owners want contractors to achieve, and bad behavior results in severe penalties. These guarantees are defined below, as well as the relationship between the bonding company and contractor.

Section 2 Bid bonds

The first guarantee required by the Miller Act is simply that a bid be honored. When a bid is turned in and opened, that's it; it will not be changed. It doesn't matter if the low bid is half of the other bids; the *bid bond* that the contractor has submitted with the bid simply means that the bid will not be changed. The student asks, "What's the big deal, why turn in a bond just to repeat what the bid is? Are you kidding me? Just go look at the bid proposal if you forget what the price is!"

Simple as it sounds, the bid bond, turned in at the same time as the bid, guarantees one of the most important and riskiest things a contractor does – deliver a firm bid that is good for a certain amount of time. It will not change, and *the contractor will sign a contract for the bid amount*, or the contractor suffers a penalty, typically five percent. If a contractor backs out of a one-million-dollar bid, it will cost him/her fifty thousand dollars. The bonding company is legally obligated to send the owner fifty thousand dollars for pulling out of the bid, but the contractor ends up paying for it. As a condition of the bonding relationship with a surety, the contractor must sign a "Master Surety Agreement", which are fancy words for a personal guarantee. If you mess up a bid, it can cost you your house! General contracting is a serious business; this is not homebuilding!

It takes a lot of work and precision estimating to determine the right price for a job. When a bid bond is required, it greatly lessens the chance that a company will hastily come up with a frivolous bid. Without a bid bond, a contractor may not spend enough time figuring the exact costs of a job and turn in an erroneous price.

If it is hard to understand why a frivolous price would be given, realize the great amount of time and effort that preparing an exacting bid takes and the expense it is for contractors.

Section 3 Payment bonds

The second guarantee required of prime contractors on government jobs is that bills will be paid. However, *the Miller Act's provisions only provide payment protection for those doing business directly for the prime contractor and first tier subcontractors and suppliers.*

Companies further down in the food chain, third tier subs and suppliers, are not protected. Payment to them is governed by their contract (or purchase order), contract law, and ethical behavior.

Section 4 Performance bonds

A performance bond is a guarantee by the surety for performance of the project, which includes timely completion. The performance bond is provided by the same insurance company providing the bid bond and payment bond. The owner must receive the bonds before the contract will be issued. If the contractor fails to do a good job and keep on track towards completion, or the completion date is reached and the job is unfinished, the owner may "call" the bond and enforce it. The insurance company backing the contractor may be forced to step in and finance the original contractor, or hire another one, or assume the role of a contractor and hire subcontractors, or pay the owner to finish the job. All of these are disastrous for the contractor, surety, the owner, and subcontractors.

Section 5 Bonding companies

Surety bonds are different from all other insurance policies. Bonding companies do not ever expect to have a claim!

The surety vets a contractor with a thorough qualification process before providing the guarantees that the bonds provide. In contrast to other insurance policies, where payments paid by the insured goes towards the inevitable expense of wrecks and fires that the insurance company has to pay for, the bonding company spends bond fees on the salaries of those banker types who are always looking at the contractor's books.

Bonding companies perform an extensive prequalification process with a new contractor, with job experience and capital being paramount. They require that the contractor have a strong relationship with a bank, and often demand that the contractor have a "line of credit", because they know there will be times that cash is needed to finance ongoing operations.

Generally, total bonding capacity at any given time is equal to about ten times net worth. If a beginning contractor wants to start out with a five-million-dollar bonding capacity, then the company will need a net worth of five hundred thousand dollars. The bonding company wants cash in the company, and insists on liquidity (*not real estate*), because they know that the contractor needs it to finance operations.

If the company is worth one million dollars, then it can be working on contracts totaling ten million. A one-million-dollar job half completed has only one half million in exposure to the bonding company, so the contractor is said to have a half million of bonding "capacity" used on this job. The sum total of bonding capacity that a contractor can have at any time is also called their total "program".

When determining whether to provide bonding to a contractor for any given job, two measures that the surety looks at are job size and the contractor's unused bonding capacity. The size of the job will be restricted to the size of prior successful projects. Increases in job size will be limited to a natural progression. If a contractor has a total bonding capacity of ten million and doesn't have any work, bidding a ten-million-dollar job will not be allowed if the largest prior job size is five million.

The contractor sends the bonding company "gross profit projections" (the surety will provide the form, thank you) every quarter that summarizes how much the contractor is earning (or losing!) to date on each job. They insist on balance sheets and income statements prepared by accountants at least once per year, with net worth and working capital being huge factors. Both bonding measures, job size and total program, are used throughout the year to evaluate the contractor's requests for bid bonds, as well as the analysis of the quarterly reports. Once a contractor establishes a bonding relationship with a surety, it is usually long term.

The liquidity and net worth requirements required by the bonding company differentiates the bonded contractor from other kinds of builders such as developers and homebuilders. The bonded contractor has to retain money in the company; it can't be taken out or bonding will cease. Where the principals of other corporations take profits out at the end of each year, bonding companies insist on growth and stability and require that the construction company maintains a net worth commensurate with the size of the bonding program.

It is a serious situation indeed for the contractor to default on any of the three bonds. If there is ever a claim, it is a big deal. It happens every once in a while, but most contractors make it through a career without defaulting on a project. If it happens, bonding companies will never deal with the contractor again. Remember, this isn't car insurance – the bonding company does not expect claims. The contractor has pledged his/her personal assets to the bonding company. The surety will not hesitate to take personal worth – cars, boats, checking accounts, everything but the family. And, that's after taking the money and assets that the company has.

The surety charges the contractor from half a percent up to two percent (of the amount of the bid) for all three bonds. There is typically no charge for a bid bond used to bid an unawarded project. When a contractor bids a job, there is no bond fee and no requirement to provide performance and payment bonds.

If an award is made, the owner will require receipt of the P & P bonds, as they are called, before issuing a contract. The cost of the bond (the invoice for the cost of the bonds comes in the mail with the original bonds) becomes one of the first job expenses for the contractor, one of the first account payables of the project.

Since the bond amount is a percentage of the total bid amount, it is the last cost shown on an estimate. The last few rows of an estimate are shown below, assuming the cost of the bond is one percent.

$9,433,962 Total costs.
$566,038 6% markup, overhead and profit.
$10,000,000 Cost and markup.
$101,000 Bond cost of 1% (.01 × 10,000,000 times 1.01).
$10,101,000 Bid amount.

Section 6 Liens

A lone counterpart to bonds, on private work, are liens. These protect payment for a project by an encumbrance attached to the real estate that the building sits on. Liens are "filed" against a nonpaying owner of real estate. As with bonds, liens help to elevate or raise the bar towards ethical behavior. For this reason, a definition of liens is given here.

By law, public property owned by the government cannot have a lien placed on it. A real estate lien, aka a property lien or mechanics lien, is a payment remedy for private projects. Liens can be created not only by those in privity with the owner (a direct relationship such as architects and builders) but by subcontractors and suppliers downline in contractual relationships. When they do so, they are said to "file" a lien.

Liens were created to "attach" to the property, not owners. A lien is a debt that the property owes and limits full entitlement to ownership, creating a "cloud" on the title of the property. Banks require settlement of liens before consummating a loan, which is why liens are often taken care of before the end of a construction project, the refinancing of an existing building, or the sale of an existing building. If there is no bank financing involved, a lien may not be cured for years. It is the transfer of the title from an owner to someone else (when they sell the property) that often triggers a lien being cleared up.

Lien laws vary from state to state and are sometimes complex and technical. Strict compliance can be required about giving notice within certain time limitations and other requirements. A "notice to owner" (NTO) is a common requirement, which is a document that a subcontractor or supplier (working for a builder) sends upline to an owner. It simply informs the owner, "Hey, I'm working on your property and you must make sure I am paid by the end of the project."

The NTO is often the first step in complying with the lien law. When done timely and on a proper NTO format, a banker and owner will carefully file these notices, which they may receive throughout the project. At the end of the job, they will make sure to receive a "release of lien" (ROL), which indicates payment, from everyone who sent NTOs. ROLs are received from subcontractors and suppliers (the same companies that sent NTOs) before final payment is advanced to the prime contractor.

If some other vendor shows up without having initially sent a notice to owner, that's too bad, the owner and banker will not be moved by tears. Hopefully, the vendor has a good subcontract or purchase order that can be enforced against the contractor, because their only recourse will be against the prime, not the owner.

PART 3
Concrete

1

INTRODUCTION

Section 1 Ruling body, the American Concrete Institute

The American Concrete Institute, founded in 1904, is a nonprofit and technical society and standards organization. It is located in Farmington Hills, Michigan. The ACI is a leading authority and resource worldwide for those involved in concrete design, construction, and materials. The ACI has over 95 chapters, 110 student chapters, and nearly 20,000 members in over 120 countries.

The ACI has many publications, including *ACI 318 Building Code Requirements for Structural Concrete*. It has a monthly magazine named *Concrete International*. Its publications are about the design and practices for concrete, from repair and grouting to pouring slabs and formwork. Their recommendations are the source of building codes and Division 3 specifications across the United States.

Section 2 Form reinforce and pour

A concrete job often begins with layout and surveying work and proceeds to earth excavation or formwork. Pouring concrete can involve four trades – earthwork, formwork, reinforcement and concrete pouring, and field work, in that order. Formwork is carpentry, and reinforcing is the fabrication and installation of rebar.

The term "form, reinforce, and pour" is the language used in the industry to mean the inclusion of an entire concrete pour – labor, material, and equipment, and the work occurs in that order. "Concrete pouring" refers to the work done on the day (and possibly the next day) of the pour, from placing and finishing to curing compound and saw-cuts. When the concrete pour is a slab, the work is sometimes referred to as "flatwork".

Section 3 Concrete takeoffs

Takeoff procedure

Before the takeoff begins, the "bid form" should be reviewed to see if there are alternates or unit prices. These require separate takeoffs and estimates.

A concrete takeoff proceeds with these steps:

1 Review the structural plans, and look for pours at the bottom elevations of the structure before higher ones. Structural plans govern over architectural if there is a conflict between them.
2 Scan the architectural, civil, and MEP plans for additional concrete. The "A" architectural plans are reviewed to see if there are pours not shown on the "S" plans, such as nonstructural concrete like patios and steps. The "C" civil plans are reviewed to see if there is site concrete (sidewalks, driveways, curbs) not shown on the other plans. The civil plans must be matched up and compared with the architectural plans – for example the matchup of sidewalks leading to the building's various entries (there are often conflicts between these two plan sets concerning sidewalks). The MEP plans often contain information about concrete pads under equipment that is not shown on other plans, both inside and outside of the building.
3 Give the specifications a first review.
4 The review of the nonstructural plans has been brief and usually no overall list of concrete pours is made, just mental notes of where the concrete is. At this point the estimator goes back to the structural plans and begins the takeoff at the lowest elevation and works up.
5 Determine the structural concrete quantities first, then the quantities of concrete on all of the other plans.
6 Determine formwork quantities next. Copy parts of the concrete takeoff to the formwork takeoff.
7 Determine earthwork quantities last. Copy parts of the concrete takeoff to the earthwork takeoff.

The takeoff does not proceed in the same order as field work. The first item that should be counted is the concrete itself, the "concrete material". Some of the same measurements initially used to count the concrete can be later reused to count the formwork and reinforcement. It is easier to understand reinforcement and formwork if concrete is taken off first. The best and fastest way to understand the layout of a building's spaces (a big key to getting everything counted!) is to understand the building's structural components, which is often concrete.

By counting the concrete material first, the "description" of the pours can be copied to the other takeoffs (formwork and excavation), as well as "reusing" units of measure. See below for the twelve column headings of the *concrete material takeoff format* used in this textbook for all types of concrete pours.

Concrete units of measure

Since a concrete plant charges for its concrete by the cubic yard, "yards" is the most important unit of measure. It is the measure needed to determine material costs on the estimate. The quantity, in cubic yardage, is found on the concrete takeoff for "like kind" pours, from concrete footings, to slabs on grade, to beams and filled cells. These various pours can require different mixes and prices of concrete, so the various pours need to be kept separate on the estimate. The unit of measure for concrete material is the cubic yard.

The pricing of labor for the various concrete pours may require a different unit of measure than the material quantity. Subcontractors are often hired to place and finish slab pours. The pouring, placing, and finishing of concrete slabs is often priced by the square foot, so this unit of measurement also needs to be shown (for slabs) on the QS.

These subcontractors own and know how to operate trowel machines and the various hand tools that it takes to place the concrete on the day of the pour. Often starting a pour at daybreak, they instruct the drivers of concrete trucks, directing them here and there as concrete is poured. By the end of a long day, the concrete is hardening and has been troweled smooth. Sawn concrete joints are sometimes the last task, which sometimes occurs the following day. However, engineers and specifications often counsel that joints should be sawn as soon as possible, preferably the same day as the pour.

The crew that pours slabs is often brought in for just a day or two of work and then they leave for another job. The way they charge for their work is by the square foot. The price varies depending on the size of the pour, whether it is a sidewalk or a large slab, whether the concrete is deposited by truck chute or concrete pump, and other factors that affect the complexity of the pour. The estimator has priced the labor cost for this pour on the estimate, either using past experience or discussing the work well ahead of time with a concrete finisher. Since their pricing language is square feet, this is the unit of measure that the takeoff has to determine.

The labor cost to pour columns, beams, and block-filled cells is often determined by the cubic yard of concrete, not measured in square feet, like a slab. The takeoff for these pours provides the arithmetic to count yardage, because that is what is needed on the estimate.

Counting concrete accessories

Two concrete accessories that are included in these chapters are vapor barriers and curing compound, items routinely assigned to the concrete trade. Vapor barrier is purchased in rolls of up to several thousand square feet (see Part 7). Liquid curing compound is usually sprayed on, but can be rolled, and is purchased by the gallon in 55-gallon drums. Its coverage is typically about 250 sf per gallon.

Since the vapor barrier lies underneath wire mesh, it is not convenient for the concrete finishing subcontractor to install it. Perhaps the general contractor does this operation, and if so the estimator determines their labor, as well as the cost of material, by the square foot.

Curing compound is brushed or sprayed onto a freshly poured concrete slab at the end of the pour. The estimator counts this product in gallons on the takeoff so that its material cost can be shown on the estimate. The labor for its placement is usually included in the square foot price of the concrete finisher.

Concrete waste factors

Contractors use different percentages of waste for various pours; there is no standard. Concrete quantities can be studied without using waste factors; however, a waste factor is always necessary, so this text uses two of them, one for concrete being poured against earth (10%), and one for concrete being poured completely within a formed surface, such as a beam (5%). See the last two right-hand columns of the concrete takeoff below.

Concrete takeoff format

313.1

Transfer these columns to the Estimate

Plus this column for flatwork

Use waste as a multiplier, not a percentage

CONCRETE TAKEOFF FORMAT

No.	Description	Detail	Qty.	L	W	SF	Ht	CF	net CY	%	act CY
1	Pour slab on grade	sht 2		10	15	150	0.333	50	1.85	1.1	2.0
2											
3											
4											

Work left to right

Use to name the detail and sheet numbers from the plans. This allows backtracking later.

Concrete pours are simply numbered in sequence on the take-off. Takeoff descriptions do not need to be cost coded until they are transfered to the Estimate.

Section 4 Formwork takeoffs

Formwork units of measure

Formwork should be counted second, after the concrete material. Formwork is not shown on the plans. However, the estimator has a list of concrete pours already figured, and the measurements for formwork are mostly already figured. There are two main units of measure to count forms.

"Linear feet" (LF) is the unit of measure for edgeforms up to 12″ high. This includes all wood edgeforms made with common lumber up to a 2 × 12. Forms for sidewalks, porches and slabs, if made of 1 × 4s or 2 × 6s (or their metal form equivalent), all are counted by the linear foot.

Linear feet would also be used to count the edgeforms of an 18″ deep monolithic edge when a 2 × 12 is used for the form and 6″ of the edge is being earthformed. What matters is the size of the form, not the depth of the concrete. Chamfer strips and recesses (of less than a foot wide) are also counted by the linear foot.

When metal or wood forms are above 12″ in height, the unit of measure to use is "SFSA" (square feet of surface area) or "SFCA" (square feet of contact area), which mean the same thing, the net area of a form in contact with concrete. This unit of measure is used for walls, columns and beams, and the bottom of suspended slabs.

Formwork takeoff format

Use the format of the ten column headings below to count most forms. Two examples of formwork activities are shown, one with a unit of measure of LF, the other of SFSA.

FORMWORK TAKEOFF FORMAT 314.1

No	Description	Detail	EA	Sides	L	Ht	%	LF	SFSA	
1	2 x 12 footing edgeforms				100		1.1	**110**		
2	Plywood wall forms				100	10			**1000**	

Ending quantities sent to the estimate are shown bolded.

Section 5 Concrete reinforcement

The reinforcement in concrete is typically straight or bent steel rebar and welded wire mesh. Reinforcement takeoffs are not covered in this introductory textbook. Rebar quantities can be too complicated for the beginner, with both lap and waste factors, and with piece after piece of rebar seeming like an endless spool of spaghetti. Rebar does appear on many of the drawings, but it is not covered on the takeoffs in this text.

The task of counting concrete reinforcement is often handed off to a fabricator that provides a quote for rebar already bent, cut, and tagged, all ready to use. Fabricators provide shop drawings with notes that correspond to the "tags" on rebar. Delivery truckloads of rebar are a concern for yard layout, with careful consideration by the contractor given for where unloading takes place. Where material is unloaded, and how it is stacked and organized, can greatly affect the cost of labor. The lay-down area for temporary storage is usually tight. Later, rebar will need to be identified and transported to where it is placed.

A price quote is provided by the fabricator as well as the "tonnage", which is the unit of measure for rebar. The general contractor or concrete trade subcontractor adds labor for reinforcement, often based on the tonnage (using dollars per ton), without ever counting the reinforcement.

"Rebar" is not referred to as "steel" because the industry uses that word to define structural steel. Rebar is rebar, and the people who do this work in the field are known as "rodbusters" who install rebar already fabricated.

Section 6 Excavation and grading

Excavation required for concrete

This term is used to define only that excavation required by the concrete trade (such as digging footings), *not* the clearing and cut and fill of a building site. Nor does it include the building pad, the earth beneath a slab on grade, except for fine grading and any trenches cut in it for thickened portions of the slab.

Note that excavation is not a part of the term "form, reinforce, and pour". In some areas of the United States, the division two contractor does all of the earthwork for the concrete contractor, and there is a strict division of labor between the two trades. In other areas, the concrete trade does some minor earthwork, involving the "excavation required for concrete". An example of this is the earth excavated in order to pour a continuous concrete footing, and takeoffs for this kind of excavation are included in this text.

Overexcavation

When pours occur several feet below grade (exactly how far down depends on regional conditions), earth has to be removed well to the side of the pour to prevent a cave-in of earth. This not only necessitates formwork to contain and shape the concrete, it requires excavation well beyond that required for the concrete pour. When this happens, the "excavation required for concrete" is called "overexcavation", discussed in later chapters.

The earthwork required for concrete is *not shown on the plans.* Whether to dig straight down (for the length and width of the concrete dimensions) or excavate beyond the footprint of the pour (overexcavate) is up to the judgment of the estimator. What is the surrounding grade going to be at the time of the pour? How deep is the pour and how much earth has to be moved out of the way vertically and horizontally? Interpreting these "conditions" is essential to defining the work.

Excavation takeoff format

Use the following format for excavation. The column headings are the same as that for concrete.

EXCAVATING AND GRADING TAKEOFF FORMAT											316.1
No	**Description**	**Detail**	**Qty**	**L**	**W**	**SF**	**Ht**	**CF**	**net CY**	**waste**	**act CY**
	Fine grade slab on grade			10	15	150					
	Hand excavate footing			100	2		2	400	14.8		

Fine grading, typically a hand operation, is making the rough grade of earth level enough to pour concrete on. A site contractor typically prepares the rough grade by machine (often to a specification of plus or minus 1/10th of a foot, or

1-1/4 inches). Fine grading, by hand using flat shovels, knocks down the highs, fills in the lows, and ends up at plus or minus 1/4 or 3/8 inch or so. The unit of measure for fine grading is square feet.

The construction sequence of excavation, formwork, and pouring is reversed when working on takeoffs, with excavation being figured last. The concrete is shown on the plans, the forms and excavation is not. It is easier to quantify formwork and earthwork after the estimator knows where all the concrete is on the plans and is counted on the takeoff. Lengths and heights already used on the concrete takeoff can be moved to the formwork and earthwork takeoffs.

Section 7 Summary

The first rule about takeoffs is to combine like kinds because they share the same unit costs. The second is to isolate differences; if pour B costs more than pour A, then it is shown separately. Both pours get a name and a quantity.

It is helpful to keep pours separate not only by type but also by area (rooms, wings, floors or elevation level). Include the area in the description of the work activity. This organization (formatting) gives the estimate clean, precise descriptions that will be useful later in purchasing, change orders, and scheduling.

2

ISOLATED CONCRETE PADS

Section 1 Drawings and photos of concrete pads

Drawing 1 Parking garage footings

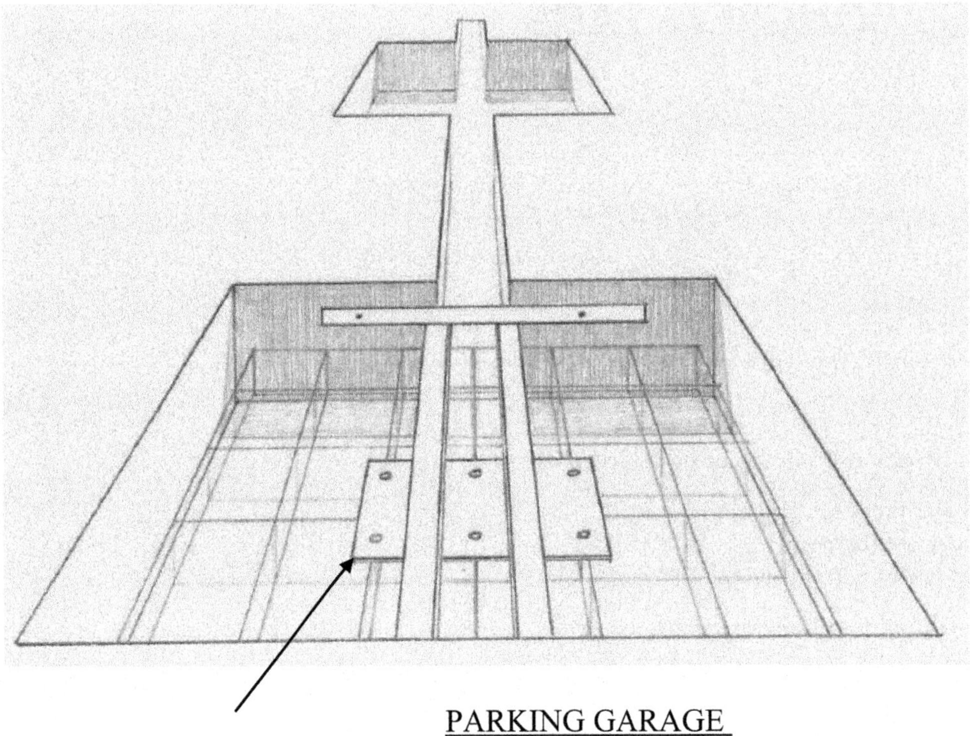

Anchor bolt template made of plywood and positioned at the top of footing (TOF).

PARKING GARAGE FOOTINGS

Drawing 1

Photo 1 Large Concrete Pads

The isolated concrete pads were formed on top of the subgrade. Look for the anchor bolts in the center of the pads. The formwork in the center of the photo is for the concrete walls of an elevator. Plastic safety caps are on top of the rebar. Steel columns will be plumbed vertical when horizontal beams are connected to them.

Drawing 1 is of large concrete pads connected with a continuous footing. It illustrates earthforming, where concrete is deposited into an excavation without the use of wood or metal forms, and plywood templates, which are for placing anchor bolts. The large pads below are similar to the layout of a parking garage or similar structure with large concentrated loads.

There are two "mats" of rebar, one bottom and one top, with rebar in both directions, or "each way". A note on the plans describing this would explain: #5s t and b 12″ o.c.e.w., which means "5/8" diameter rebar located top and bottom 12″ on center each way.

Steel anchor bolts approximately 12″–16″ long are to be embedded in the concrete. They are supported with a plywood template. A steel column will later stand at the center of the pad, its base plate (with drilled holes) held down by the anchor bolts. Consider the accuracy required here in the placement of the wood template. Steel columns are placed at fixed locations and the bolt locations have to be accurate.

The bracing for the plywood template can be accomplished in a number of ways. The earth in this drawing must be firm and hard as it appears that 2 × 4 horizontal braces are somehow attached to it (this wouldn't work in Florida!). An alternate method is to position 4 × 4 or larger timbers on top of the ground stretching across the entire pad. Plywood templates can then be suspended overhead from the timbers with some wood bracing.

This formwork requires cooperation between the trades. The Division 5 steel subcontractor provides the bolt pattern spacing for the steel column base plate (by either a shop drawing or a sample base plate) to the prime contractor, where it is forwarded to the Division 3 concrete trade subcontractor.

Photo 2 Steel Column Base Plate And Bolts

Vertical anchor bolts with leveling nuts under and top of steel base plate. Notice the weld at the bottom of the column to the base plate. These separate pieces, the column and the base plate, were fabricated (welded together) in a local shop, which is known as "fabrication".

Section 2 Earthforming concrete pads

Introduction

Isolated concrete pads are individual concrete footings as opposed to continuous footings. They might be 2′ square for wood posts or 15′ square for a parking garage. Formwork is not shown on the plans. The judgment of whether forms are required (at the bidding stage) is up to the estimator.

When concrete can be poured into shallow continuous trenches or into individual shallow excavations, forms may not be needed. Earthforming is the term used to describe this absence of wood or metal forms, where concrete is deposited against earth that stands firm.

At a certain depth of excavation, it becomes dangerous to work in a trench because the earth will cave in. Determining this depth depends on highly local factors, including type of earth, time of year, the water table, etc. Geotechnical reports, giving the soil characteristics from borings taken onsite, are often provided by the owner to bidders. These reports can be used to determine the stability of soil.

The estimator can usually make a decision about whether earthforming is feasible by being aware of typical conditions in geographical areas that his or her company has worked in many times. In mid Florida, when the bottom of concrete footings are deeper than about three feet below surrounding grade, the estimator figures the earth "won't stand" and is going to be widened beyond the width of the concrete (overexcavated). Forms will be needed against the sides of footings.

Sometimes the decision about earthforming has to wait until the job is awarded and the site contractor is known. After the job is let, the concrete and site contractors can plan out whether the subgrade will be provided at the bottom of footings (requiring edgeforms) or to the top of footings (requiring excavation). For the estimator, who is charged with "getting the costs covered", choosing either forms or excavation is sometimes an academic one. The exact means and methods will be determined later.

The isolated concrete pads in this chapter are shallow and judged not to require forms.

Shed on piers foundation

See the online resources for diagrams 322.1 & 322.2

Plan interpretation

There are eight each concrete pads and piers. From Section A, all of the pads measure 3′ square and 1′ high. The top of the footing is a minimum of 8″ below grade.

Sketch 1 is not a part of the plans and must be visualized by the estimator.

Scope of the work

Include:

Form, reinforce, and pour the footing pads.
Excavate earth required for concrete.
Include concrete material, labor, and equipment.

Exclude:

Vertical piers.

Construction techniques

A site visit reveals the ground is up and down two inches or so over the footprint of the site. The concrete will be "earth-formed"; metal or wood forms are not necessary. The earth will be excavated by hand after the layout is completed. Excess earth will be spread onsite.

The concrete will be delivered by a ready mix truck, and the truck chute will be used to pour concrete directly into the excavations. Wheelbarrows or concrete buggies might be used if the chute doesn't reach all of the footings.

Concrete takeoff

There is only one concrete item that needs describing on the takeoff in order to price this work, the group of "like kind" concrete pads. The purpose of the takeoff is to determine this single quantity, which will be transferred to the estimate. The estimate prices both labor and material, both a consequence of one quantity, the amount of concrete in the pads.

The unit of measure for concrete footing pads is the cubic yard. Using the standard concrete format as shown in the introduction, this yardage is determined as follows:

322.3	Concrete Takeoff Shed On Piers							abce	f / 27		gh
			a	b	c	d	e	f	g	h	i
No.	**Description**	**Detail**	**Qty.**	**L**	**W**	**SF**	**Ht**	**CF**	**net CY**	**%**	**act CY**
1	Pour concrete pads	A	8	3	3		1	72	2.67	1.1	**2.93**

Like Kind Pour, all poured at same time, one description.

From the plan and section A

Use waste as a multiplier
Use 1.1 waste if earth formed
Use 1.05 with wood or steel forms.

This is the result, the answer to how much concrete incl. waste it takes to pour the pads.

Since concrete is measured at the plant in quarter yards, the quantity would be rounded up to 3.0 cy when transferred to the Estimate.

Use the waste column as a multiplier greater than one. If the waste factor is 10%, use 1.1, not 10%; it saves a step.

Earthwork takeoff

The unit of measure for excavation, both labor and material, is the cubic yard. One number is needed for this project, the amount of earth that needs to be excavated in order to perform the concrete work. This quantity is needed for the estimate, and it is the duty of the takeoff to provide it.

When concrete is earthformed, the length and width of concrete and earth excavation are the same, but the excavation "height" is taller (see Sketch 1) than the concrete height.

Review Section A and the 8″ below grade requirement. What is the average depth of excavation given all footing bottoms are at the same elevation?

Rounding off is common when figuring earthwork. For the depth of excavation, Section A shows that it will be a minimum of 20″. There are two factors adding to this measurement, both involving *top of grade*. One is the relatively minor 2″ up and down unevenness seen at the site visit. The other is the allowance of a couple of inches to maintain the requirement of 8″ minimum below grade. The estimator realizes that the superintendent will use an average excavation depth of 10″ or so to factor in a margin of error. So, two feet is used for the depth of excavation, a nice round number. Always round up when figuring earthwork by hand.

When figuring hand excavation, use the format of 12 column headings used for concrete.

EXCAVATION AND GRADING TAKEOFF SHED ON PIERS											322.4
No.	**Description**	**Det.**	**Qty.**	**L**	**W**	**SF**	**Ht**	**CF**	**net CY**	**%**	**act CY**
1	Hand excavate pads	A	8	3	3		2	144	**5.33**		

The 2' ht is the only difference between earthwork and concrete measure.

This is the answer, how much (compacted) earth must be excavated.

"Hauling away" this same earth by truck (cubic yards), now in a ***loose*** *state, would require a multiplier because the earth was compacted before it was excavated.*

The 5.33 cy of excavated earth will be spread around the site and "lost", not hauled away.

Summary

Counting excavation required for concrete is perplexing because, like formwork, it is not directly shown on the plans and requires a good knowledge of construction techniques and sequencing.

Excavation labor is a risky task, whether by machine or hand or a combination of both. Unforeseen conditions are often encountered. There are two areas of construction work in which unforeseen conditions are found more often than anywhere else, and earthwork is one of them. The other is the existing enclosed space (attic, ceiling plenum, concealed walls, etc.) of a building renovation.

The risk is that the work can be quite different from estimated. Having too much rain while excavations are open can cause earth to slide to the bottom, requiring its removal a second time. Groundwater levels that fluctuate can make dewatering an unanticipated requirement. Boulders can impede progress. Nowhere else does estimating approach gambling any closer!

Section 3 Overexcavation and edgeform concrete pads up to 12″ high

Introduction

When concrete is to be placed several feet beneath grade, room will be needed around the pour for people to work and set forms. This is *overexcavation*. Often there are many levels to a building foundation, and there are various "subgrades" on which the work takes place, some or all of it requiring overexcavation. The sides of the dig are laid back at an "angle of repose" that will not cave in. In this textbook, "overexcavation" is that amount of earth removal beyond "straight down" excavation.

The term "excavation required for concrete" is an oft-used industry term. It means earth that has to be moved out of the way in order to accomplish concrete work. It is usually up to the concrete trade to provide this part of earthwork, as contrasted with the mass excavation and earthwork fill provided by the site contractor.

Overexcavation creates a "staging area" for workers and material. The amount of room needed depends on the construction techniques being employed and is a determination of the estimator. Overexcavation is *not* shown on the plans.

Concrete *edgeforms* are used to confine concrete at the sides of the pour. Forms less than 12″ in height can be made of wood or metal. Often, wood boards sized from 2 × 4 through 2 × 12 are used, both being 1-1/2″ thick. However, 3/4″ thick boards are used also, 1 × 4 through 1 × 12, especially if a curve needs to be made. Vertical wood stakes are used to hold the form steady. Steel forms are sometimes made from C channels (see Part 5, Chapter 1 Structural steel), modified by having holes drilled in the top and bottom of the C to allow passage of steel rods used for stakes to hold the form rigid.

Using common lumber, a 2 × 12 is the largest size, having an actual side length of 11-1/4″, with this length being used as the height of the form. Forms of this size and smaller are called edgeforms in this text.

Elevator pad

See the online resources for diagrams 323.1 & 323.2

Plan interpretation

There are two slabs measuring 11′-4″ × 12′. Section A shows the slabs to be 1′ thick and 4′ below the top of the main slab. A 3′-8″ high concrete wall is to be built around the perimeter of the slabs. The dotted line on the foundation plan is where the inside face of the concrete wall will be placed. There are two mats of rebar.

Scope of the work

Include:

Form, reinforce, and pour the elevator pads.
Excavation required for concrete.
Formwork design by trade contractor. Provide edgeforms and bracing to maintain tolerances.
Tolerance is 1/4″ horizontal, 1/8″ vertical.
Include concrete labor and material, pumping and equipment, place and finish.
Employ an independent testing service to take 4 cylinders per 50 cy of concrete.
Test one cylinder at 7 days, one at 14 days and one at 28 days, hold one in reserve.

Exclude:

The building pad. (Division 2 is providing the subgrade at −4″.)
Concrete walls.

Construction techniques

Earthforming is ruled out for safety and practical reasons as it would require excavating "straight down" and leave earth side walls 4′-8″ high (too deep for soil to "stand").

How much earth needs to be moved out of the way in order to form and pour the pads? The geotechnical report reveals sandy soil that probably wouldn't stand more than a couple of feet. *It would be too dangerous to attempt because of a potential cave-in.*

Overexcavation beyond the footprint of the pads will be required to place the forms and concrete. Assume this "walk around" area will need to be about 3′ beyond the length and width of the pad.

Sketches 1 and 2 are shown in order to visualize the "overexcavation" required to pour the two elevator pads. Sketches and notes like these would not be a part of the plans, but estimators have to visualize this geometry in order to determine earthwork quantity

See the online resources for diagrams 323.3 & 323.4

Use 2 × 12 lumber for the edgeforms, sold in 2′ increments. Typical lengths are 8′ to 16′. Hold edgeforms in place with vertical wood stakes.

A concrete truck will bring concrete to the edge of the large main slab. From there it will be pumped to the two elevator pads, then placed and finished within the forms.

Concrete takeoff

The *unit of measure is the cubic yard.* For slabs, also determine square footage.

											323.5
						abc		de	f / 27		gh
			a	b	c	d	e	f	g	h	i
CONCRETE TAKEOFF ELEVATOR PAD											
No.	Description	Detail	Qty.	L	W	SF	Ht	CF	net CY	%	act CY
1	Elevator Pads	A	2	12	11.33	**272**	1	272	10.071	1.05	**10.57**

Work left to right

There are 2 elevators

For slabs, two units of measure are important - SF and CY.

Use 1.05 conc waste since forms are wood

When sent to the estimate, this ending quantity of 10.57 would be rounded up to the nearest 1/4 yard, or 10.75 cy, because a quarter yard is the smallest unit of measurement used by concrete plants.

Formwork takeoff

"Linear feet" is the unit of measure for counting the edgeforms of the elevator pad. The estimate will need the total linear footage in order to price the labor and material. The purpose of the takeoff is to calculate this quantity.

Two ways are shown to count the length of the edgeforms, Methods 1 and 2. Method 1 is used for adding up the exact lengths of forms in contact with the concrete and simply counts the perimeter of the concrete. Then a standard waste factor, say 1.1, is multiplied by the perimeter.

Method 2 is used when the length of boards that actually need to be used can be readily observed from the plans. In this example, it is apparent that 12′ boards will be needed for two sides and 14′ boards for the other two sides.

Method 2 produces two items of description instead of one. This distinction can be important because the estimate can benefit from the knowledge that both 12′ and 14′ lengths are needed to properly price the work. The material cost of lumber is not proportionate, and a 12′ long board is not the same price per linear foot as a 14′ length. If there are hundreds or thousands of feet of forms, the difference can be significant. The estimate will be more accurate if the takeoff provides a more precise *description of the work* and the price for both lengths are used to cost the material (instead of a blended price).

The labor per linear foot is the same whether a board is 12′ or 14′ in length.

In method 2, instead of a single description of "2 × 12 edgeforms", the work items become "2 × 12 × 12 edgeforms" and "2 × 12 × 14 edgeforms". This is a huge distinction in the takeoff world. When just 2 × 12s are used, without a length, it can mean that the length is unimportant and that any length will suffice or that the length may be important but the estimator has left it up to the project manager or superintendent to figure it out later.

When the length is shown as part of the description of lumber, it completes a full description of the material needed in order to accomplish the work. The estimator should use specific lengths whenever they are important. Method 2 is better than Method 1. The material price will be more accurate, and the task of determining lengths is not passed on to someone else. After all, the estimator is already counting the material and, if specific lengths are beneficial, they should be described.

When lumber is shown not to have a "length" in this textbook, then lengths are not important. If the length is important, it is shown as a part of the work description.

323.6

Form length counted by METHOD 1 Exact Count

FORMWORK TAKEOFF ELEVATOR PAD

No	Description	Detail	EA	Sides	L	Ht	%	LF	SFSA
1	2 x 12 edgeforms	A	2	2	12		1.1	53	
2	2 x 12 edgeforms	A	2	2	11.33		1.1	50	
								103	

12" is a description, not a unit of measure! Ht. column not used.

Use 10-20% waste for common lumber

The ending quantity of 103 lf would be rounded up to an even number because lumber is made in 2' increments.

323.7

Formwork counted by Method 2 Visual Observation adds length to the description

FORMWORK TAKEOFF ELEVATOR PAD

No	Description	Detail	EA	Sides	L	Ht	%	LF	SFSA
1	2 x 12 x 14 edgeforms	A	2	2	14			56	
2	2 x 12 x 14 edgeforms	A	2	2	12			48	
								104	

Method 2 is used when you can write the answer down directly without doing any arithmetic (by simply looking at the plans and observing the lengths needed).

The lengths of 12′ on two sides of the elevator pad will require 14′ long 2 × 12s. The 11′-4″ sides will require 12′ long boards.

In Method 2, the "description" of the item changes from just "2 × 12s" of unspecified length to "2 × 12 × 12s" and "2 × 12 × 14s".

Earthwork takeoff

The division 2 subcontractor is leaving the earth subgrade at minus 4″, the bottom of slab, see above scope of work. The excavation depth for the concrete contractor is from minus 4″ to the bottom of the elevator slab, which is 5′ below the finished floor, so the concrete contractor excavates from a depth of minus 4″ to minus 5′, which is 4′-8″.

There are two units of measure needed on the takeoff to describe the work (well, three, but backfill will be discussed later). They are excavation and fine grading. Both will be needed on the estimate for pricing labor; there is no material. The first one is digging the pit, getting it down to a more or less level surface at the bottom, with enough room around the pour to set forms. This work would be done by machine and an operator, or by hand, or both. Regardless, the unit of measure is cubic yardage of removed earth, and there will be labor and perhaps equipment, but not material.

The second earthwork operation is by hand and involves using flat shovels to level the "rough grade" left by the initial excavation. This is called "fine grading" and occurs on the ground that will be in contact with the bottom of the elevator slab.

A third earthwork operation is the replacement of the "overexcavated" earth that was only removed in order to walk around the slabs and accomplish the work. This is called backfill.

The height (depth of excavation) can be determined from Section A (4′-8″), but the length and width requires judgment. An assumed "walk around" space of 3′ shown in Sketch 1 determines the size of the "pit". Note this 3′ area could just as well be 2′ or 4′, and the estimator uses experience to decide.

An assumption that has to be made is the "angle" that the earth is going to be at on the sides. This is called the angle of repose, and for estimating work, a 45-degree angle is handy to use and often fairly accurate for figuring excavation required for concrete. Always be liberal in the count, after all it might rain and the earth may cave in. Earthwork dimensions can be arbitrary!

The volume of the pit, using 3′ walk around room, is 6′ longer and 6′ wider than the concrete pad, *plus the angle of repose at each end.*

Convert excavation to an equivalent rectangular length and width, which is easy to do using 45 degrees. See Sketches 1 and 2, diagrams 323.3 and 323.4.

323.8

See Sketch 2

This is a lot of earth to remove just to pour 10 cy of concrete!

EXCAVATION AND GRADING TAKEOFF ELEVATOR PAD

No.	Description	Det.	Qty.	L	W	SF	Ht	CF	CY	%	CY
1	Exc. Pads, hand and machine	A	2	22.67	22	997	4.67	4658	**173**		
2	Fine grading		2	12	11.33	**272**					

Section 4 Plywood forms for concrete pads over 12″ high

Sideforms

Concrete forms greater than 12″ high on the sides of a pour are typically called sideforms, while those less than 12″ described in the last section were called edgeforms. Forms higher than 12″ have to withstand increasing pressure, and one way to make them is with plywood braced with 2 x lumber. The plywood industry makes a special plywood used for formwork called "plyform". This plywood, made in 4′ × 8′ sheets, is 3/4″ thick and has wax, yes that's wax, on one side to help provide a smooth concrete finish. This plywood and its bracing, which can be complicated, is called form design.

Schoolhouse slab and footing

See the online resources for diagrams 324.1, 324.2, 324.3, & 324.4

Plan interpretation

The plan shows the outline of a 6″ concrete slab above various footing pads. There are eight each type A footings, six each type B, and four each type C. All three have different dimensions.

The perimeter of the slab rests on type A footings, which have steel anchor bolts embedded in them. The other two footings support columns, which in turn support the slab.

Type C pads are the lowest at an elevation of minus 9′-8″. Type A and B pads are at minus 3′.

This "plan view" of a school and its dimensioning illustrate some important aspects about architectural and engineering drawings. A plan view, by definition, is a drawing that shows a horizontal surface at a specific elevation. Sometimes plans are combined, such as floor plans and foundation/footing plans. Often the name of the drawing describes its dual purpose; such as being called a "slab and footing plan". When drawings depict horizontal surfaces at multiple elevations, dashed lines and notes are used to point them out.

Review how the dimensions are presented on this plan. Only the center grid lines are shown. The size of isolated footings A, B, and C are not shown on the plan. Neither is the length and width of the slab. The size of the footings and the overall slab dimensions are left up to a review of the sections. This may seem arbitrary and lazy on the part of the architect because the plan view could show more, there is room on the sheet, and the contractor has to search the remainder of the drawings for dimensioning. However, many drawings use a grid like this, and there are reasons for how the concrete sizes are presented by the architect to the contractor.

Understanding how an architect and engineer work together can explain how some dimensions are presented. Architectural drawings may be completed or almost completed before they are sent to an engineer. By using centerline dimensioning for a foundation, the architect can get a drawing done *even though he or she doesn't know what size the footing pads will be.* The footings sizes may be shown as a generic size.

The architect, who is being paid by the owner for a full set of plans, furnishes the architectural plans to the engineer, who is hired to size the isolated footings. The engineer does the load calculations and sizes the footings, gives the information to the architect, who then has options as to how the footing sizes are shown.

The plan can be revised and generic footing sizes changed to scale properly and more dimensioning and notes given. Or, the footing sizes may be shown in a "footing schedule" and placed on drawings by the architect or engineer. A schedule is a handy way for them to be shown because often there are many sizes of footings, and accessory notes about reinforcement can be presented in the same table or schedule. In the drawings presented here, the footing sizes are shown in the section details and the drawing plan is to scale.

Another point about the interpretation of drawings concerns the number of times the letters (A, B, and C) of the sections are shown on the slab plan. The architect figures s/he has shown them enough times on the schoolhouse plan for it to be clear what the other (unnamed) footing types are.

Scope of the work

Include:

Form, reinforce, and pour the concrete pads.
Install anchors bolts furnished by others.
Formwork design by trade contractor.
Tolerance 1/4″ horizontal, 1/8″ vertical.
Include concrete pumping and concrete material and labor.

Exclude:

Sitework. The earthwork subcontractor will provide the subgrade along the east side of the site at minus 9′-8″, where the lower C pads occur. After the C footings and columns are poured, the grading sub will bring the pad up to minus 3′, the elevation of pads A and B.
Structural steel.
Reinforcing steel.

Construction techniques

There is no excavation required by the concrete trade.

Sideforms will be necessary at all of the pads because grade is being provided at the BOF (bottom of footing). Pads A and C will require plywood edgeforming because they are over 1′ high. Pad B will require 2 × 12 edgeforms.

The concrete will have to be pumped in order to access the interior pads. There will be two pours, one for the lower C pads and a second for pads A and B.

The anchor bolts at Section A will have to be held in place with an anchor bolt template, a common type of form, while the pour occurs. The purpose of an anchor bolt template is to support and hold steel anchor bolts in very specific locations while concrete is poured around all but the top few inches of the bolts, which will receive the base plate of a column. See Sketch 1, a plan view of the anchor bolt template on top of sideforms.

See the online resources for diagram 324.5

Concrete takeoff

Two quantities are needed since there are two pours; see rows 1 and 2. There is no problem in combining the type A and B pads for the second pour, because they have the same labor and material pricing on the estimate. The takeoff only separates them because of the arithmetic needed to count their quantities.

The quantity of concrete pads is shown on the plan and their L/W/Ht is shown in the sections.

There might be hundreds of pads, inviting confusion. They should be counted the same way every time. The plans can be used to write on, collecting and grouping like kind items, the sum of which is transferred to the takeoff. Often there are handy footing schedules on the plans beside which quantities can be written.

Add these separate footing types for a total of 18 each. Then, *count all of the footing pads* on the plans *without regard to type* in order to verify the *total of 18*. This is a simple procedure and a good cross-check. This is the best way to consistently get it right – not only should pads be counted by type, the total of all types should be checked.

Always check subtotals against the total!

Count the same way every time and follow a procedure that maybe takes a little time but helps get it right. Use the plans to write on – it leaves a record that can be referred to later when an extra footing is found on sheet 50!

CONCRETE TAKEOFF SCHOOLHOUSE											324.6
No.	**Description**	**Detail**	**Qty.**	**L**	**W**	**SF**	**Ht**	**CF**	**net CY**	**%**	**act CY**
1	1st Pour, C pads	C	4	6	6		2	288	10.7	1.05	**11.20**
2	Pour A pads	A	8	4	4		1.5	192			
3	Pour B pads	B	6	2	2		1	24			
4	2nd pour		18					216	8.0	1.05	**8.4**

Formwork takeoff

The unit of measurement for the 2 × 12 edgeform at Section B is the linear foot. Any length of board can be used to cut the small lengths needed here for these 2′ sides, so the linear foot unit of measure is simply described as "2 × 12 edgeforms", no board length shown.

The unit of measurement for the plywood edgeforms at Sections A and C is SFSA (square feet of surface area) or SFCA (square feet of contact area). The formwork measurement is a geometric one; count the sides of the concrete in contact with forms.

The unit of measure for small anchor bolt templates is usually "each" as they often involve just a few dollars of material and can be made with either a board or plywood, or a combination of both. If the template is more elaborate, it can be described more appropriately.

Counting the 2 × 12 edgeform for type B footings is similar to counting the perimeter of the elevator pad in the last chapter. Only 3 numbers are needed – 324.7

6 ea *4 sides* *2′ long per side*

Not used. 12″ is a description, not a unit of measure.

FORMWORK TAKEOFF							SCHOOLHOUSE			
No	**Description**	**Detail**	**EA**	**Sides**	**L**	**Ht**	**%**	**LF**	**SFSA**	
1	12″ edgeforms type B ftgs	B	6	4	2		1.1	**53**		
2	Plywood form type A ftgs	A	8	4	4	1.5			**192**	
3	Plywood form type C ftgs	C	4	4	6	2			**192**	
3	Anchor bolt templates		**8**							

The unit of measure for anchor bolt templates is EACH.

Use plywood forms for heights over 12″.

Net area of concrete in contact with plywood.

Summary

Edgeforms of commonly sized lumber, counted by the linear foot with exact plan lengths, should have a waste factor of 10%–20% added. If the takeoff has 1,000 LF of 2 × 12 edgeforms including the waste factor, that's how much is placed on the estimate and how much will be purchased later. (This assumes no reuse, and note 2 × 4 × 4′ vertical stakes and nails are not counted here). Counting edgeforms up to a 2 × 12 in size is pretty simple for the estimator. It's by the running foot.

By definition, the measurement of plywood sideforms does not use a waste factor; it includes only the area in contact with concrete. This formwork can be elaborate, and the square footage of contact area does not explain how much plywood or 2 x bracing is going to be used. Though there is a lot of bracing in order to hold the plywood in place, this lumber is not counted on the takeoff sheet.

Often there is a lot of "formwork design" that is not known at the estimating stage because it is complicated and takes a lot of time. Imagine stadium steps, walls of various heights with multiple openings, and columns with embeds. Forms, and formwork drawings (supplied later if they get the job), may be quoted by a supplier who "rents" all of the parts and pieces to the contractor. The rental is often simply a monthly lump sum fee plus delivery and pickup. That is material costing, and it goes straight to the estimate while labor is figured by contact area furnished by the takeoff.

3

CONTINUOUS CONCRETE FOOTINGS

Section 1 Photos and drawing(s)

Photo 1 Earthformed Continuous Footing

Imagine what a hard rain would do to it! And, it does not look like the sides of this trench would "stand" if the trench were dug much deeper.

Rebar will be set inside the trench after compaction is completed. On commercial projects, the earth at the bottom of the footing will be tested for compression by a testing company.

Photo 2 Poured Continuous Footing

The worker at the end of the chute is depositing concrete from a ready-mix truck. The truck in the background has already deposited its concrete and its chute is being washed out. All of the earth appears to be imported sand. A few vertical rebar dowels extend from the concrete. Note the stepdown in the footing, which is probably 8", a block dimension.

Photo 3 Formed Continuous Footing

Continuous footing with plywood sideforms bearing on the ground. The sideforms are in 8' long sections and are reused. Vertical 2 x 4 stakes (against the forms) are diagonally braced with wood "kickers" which are in turn braced with another vertical stake.

The vertical rebar is positioned to be within the cells of concrete block which will be filled with concrete.

Section 2 Centerlines and rectangles

Counting the length of concrete footings

There are two primary methods to count the linear footage of continuous concrete footings. One is by following the footing centerline and the other is to count the length of rectangular sections, counting the ends once with no overlap. Regardless of the method, each differently sized (a change in width or height) rectangular continuous footing has its own length.

The centerline and rectangular methods will always provide the same correct length. Consider some characteristics of continuous footings:

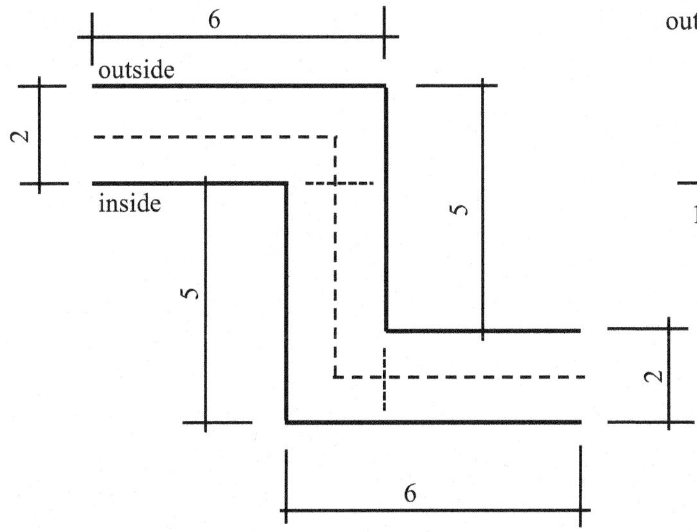

outside	inside	centerline	rectangle
6	4	5	6
5	5	5	5
4	6	5	4
15	15	15	15

When there are the same number of left and right turns in a zig-zag pattern, the inside and outside edge lengths equal the centerline and rectangular lengths.

Sketch 1 How to Count Footing Zig-Zags

Garage foundation plans

See the online resources for diagrams 332.2, 332.3, & 332.4

Plan interpretation

A garage slab is surrounded by a concrete stem wall, which is bearing on continuous footing A. The wall has an opening for the garage door per Section B. The dotted line on the plan indicates a continuous footing 8″ below grade. The footing is 2′ wide and 1′ high, and continues under the garage door per Section B.

Scope of the work

Include:

Form, reinforce, and pour the concrete footing.
Excavate earth required for concrete.
Include concrete material and labor.

Exclude:

Concrete slab and stem wall.

Construction techniques

The footing will be earthformed, and the footing trench will be at least 20″ deep. There is no formwork. The site is level. The concrete pour will be by ready mix truck.

Concrete takeoff

Only one quantity is needed on the takeoff and estimate in order to count the concrete material and labor. The unit of measure is cubic yards.

Determine the length of the footing from the plan view. Find the width and height of the footing in Sections A and B. Continuous like kind footings *have the same width and height.* Count them by the *rectangular method* or the *centerline method.* Count the length of footings:

See the online resources for diagram 332.5

Only one number is needed, the cubic yardage of the concrete material. Both labor and material will be accounted for on the estimate with this quantity.

CONCRETE TAKEOFF GARAGE FOOTING											332.6
job: Garage					job number				date		
No.	**Description**	**Det**	**Qty**	**L**	**W**	**SF**	**Ht**	**CF**	**net CY**	**%**	**act. CY**
1	Conc continuous ftg	A, B		85.32	2		1	171	6.32	1.1	**6.95**

Start here after the length is counted on the plans (L)

Section A and B (W, Ht)

Round to 7.0 cy (act. CY)

Earthwork takeoff

The total cubic yardage of excavated earth is the only description that needs to be quantified.

EARTHWORK TAKEOFF GARAGE FOOTING											332.7
job: Garage				number					date		
No.	**Description**	**Det.**	**Qty.**	**L**	**W**	**SF**	**Ht.**	**CF**	**net CY**	**%**	**act CY**
1	Hand excavate ftg	A, B		85.32	2		2	341	**12.6**		

The site is level and the top of the footing is 8" below grade. Since the min. depth of excavation is 1'-8", the estimator uses 2' as an average for takeoff purposes.

Section 3 Footings and backfill

Livingston Square foundation plans

These plans illustrate various characteristics of plan sheets. The way that the drafter presents information is highly individualistic.

Notice the dotted lines at the left and bottom of the plan sheet. Section A is cut through them, and this detail is of a 30″ wide footing 16″ high. Section A is also shown to cut through the other three exterior walls, although the dotted lines representing them are not shown. The estimator interprets this to mean that the 30″ footing is present for all four sides of the building, which is correct. It seems clear in this example, but what if Section A only cut through two of the four walls? The interpretation would be the same, and no other conclusion is possible; there are only two sections shown, one for the exterior and one for the interior.

There is more interpretation involved with larger buildings, and more conditions are all shown with a minimum amount of detail. The architect is not obligated to have an abundance of sections all over the place. Placing all of the information needed to build a building on a set of plans is a complicated task. It is a one-time occurrence; sometimes details are drawn to an awkward scale, or a better place for a section should have been chosen.

A point to remember about all plans is that different designers are going to draw the same structure with different details and sections. Estimators see hundreds of sets of plans; their interpretation skills get a lot of practice. *When there is a conflict(s) in the plans, there still is a singular interpretation.* It may not be what the drafter thinks it is; the legal reasoning for what the plans mean depends on the preponderance of "written specifications governing over drawings", "large details ruling over small ones", and "precise information from drawings or specifications meaning more than general instructions".

The estimator should never judge the drawings based on "what makes sense". That is the "construction techniques" bias; which can happen if "how it should be done" is substituted for what is strictly drawn. The estimator should stick to the facts.

See the online resources for diagrams 333.1 & 333.2

Plan interpretation

The wall structure is block with 8″ wide exterior walls and 6″ interior walls. Footing A at the perimeter is 30″ wide and TOF (top of footing) is minus 6′. Interior B footings are only 16″ wide and TOF is minus 16″.

The documents provided by the owner include a geotechnical report with information about borings taken at the site. They reveal that trash and road debris was left there long ago, and this is why the site is being excavated to minus 7′-4″.

Scope of the work

Include:

Form, reinforce, and pour the exterior and interior footings.
Concrete trade provides hand excavation of the building pad for interior footing B.
Provide edgeforms for footing A.
Include concrete material, labor, and pumping.

Exclude:

Division 2 Sitework subcontractor will excavate the entire building area to minus 7′-4″. Division 2 will provide the earth building pad after the exterior block walls are built.
Division 4 Masonry

Construction techniques

There is no hand excavation for footing A. Form the sides of footing A with plywood. After the perimeter foundation walls are built, sand will be imported as building fill dirt to minus 4″. Some of the foundation wall will not be completed to leave access for dump trucks with fill dirt. After the building pad is imported, the remaining 8″ block wall will be "buttoned up".

Footing B is then excavated and poured. After the 6″ block wall is built, dirt will need to be backfilled; see Sketch 2.

Concrete takeoff

Footings A and B are counted separately. They are poured at separate times and are not "like kind" pours. The perimeter footing pour will be easier than footing pour B. Also, the excavation for footing A, into undisturbed soil, will have a different production rate than digging the imported sand of footing B.

The centerline method is used here to determine the length of the footings.

333.3

Begin lower left corner proceed clockwise →

Review the foundation plan until you get these same lengths.

	Footing A Centerline		*Footing B Centerline*
	60		*6.33*
	58.67		*3.33*
	53.33		*53.33*
	3.33		*63*
	6.67		
	55.33		
	237.33	lf	

Since the length has 3 digits left of the decimal, use 3 digits to the right of the decimal when multiplying.

See Section A

CONCRETE FOOTING TAKEOFF											
job: Livingston Square Condos				job number				date			
No.	**Description**	**Det**	**Qty**	**L**	**W**	**SF**	**Ht**	**CF**	**net CY**	**%**	**act. CY**
1	Pour footing A			237.33	2.5		1.333	791	29.29	1.05	**30.76**
2	Pour footing B		2	63.00	1.33		0.67	112	4.16	1.1	**4.6**

See Section B

Round to 30.75 cy.

Round up to 4.75 cy.

Formwork takeoff

There is a plywood sideform at footing A (it is taller than 12″), and the unit of measure will be square feet of surface area. Since footing B requires excavating down into the building pad, the concrete there is earthformed.

FORMWORK TAKEOFF									333.4
job: Livingston Square Condos		job number					date		
No.	**Description**	**Det.**	**Each**	**Sides**	**L**	**Ht.**	**%**	**LF**	**SFSA**
1	Plywood edgeform ftg A	A		2	240	1.333		**640**	

Count the perimeter not the centerline

Earthwork takeoff

See Sketch 1 and 2 for footing B excavation and backfill.

See the online resources for diagrams 333.5 & 333.6

FOOTING EXCAVATION TAKEOFF											333.7
job: Livingston Square Condos				job number					date		
No.	**Description**	**Det.**	**Qty.**	**L**	**W**	**SF**	**Ht.**	**CF**	**net CY**	**%**	**act CY**
1	Excavate footing B	B		126	1.333		1.67	280	**10.39**		
2	Backfill at ftg B	B		126	0.833		1	105	**3.89**		

See Row 2, the width of 10″ or 0.833. Note that after the footing is poured and the blocks are laid, there are two rectangles to fill, one on each side of the block. They are each 5″ wide. Row 2 sums these two widths for a combined width of 10″. It would also be correct to place a quantity of 2 in the quantity column and a width of 5″ in the width column.

Section 4 Footing stepdowns

Introduction

A "footing stepdown" is an elevation change of the top surface of a footing. This often occurs when the ground elevation, across the footprint of a structure, changes by more than a foot or two. A footing stepdown does not have to be on the plans; the depth of footings below grade are governed by building codes, and a footing can step up and down across a site without any indication on the plans.

Some drawings have topographic information on them and show where footing elevations change and present a detail of the shape of the stepdown. Stepdowns add to the concrete yardage of a pour because the footing is thickened. The shape of a stepdown usually involves triangles, and estimators often look for a quick way of accounting for them without being one hundred percent correct. Unless there are many of them, it won't matter. Sometimes a couple of linear feet per stepdown is added to the length of the footing. However, when there are numerous stepdowns, they should be taken off accurately with multiplication of the takeoff.

The drawings for this chapter illustrate the difficulty in measuring footing lengths. This task is often made burdensome by not having enough dimensioning shown on the plans. When the estimator finds him- or herself being confounded by the lack of information, scratching in dimensions on the plans only to find oneself doing it again in another location, it may be time to pull another line(s) of dimensions all the way through the plan.

Foundation plans often dimension to the edge of the foundation wall, not the footing. A handy dimension for the estimator to keep in mind, when counting footing lengths, is the distance from the edge of the foundation wall to the edge of the footing. This distance is used when footings are counted by rectangles and corners are turned.

As said in the introduction, as long as drawings are to scale, it is close enough to count with the use of a scale or mouse. However, if the student does not practice adding plan dimensions (whether it be footings or block walls), then there will be mistakes made when using devices that help to count. Only by understanding *exactly* how a footing or wall turns a corner will the student be able to always get the right answer, even with the plans on a screen and with the aid of a mouse.

Mad Hatter foundation plan

See the online resources for diagrams 334.1, 334.2, 334.3, 334.4, 334.5, 334.6, & 334.7

Plan interpretation

The property slopes down from west to east; see the TOF (top of footing) elevations of footing pads A and B from the footing pad schedule, or Sections A and B.

There are six footing pads at the corners of the house 4′ square and 2′ deep. See the footing pad schedule and the outline of a pad in Section A.

Pads 2, 3, and 4 on the low east side are at minus 5′ TOF. Pads 1, 5, and 6 on the high west side are at minus 2′ TOF.

Continuous footing B, on the low east side, is 30″ wide, 16″ deep, and minus 5′ TOF. Continuous footing A, on the high west side, is 2′ wide, 1′ deep, and minus 2′ TOF.

The continuous north/south C footings will have to "step up" in elevation from minus 5′ low side to minus 2′ high side. There are eight pilasters with "bumpouts" in footing C. The bumpouts are where the footing is widened at the pilasters.

Each stepdown is limited to a 12″ drop, so at least three are required both north and south. Their exact location will be field determined.

Footing D is an interior footing and is 16″ wide and minus 16″ TOF.

The pads and continuous footings must be poured at the same time in a continuous pour as instructed by the foundation plan.

Scope of the work

Include:

Form, reinforce, and pour concrete footings and pads.
Excavate earth required for concrete.
Concrete trade to provide excavation for footing C.
Concrete trade to provide partial excavation of footing pads one through six, which are 2′ deep (deeper than the grade provided by Division 2).
Formwork by concrete trade contractor. Tolerance 1/4″ horizontal, 1/8″ vertical.
Include concrete pumping and concrete material and labor.
Employ an independent testing service to take four cylinders per 50 cy of concrete.
Test one cylinder at 7 days, one at 14 days, and one at 28 days, hold one in reserve.

Exclude:

Division 2 subcontractor to provide grade at the bottom of the footing on the low east side (−6.33′) and the high west side (−3′).

Construction techniques

A site visit and grades on the plans show a three-foot downward slope across the property from west to east. Since bottom of footing grade is provided for the low footing B and the high footing A, forms can be built for these while hand excavation begins on footing C. See the footing stepdowns shown in Sketch 3. The footing pads at the corners of the house, being 2′ deep (see Sketch 1), all drop a foot lower than the adjacent footings and the grade provided by others. A decision is made to earthform the bottom half of the pads; see Sketch 1.

Provide formwork for footings A and B, and for footings C and D if required.

See the online resources for diagram 334.8

The contractor figures s/he is already too busy with pouring the perimeter footings and decides to pour footing D separately.

Concrete takeoff

There are two pours and two quantities needed on the takeoff. The first pour is made up of several rectangles. On the takeoff they are simply added down and then summed for a total cubic yardage.

MAD HATTER CONCRETE FOOTING TAKEOFF											334.9
No.	**Description**	**DET**	**Qty**	**L**	**W**	**SF**	**Ht**	**CF**	**net CY**	**Waste**	**act CY**
	1st pour										
1	Footing pads 1-6	schedule	6	4	4		2	192			
2	Footing A west side	A		38	2		1	76	see sketch 2		
3	Footing C no & so	C	**2**	28	3		1	168	see sketch 3		
4	add stepped ftg's		6	1.5	3		1	27	see sketch 3		
5	add bumpouts	C	8	0.5	1.67		1	6.7	see sketch 3		
6	Ftg B east side	B		38.25	2.5		1.33	127			
								596.9	22.11	1.1	**24.32**
	2nd pour										
1	Interior ftg pad D			43.17	1.33		1	57.42	2.13	1.1	**2.34**
									see sketch 4		

See the online resources for diagrams 334.10, 334.11, & 334.12

Formwork takeoff

The 12″ edgeforms for the continuous footing will extend around the top half of the six footing pads; see Sketch 1. The unit of measure for stepdowns is "each".

Footing B has sides greater than 12″ in height, as have pads 1–6. These will take plywood forms, and their unit of measure is "square feet surface area."

Footing A is formed with a 2 × 12 because it is 12″ high. The best length to use is not readily apparent, so the exact linear feet are determined and 10% is added.

FORMWORK TAKEOFF MAD HATTER FOOTINGS										334.13
No.	**Description**	**DET**	**EA**	**Sides**	**L**	**HT**	**%**	**LF**	**SFSA**	
1	24" Plywood form pads 1-6	sched	6	4	4	2			**192**	
2	12" Edgeform ftg A	A		2	38		1.1	**84**		
3	16" Plywood form ftg B	B		2	38	1.33			**101**	
4	Form all ftg 12" stepdowns	see typ	6		3		1.2	**22**		

Earthwork takeoff

Six footing pads will be hand excavated to a depth of one foot below the provided grade. Some estimators always show footing pad excavation separate from continuous footing excavation, because the small footprint of the pads requires more labor per cubic yard than a straight continuous footing. However, on this job, with the "bumpouts" considered, which are small in area, the relatively short continuous footings, plus the stepdowns, all are slow to deal with and placed in one overall quantity.

MAD HATTER EXCAVATION TAKEOFF											334.14
No.	**Description**	**DET**	**Qty**	**L**	**W**	**SF**	**Ht**	**CF**	**net CY**	**Waste**	**act CY**
	1st pour										
1	Footing pads 1-6	schedule	6	4	4		1	96	see sketch 1		
3	Footing C no & so	C	2	28	3		2.5	420			
4	add stepped ftg's		6	1.5	3		2.5	68			
5	add bumpouts	C	8	0.5	1.67		2.5	17			
								600.2	**22.23**		
	Footing A and B	NO EARTHWORK									
	2nd pour										
1	Interior ftg pad D			43.17	1.33		2	114.8	**4.25**		

4

SLABS ON GRADE

Section 1 Photos and drawing(s)

Photo 1 Sidewalk Edgeforms

1 x 4 edgeforms for a sidewalk to be poured on top of sand. The vertical stakes are minimal 1 x 2 grade stakes, or possibly 1 x 4's. There are no crossway (transverse) perpendicular forms to set expansion joint against. Expansion joints will probably be placed on the day of the pour, and/or sawn control joints will be made late in the day or the following morning.

Photo 2 Slab Prep With Column Blockouts

Column blockout forms surround the steel columns on the left and center

This is of one of several slab pours to be made for a four story apartment complex. Note the wood "column blockout" on the left side of the photo. At the center column a short piece of a round column form is being used to blockout the concrete slab.

The small piece of equipment is a walk behind gas compactor. A compaction test will occur, and the the area will be sprayed with "soil poison". Vapor barrier and wire mesh will be placed on the ground, and after inspection the slab will be poured.

Photo 3 Slap Prep With Header Block, aka Shoe Block

Split faced header block at perimeter | Column surround with anchor bolts | Regular Block first two courses

Note the wood "column surround" form, and the anchor bolts ready for the steel column at the base of the picture.

At the border of the slab are split faced shoe block at the slab elevation. They are more expensive than regular block so they must be visible upon completion of the building (why use them otherwise?). The contractor has used regular block on the first two courses.

The dirt appears to be compacted. After it receives a compaction test by an independent engineering company it will be sprayed for termite prevention. A vapor barrier and wire mesh will be placed, and then the pour will be ready, which will happen early in the morning.

Photo 4 Slab Sections

Photo 5 Slab On Grade Pour

Believe it or not, these are called "come-alongs" , which you may not have heard of, but rhymes with thongs, which no doubt you have…they are for pulling the concrete down to level.

Concrete is placed somewhat high, or above elevation, and raked down to level. Surface, or "bleed" water, is present on top of the slab.

Photo 5 - 6 x 6 Welded Wire Mesh

Welded wire mesh in sheets of 6' x 20' shown here, 6" square grid. It is also made in rolls 5' wide x 150'L, or 750sf, and 6' wide x 100'L, 900sf. A typical plan note reads, "6 x 6 W1.4 welded wire mesh". Wire diameter varies with the smallest being designated W1.4, with others being W2.1, W2.9, etc.

Photo 6 Thickened Slab

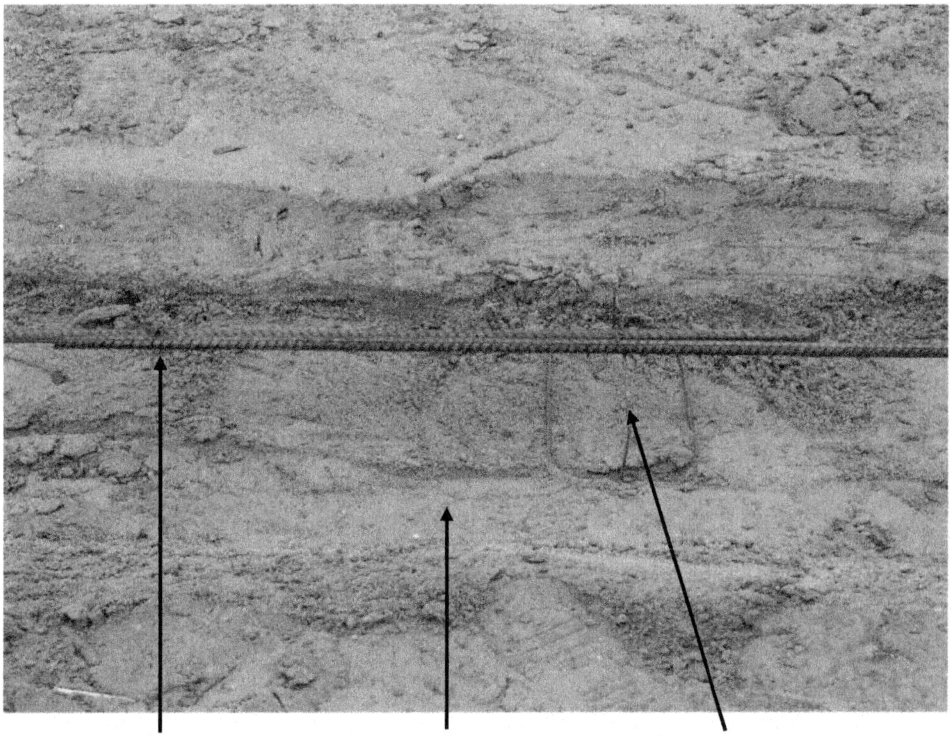

Note the rough cut of the earth. Since width and depth of a thickened slab (TS) from the drawings are "minimums", a large waste factor should be used for the takeoff of both earth excavation and concrete. The photo was taken in Florida; in other locations the earth is firmer, harder to excavate, and "stands" up better (deeper excavations can be made before cave-ins occur).

Section 2 Fill dirt and slab thickened edges

Introduction

The perimeter edge of a concrete slab is often thickened. A 4″ sidewalk may be six or eight inches thick at the sides while the edge of a monolithic slab can be two feet deep or more. The purpose at the sidewalk is simply to make it stronger, while the mono slab may support the exterior walls of a structure. Thickened edges are a common occurrence on slab plans. In this text, they are sometimes referred to as a TE.

The quantity of concrete for the TE is added to the flat portion of the main slab to obtain the cubic yardage. The quantity of excavation is the same amount as the TE figured for concrete. The square footage of the slab will also be needed for placing and finishing concrete and for fine grading.

The site contractor typically provides the rough grade of a building pad made of earth and levels it to within plus or minus 1/10th of a foot. "Fine grading" is the term used to describe the labor of using flat shovels to level the imported fill from rough grade to finish grade. The highs of the rough grade may equal the lows and often no fill dirt enters or leaves the slab area, it is just moved around by the leveling process. Finish grade is met when the dirt is 1/4″ or 3/8″ within level all across a 4″ slab.

The unit of measure for digging the earth from the edge of a slab is cubic yardage. This hand work (or backhoe work) is accomplished by the concrete subcontractor who is charged with providing all earthwork excavation "required for concrete". Since the earth has to be excavated *beneath the flat building pad provided by others*, it is excavation required for the concrete trade.

The concrete trade has to provide both "fine grading" by the square foot and the "excavation of thickened edges" by the cubic yard.

Garage plans

See the online resources for diagrams 342.1, 342.2, 342.3, & 342.4

Plan interpretation

A 4″ slab on grade is surrounded on three sides by expansion joint against an 8″ stem wall. There is a thickened edge (TE) at the garage door, recessed to keep out water. The slab slopes 4″ west to east.

Scope of the work

Include:

Form, reinforce, and pour the garage slab.
Excavate earth required for concrete.
Provide fill dirt as required for concrete. Fine grade and compact the earth.
Furnish and install 8″ edgeform at the garage thickened edge including recess.
Furnish and install expansion joint.
Include concrete material and labor to pour and finish.
Include curing compound and 6 mil Visqueen.

Exclude:

Concrete footing and stem wall.

Construction techniques

Import fill dirt, spread and level by hand, and compact. Excavate garage thickened edge. Use 2 × 8 (actual height 7-1/4″) lumber to form the thickened edge. Use a flat 1 × 12 sawn to a width of 9-1/2″ to form the recess. Nail the expansion joint to the wall.

Concrete pour by ready mix truck.

Concrete takeoff

See Sketch 1 for the three geometric shapes of concrete that have to be counted. There is one for the main 4″ slab and two for the thickened edge.

There are two concrete quantities shown. One is the square footage of the slab, which is needed to price the labor (or subcontract) of finishing operations that take up the day of the pour. This is the cost of "placing and finishing". The other concrete quantity needed is the cubic yardage, which is needed to price the material cost on the estimate.

The two concrete "accessories" of curing compound and vapor barrier are also shown.

CONCRETE TAKEOFF											342.5
job: Garage				number				date			
No.	**Description**	**Det**	**Qty**	**L**	**W**	**SF**	**Ht**	**CF**	**net CY**	**%**	**act. CY**
1	4" garage slab			22.67	18.67	423	0.33	141			
2	Thicken edge garage			12.00	0.67	8	0.60	5	(12' x 8" x 7 1/4")		
3	T.E. triangle			12.00	0.17		0.33	0	(12' x 2" x 4")		
4	Place and finish SOG					**431**					
5	Concrete material							146	5.4	1.1	**5.9**
(finish area divided by product coverage of 250 sf)											
							Finish area				
6	Curing compound		**2**	**gal**							
7	Visqueen (a x b = c)		474			**431**				1.1	
			c			**a**				**b**	

Formwork takeoff

The garage door form consists of two pieces, see Sketch 1, and the quantity of it is shown on rows 1 and 2 below.

See Section A for the expansion joint, which is counted as a form. Note how it is depicted on the floor plan in shorthand form.

FORMWORK TAKEOFF									342.6
job: Garage		number					date		
No.	**Description**	**Det.**	**Each**	**Sides**	**L**	**Ht.**	**%**	**LF**	**SFSA**
1	Edgeform 2 x 8 gar dr	B		1	14		1.1	**14**	
2	1 x 12 horiz. form recess	B		1	14		1.1	**14**	
3	Expansion jt 4"				73.33		1.1	**81**	

The form will need to be several inches longer than the 12' exact length of the opening. Choose a 14' long 2 x 8. Some estimators would combine the 2 x 8 and the 1 x 12 into one form with one description and quantity.

Do you get this total?

Round off to 80lf when transferring to Estimate. Expansion joint comes in 10' lengths.

Earthwork takeoff

There are three total earthwork operations that need to be quantified in order to describe the work for this project. See Sketch 2 for the cubic yardage of excavation needed at the bottom of the garage door. The floor plan dimensions provide the square feet of fine grading.

Fill dirt is a positive number and has to be imported. However, the excavation, which can be viewed as negative, can be used as fill dirt if it is "suitable soil" as defined by the specifications. Onsite excavation is often used for fill dirt. In the following earthwork takeoff the TE under the garage door, see row 4, is subtracted from the total fill dirt required. Sketch 2 is provided to visualize the geometry of the fill.

See the online resources for diagram 342.7

EARTHWORK AND GRADING TAKEOFF											342.8
job: Garage				number					date		
No.	**Description**	**Det.**	**Qty.**	**L**	**W**	**SF**	**Ht.**	**CF**	**net CY**	**%**	**act CY**
1	Hand excavate TE	B		12	1.5		0.33	5.94			
2	Import fill dirt garage					423	0.167	70.64			
3	less T.E. fill onsite							5.94			
4	Fill dirt required							64.70	2.40	1.3	**3.1**
5	Fine grade and compact					**423**					

30% difference between loose and compacted earth.
30% is a factor used in central Florida. Compaction factors for turning loose fill into packed earth varies by region.

Section 3 Thickened slabs

Concrete slabs are often thickened to provide bearing for interior walls. Sometimes concrete slab thickness is added to provide bearing for heavy objects (point loads) or simply to provide some concrete for the legs of bike racks that need to be embedded deeper than a typical 4″ slab depth.

The earth excavated at a thickened slab (TS) is "excavation required for concrete" and is often hand excavation estimated by the cubic yard. After excavation, this same area is fine graded and its unit of measure is square feet (SF).

Mad Hatter slab plans

See the online resources for diagrams 343.1, 343.2, 343.3, 343.4, & 343.5

Plan interpretation

The main slab on grade is 4″ thick and irregular in shape. At Section A the slab extends over the top of the foundation wall. A small area of the slab, approximately 5′ × 20′, is 4″ lower than the main slab and butts into the foundation wall. The slab is thickened at Section C to support a wall and at Section B adjacent to the recessed slab. The recessed slab at Section B is surrounded by expansion joint and slopes to a drain.

There are six each 24″ square block columns at the corners of the slab.

Scope of the work

Include:

Form, reinforce, and pour the slab on grade.
Excavate earth required for concrete.
Fine grade and compaction.
Provide all forms necessary for the concrete work.
Include expansion joint, perimeter edgeform, and slab recesses.
Include concrete material and labor.
Furnish and install curing compound and Visqueen.

Exclude:

The Division 2 subcontractor will provide the building pad at −4″ to a tolerance of plus or minus 1/10th of a foot.
Division 4 Masonry

Construction techniques

Since the building pad elevation is being provided at the bottom of the main slab, the entire recessed area, see Section B, will need to be excavated. As part of the preparation of the main slab, the TE shown at Section B (on the left side of the recessed slab) will be formed and excavated. A 10″ high form will be used, and the earth, which is at the bottom of slab depth, will be excavated 6″ further down.

Form the perimeter edge of footing A with a 2 × 6 (5-1/2″ actual height) or larger. Use a 2 × 10 to form the 10″ thickened edge at Section B.

The location of the TS at Section C will be determined with layout strings. The earth is excavated here to a depth of 8″ to support a 6″ thick wall.

The contractor decides two pours will be made, the main and recessed slabs, due to the formwork and slopes at the recessed area. An argument could be made for pouring either slab first, or for them to be poured at the same time. In this example, the contractor chooses the larger pour to be completed first, the main slab. The takeoff is unaffected by this sequence; both quantities are counted separately.

The earth at the recessed slab area will be excavated after the first pour is made. Both slabs require fine grading.

At the recessed slab, some means of "screeding" will need to take place to level the concrete as it is placed. The expansion joint around the perimeter, if placed carefully to a horizontal line, will suffice for this leveling operation. The expansion joint is nailed to the first slab pour and wall (the wall has to be in place) in a straight line prior to the recessed pour. Another form will have to be placed from the four corners of the recessed slab to the drain. An easy product to use for this type of location (which may or may not be shown on the plans) is a thin 25-gauge 4″ metal keyway (a standard product) that remains in place. The expansion joint and metal keyway will be used to *screed against*.

Concrete takeoff

Six quantities are needed on the takeoff. Two are cubic yardages of concrete, the main slab and the recessed slab; see rows 4 and 7. The placing and finishing square footages are two more; see rows 5 and 8. The last two concrete takeoff items are the accessories of curing compound and a vapor barrier; see rows 10 and 11.

One of the two interior block walls extends through the slab (see Section B), and its footprint is not counted in the area of the slab.

Use the plans to do a bit of figuring on. The preliminary arithmetic leading up to the total square footage of a slab is not needed on the takeoff, only the total slab area.

See Sketch 1 for the slab dimensions that an estimator needs for counting the individual rectangles of the slab.

See the online resources for diagram 343.6

The slab areas are calculated separately but added together below in slabs first pour, perhaps written on the plans, before being sent to the takeoff.

343.7

SLAB AREAS

See Sketch 1 for these slabs first pour

	SLABS FIRST POUR					
	Feet			Decimal		
Area	L	W		L	W	SF
1	10-7	14		10.58	14.00	148.1
2	5-7	6-0		5.58	6.00	33.5
3	11	26-8		11.00	26.67	293.4
5	16-1	40-8		16.08	40.67	654.0
6	6	20-8		6.00	20.66	124.0

Area of Pour 1 ⟶ 1252.9

Less square corners 6 ea 16" sq ⟶ -10.6

Net Area of Pour 1 ⟶ **1242.3** See Takeoff below for this **4" Slab**

Area of recessed slab 2nd pour

	SLABS SECOND POUR					
Area	L	W		L	W	SF
4	5-1	20		5.08	20.00	**102.0**

See Takeoff below for this **6" Slab**

TOTAL SLAB AREAS POUR 1 AND POUR 2 **1344.3**

CHECK : Start with the area of the BIG SQUARE (out to out dimensions).

Then deduct the rectangles that are NOT a part of the slab.

Total area big square	East/West 38'-8"	X	North/South 40'-8" equals		1572.7
Less area A	6	X	20	=	120.0
Less area B	6	X	14	=	84.0
Less area C	20	X	0.5	=	10.0
Less area D	5.08	X	0.67	=	3.4
Less square corners	6 ea X 16"	X	16"	=	10.6
TOTAL SLAB AREAS					**1344.7**

The quantities for the thickened edge B and thickened slab C are shown on rows 2 and 3 of the concrete takeoff. The shape of the thickened slab is converted to an equivalent rectangle 12″ wide and 4″ high.

CONCRETE TAKEOFF											343.8
job: Mad Hatter					number				date		
No.	**Description**	**Det**	**Qty**	**L**	**W**	**SF**	**Ht**	**CF**	**net CY**	**%**	**act. CY**
1	Slab first pour					1242	0.3333	414			
2	T.S. against recess	B		23.75	0.667		0.5	8			
3	T.S. under wall	C		23.25	1		0.33	8			
4	Concrete material							430	15.91	1.1	**17.50**
5	Place and finish					**1242**					
6											
7	Recessed slab 2nd pour					102	0.5	51	1.89	1.1	**2.08**
8	Place and finish					**102**					
9											
10	All curing compound 250 sf/gal		1344	sf	**say 6 gallons**						
11	All visqueen		1	**roll (2000sf)**							

Formwork takeoff

FORMWORK TAKEOFF									343.9
job: Mad Hatter		number					date		
No.	**Description**	**Det.**	**Each**	**Sides**	**L**	**Ht.**	**%**	**LF**	**SFSA**
1	2 X 6 perim. main slab	A			143		1.1	**157**	
2	10" edgeform main slab at recess	B			25		1.1	**28**	
3	1/2" x 4" exp. Jt @ recess	B	3		20		1.1	66	
			1		5		1.1	6	
								72	

Earthwork takeoff

There are four earthwork operations required of the concrete contractor. Two of them are fine grading, one for each pour. These quantities are not combined because they are not "like kind" labor operations. Both quantities need to be sent to the estimate separately because the grading at the recessed slab will be slower than the main slab, due to the inclined surface and smaller area. The estimate will need them separately, so they are counted separately on the takeoff.

Two excavation operations will be done at the main slab. See Sketch 2, where the average depth of excavation for the recessed slab is shown to be 7″. See rows 2 and 3 of the earthwork takeoff.

Both slabs will require fine grading, the work occurring at different times.

EARTHWORK TAKEOFF											343.10
job: Mad Hatter				number					date		
No.	**Description**	**Det.**	**Qty.**	**L**	**W**	**SF**	**Ht.**	**CF**	**net CY**	**%**	**act CY**
1	Fine grade main slab					**1242**					
2	Hand exc TE at recess	B		25.75	0.67		0.5	9	**0.32**		
3	Hand exc recess slab					102	0.58	59.23	**2.19**		
4	Fine grade recessed slab					**102**					

See the online resources for diagram 343.11

Section 4 Embeds

Introduction

Steel is often embedded flush to the top or side of cast-in-place concrete to provide a welding surface for a steel component. Sometimes the embedded steel has anchorage, called "studs", attached to it that extend into the concrete. This steel is furnished to the concrete trade by the steel subcontractor to be installed with the concrete formwork.

Contractors refer to an individual embedded piece of steel as an "embed", used as a noun. It is also used as a verb, for example, "Herb, go embed these steel weld plates into the top of that tie beam."

The form material for an embed is usually wood with the embed item attached to it. When wood and a steel embed are used together as a form, it is important for the work description on the takeoff to include the word "embed" as an alert for the labor to be properly appraised on the estimate. The bit of lumber may not cost much, and the steel embed is provided by others, but the labor can be a lot compared to the material. Often the takeoff does not need to name the small pieces of lumber required to position the embed unless it is elaborate; the unit of measure for the formwork quantity of an embed can be simply "each", or "lump sum".

Perhaps the most common embed is used for bar joists to weld to on the top of block or concrete walls. These are typically 4″ or so square *plates* (plate steel is made in sheets like plywood and can be cut into small pieces) placed on top of block bond beams (block filled with concrete) or concrete walls or beams. These weld plates often do not require formwork, because the embeds only need to be positioned into the freshly poured concrete in an on-center basis. The pattern is laid out before the pour. In this case, the unit of measure is "each" and the work description is "Embed bar joist weld plates". This requires labor but no material cost because it is furnished by others. See Sketch 2.

Before the plans for this chapter are presented, two examples of steel embeds are shown in the sketches below.

See the online resources for diagrams 344.1 & 344.2

Water treatment plant plans

See the online resources for diagrams 344.3, 344.4, 344.5, & 344.6

Plan interpretation

The main 6″ slab butts into the side of a concrete foundation wall on three sides, see Sections A and B. The north end of the slab is 1′ thick and sits on top of the foundation wall, see Section C.

The slab is interrupted by an open pit area with bar grating at the floor level. The grating rests on 4 × 4 steel angles embedded into the concrete. See Sections A and B for the thickened edge adjacent to the pit. The sides of the pit are made of 8″ block.

There are four equipment pads on top of the main slab.

At the edge of the pit, the main slab thickens to a 10″ TE shown in Sections A and B, then drops to 22″ deep at six each column pad locations shown in Section A. There are four anchor bolts that get embedded at six column locations.

Scope of the work

Include:

Form, reinforce, and pour the slab on grade.
Embed 4 × 4 steel angle furnished by others.
Excavate earth required for concrete.
Import fill dirt as required.
Provide all forms necessary for the concrete work including expansion joint.

Include concrete pumping and concrete material and labor.
Furnish and install curing compound and Visqueen.

Exclude:

Division 2 will place dirt at minus 12″ at the north end and minus 10″ at the remainder (which is the same level as the top of the pit wall).
Division 4 Masonry
Division 5 Structural steel

Construction techniques

The concrete foundation stem wall, the block walls at the pit and Division 2 earthwork will be in place before concrete work begins. It would not be practical for the site contractor to provide subgrade earth any higher than the interior block walls of the pit at −10″. Earth would fall into the pit. See Section B.

Four inches of imported fill dirt will be imported for the slab by the concrete trade (except at the north end). This earth will need to be spread, leveled, and compacted. Excavation is required at the slab edge along the edge of the pit, both for the continuous TE and the deeper column pads.

The north end of the slab edge will be formed with plywood about 16″–18″ high, overlapping the 8″ wall below; see Sketch 5. Expansion joint will be nailed 6″ high on the other three sides of the main slab.

Form the thickened edge at the columns with plywood approximately 2′ high; see Sketch 3. A board, say a 2 × 12, or some plywood can be used to form the continuous TE, see Sketch 4. Sketches 3 and 4 are after the Formwork takeoff.

Concrete takeoff

Should the equipment pads be poured with the main slab? If so, consider the formwork. The stakes supporting the sideforms will be within the slab pour and pulled out after the concrete partially sets. Fresh concrete will then have to be deposited into the holes left from the stakes. The sideforms of the mechanical pads will have to be removed so that the concrete edges can be rubbed and finished, causing traffic on top of the main slab while the concrete sets.

Because of the difficulties listed above, the takeoff below assumes a second pour for these equipment slabs, known as "housekeeping pads", and are shown separately. No vertical stakes are needed, only sideforms, and the newly poured slab allows easy access and staging around the work.

The concrete accessories of curing compound and vapor barrier are the other two items on the concrete takeoff.

CONCRETE TAKEOFF											344.7
job: Water Treatment Plant					number				date		
No.	**Description**	**Det**	**Qty**	**L**	**W**	**SF**	**Ht**	**CF**	**net CY**	**%**	**act. CY**
1	Main 6" slab	plan		50.50	24.67	1246					
2	less pit open area	plan	-1	42.00	5	-210					
3	Area and CF of slab					1036	0.5	518			
4	less out of 4 x 4 recess	B	-1	94.00	0.33		0.33	-10			
5	Add CF of 12" slab			26.00	4.583	119	1	119			
6	Continous TE @ pit			80.66	0.833		0.33	22			
7	Column pads		6	2.67	2		1.33	43			
8	Concrete SOG					1155		692	25.6	1.1	**28.2**
9	Equipment pads		4	7.00	4	112	0.33	37.296	1.4	1.05	**1.5**
10	Curing compound		**6**	**gal**		1267					
11	Visqueen		1	**roll 1500 sf**		1036				1.1	1139
			d			a				b	c

Row 1: The length is 56′ less 8″ less 4′-10″.

Row 2: The minus 1 shown for the quantity is used to get a negative number for the SF.

Row 4: This is the volume of the void at the 4 × 4 angle. The minus 1 shown for the quantity is used to get a negative number for the CF.

Row 5: The W of 4′-7″ is determined by equivalent rectangles; 3″ is added to the 4′-4″ dimension from Section C.

Row 6: The L is 6.33′ × 6′ plus 10.67′ × 4′.

Row 8: The waste factor is 1.1 because the slab on grade (SOG) is poured on earth, an uneven surface.

Row 9: The waste factor is 1.05 because the concrete is poured within forms.

Formwork takeoff

Although the form at the north end of the slab will be 16″–18″ high, only the surface area is counted on the takeoff. See Sketch 5.

Only one side of the column pads is formed, the side running with the block pit wall; the other three are earthformed. See Sketch 3. This is a plywood form and is counted by the SF, not running foot.

The TE shown in Section B can be formed with a 2 × 12 board, see Sketch 4. It will have a steel angle attached to it, secured with wood blocking. Sketch 4 shows two ways to "hang" the steel angle from wood forms.

FORMWORK TAKEOFF									344.8
job: Water Treatment Plant		number					date		
No.	**Description**	**Det.**	**Each**	**Sides**	**L**	**Ht.**	**%**	**LF**	**SFSA**
1	Form 12" slab edge	Sketch 5			34.67	1			**35**
2	Form sides of col pads	Sketch 3	6	1	2.67	2			**32**
	(incl embed stl angle)								
3	2 x 12 and 2 x 4 edgeforms @ pit	Sketch 4			78		1.1	**86**	
	(incl embed stl angle)								
4	Anchor bolt templates	A	6						
5	2 x 4 edgeform equip pads		4		22		1.1	**97**	
6	1/2" x 6" exp jt	A, B			122.01		1.1	**134**	

Row 1: The L is 26′ plus 8.67′.

Row 3: The L is 10 plus 6.33′ × 4′ plus 10.67′ × 4′.

Row 6: The L is 24.67′ plus 48.67′ × 2′. This includes the 8″ × 2 at the north ends of the foundation wall.

See the online resources for diagrams 344.9, 344.10, & 344.11

Earthwork takeoff

The earth pad is being provided at −12″ at the north end where the slab is 12″ deep; no fill is needed in the area of Section C. To the estimator this means rough grading is unnecessary but fine grading remains.

The remainder of the building pad is being left at the top of the walls of the pit, which is −10″ (see scope of work). Since the main part of the slab is 6″ thick, *4″ of imported fill dirt will be needed*, except at the north end. This earth is rough graded by the CY and fine graded by the SF. Another earthwork operation is hand excavation of the column pads and TE.

The quantity for fine grading is often taken from the concrete takeoff. Note that in this case the exact amount of fine grading does not technically include the top of the block wall at the pit. This quantity is deducted on the earthwork takeoff below. However, many estimators would consider subtracting this (on the earthwork takeoff) an academic exercise not worth going into. They would use row 2 without row 3.

EARTHWORK TAKEOFF											344.12
job: Water Treatment Plant				number					date		
No.	**Description**	**Det.**	**Qty.**	**L**	**W**	**SF**	**Ht.**	**CF**	**net CY**	**%**	**act CY**
1	Spread and level by	A, B				1036	0.333	345	12.78	1.3	**16.6**
	hand add 4" of fill										
2	Fine grade slab					1155					
3	less top of blk at pit		-1	96.66	0.667	-64					
4	Fine grade SOG					**1091**					
5	Hand exc. Col pads	A					1.33	32	**1.19**		
6	Hand exc. TE	B					0.333	32	**0.83**		

5

MONOLITHIC SLABS

Section 1 Photos and drawing(s)

Monolithic slabs defined

Monolithic slabs are slabs on grade combined with a TE (thickened edge). "Mono slabs", as they are called, are technically those slabs thick enough at the edge to support a structural load. However, common usage has blurred this distinction and often any slab, such as a basketball court made of a 4″ slab with a shallow edge thickened to 8″–12″, is also referred to as a monolithic slab. However, monolithic slab edges that support several floors of a building thicken to several feet in depth and width. A common use for monolithic slabs are the foundations for metal buildings, with a monolithic edge all around, becoming wider and thicker at the columns.

Photo 1 Monolithic Edge Overexcavated And Braced

The following photos are of basketball courts at the Jonesville Park in Alachua County, Florida. The courts consist of two monolithic slabs 1′ high at the edge.

Photo 2 Basketball Courts, Building Pad And Edgeforms

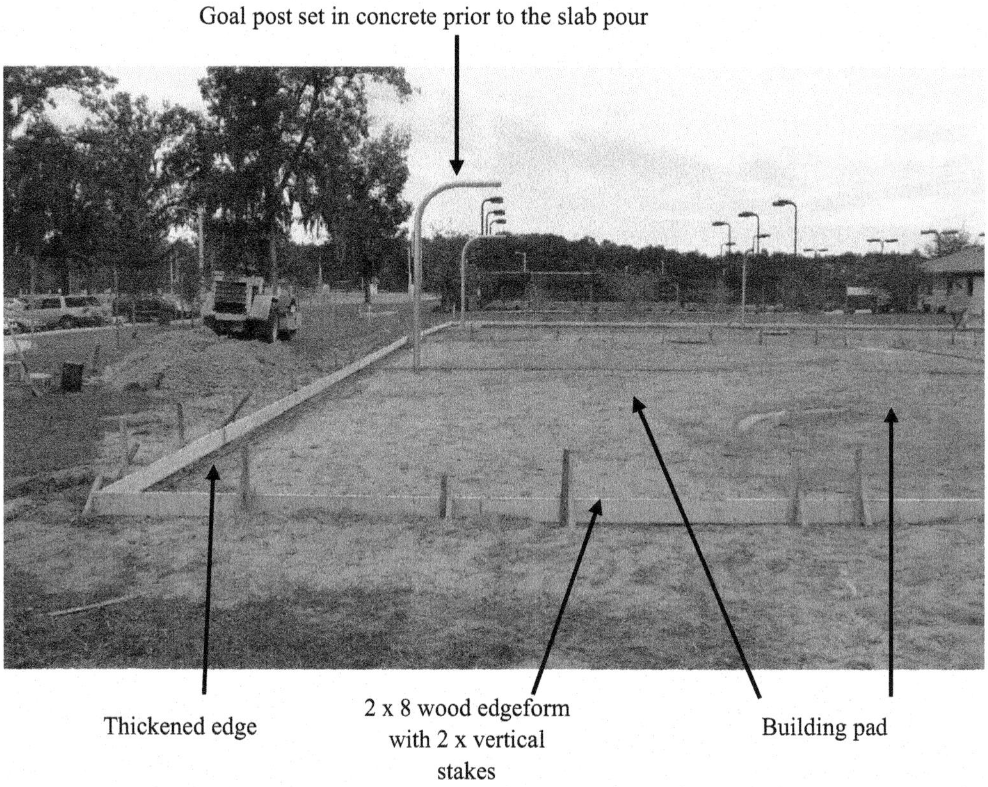

Photo 3 Formwork Sequence Between Courts 1 And 2

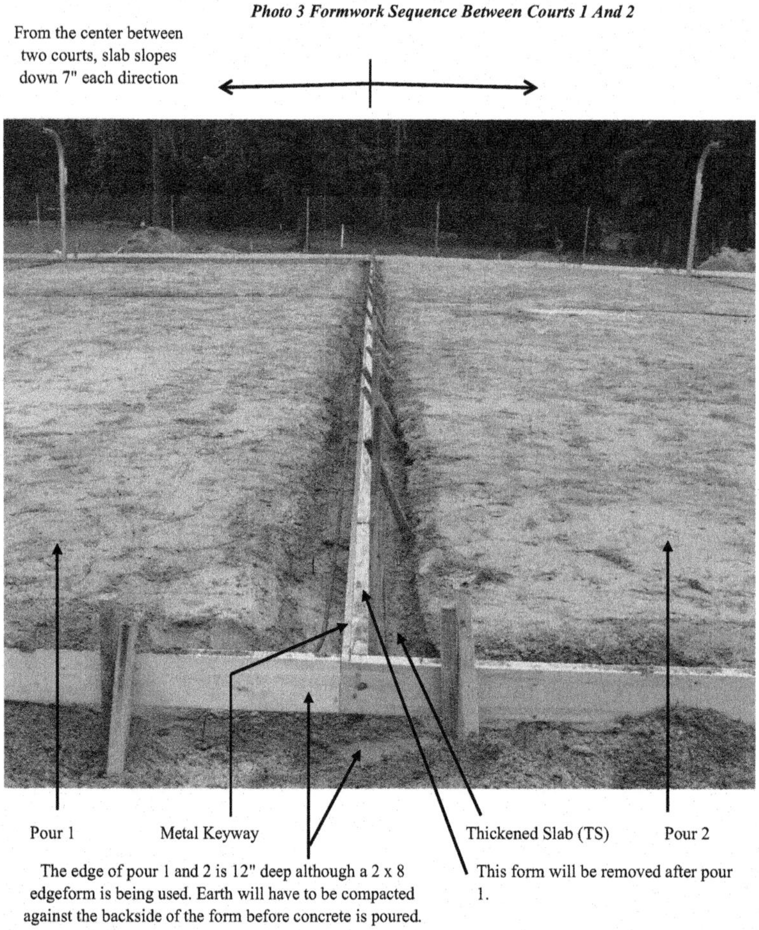

Photo 4 Shallow Monolithic Edgeforms

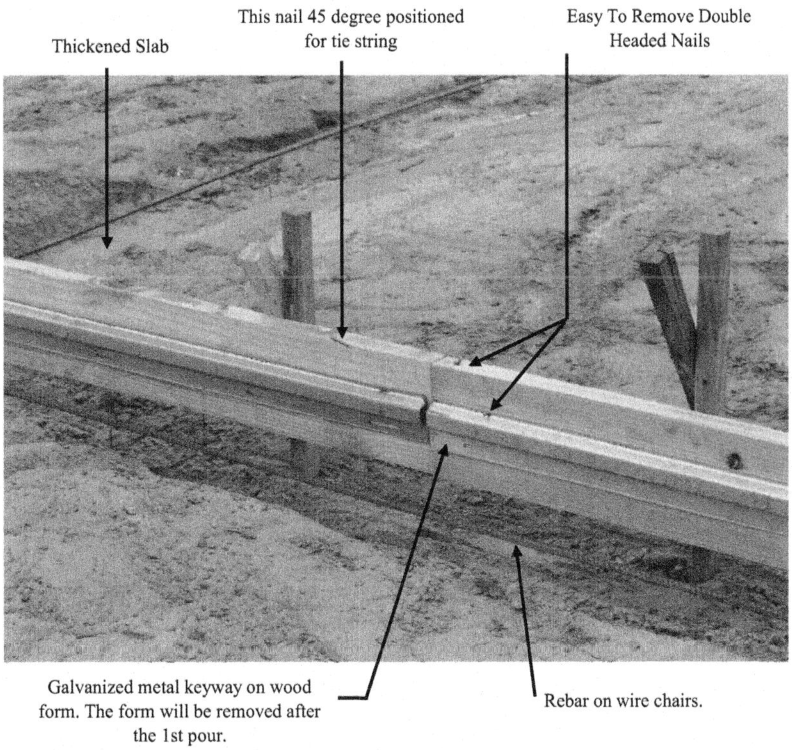

Section 2 Changing triangles into rectangles

Ticket booth plans

Review Section A. The inside edge of a mono slab is usually angled, not vertical, creating a triangle beside a rectangle. The estimator should keep things simple and convert the entire edge to an *equivalent rectangle*. If the triangle is not dimensioned, use a scale, and assume the angle is at 45% if it looks close.

See the online resources for diagrams 352.1 & 352.2

Plan interpretation

Block walls sit on the edge of a monolithic slab 4″ above grade. The concrete edge is 20″ deep and 16″ wide (at the bottom). The interior side of the mono edge is angled; it is not at a 45-degree angle.

There is also an interior thickened slab; see Section A.

Scope of the work

Include:

Form, reinforce, and pour the slab on grade.
Excavate earth required for concrete.
Provide edgeforms.
Include concrete pumping and concrete material and labor.
Furnish and install curing compound and 6 mil Visqueen.

Exclude:

Walls.

Construction techniques

A site visit reveals the ground is near level. An area beyond the perimeter of the slab will have to be overexcavated in order to accomplish the work. See the "walk around" room in Sketch 1 required to set the plywood sideforms.

The interior TE will be hand excavated.

Concrete takeoff

To figure the quantity of concrete material, count the 4″ depth of the entire slab on the first two rows of the takeoff, then count the mono edge on row 3, and the interior TS on the following row 4. The "place and finish" quantity is the entire square footage of the top surface of the slab.

Curing compound and vapor barrier are figured in gallons and square feet.

See Sketch 1 for how to form a mono edge.

See the online resources for diagrams 352.3 & 352.4

CONCRETE TAKEOFF											352.5
job: Ticket Booth					number				date		
No.	**Description**	**Det**	**Qty**	**L**	**W**	**SF**	**Ht**	**CF**	**net CY**	**%**	**act. CY**
1	Top 4" slab	A		17.67	10	177	0.333	59			
2	Top 4" slab			5.67	6.33	36	0.333	12			
3	Thicken mono edge	A		59.33	1.83		1.33	145	See Sketch 2		
4	Thicken interior	A		6.33	1		0.33	2			
5	Place and finish conc					213					
6	Concrete material							218	8.06	1.1	8.86
7	Curing compound		1	gal							
8	Visqueen		250	sf							

Row 3:

See Sketch 2 for L of the edge. For the width, use an equivalent rectangle. To 16" add half of the triangle, or 6", to get 22" W, or 1.83.

Row 4:

See Sketch 2 for L of the interior footing. For the width, use an equivalent triangle. To 8" add 2" on each side to get an average width of 1'.

Row 3: See Sketch 2 for the length of the edge. For the width, use an equivalent rectangle. To 16″, add half of the triangle, or 6″, to get 22″ W.

Row 4: See Sketch 2 for the length of the interior footing. For the width, use an equivalent triangle. To 8″, add 2″ on each side to get an average width of 1′.

Formwork takeoff

Only one quantity is needed to define the work, which is the surface area of forms against concrete at the perimeter.

FORMWORK TAKEOFF									352.6
job: Ticket Booth		number					date		
No.	**Description**	**Det.**	**Each**	**Sides**	**L**	**Ht.**	**%**	**LF**	**SFSA**
1	20" h plywood edgeform	A		1	66.67	1.67		111	

Perimeter, not length of TE

Earthwork takeoff

There are two areas of excavation, the perimeter and the interior, both shown in Section A. These operations are done at the same time and are not "different" enough to be shown separately.

EARTHWORK TAKEOFF											352.7
job: Ticket Booth				number					date		
No.	**Description**	**Det.**	**Qty.**	**L**	**W**	**SF**	**Ht.**	**CF**	**net CY**	**%**	**act CY**
1	Fine grade top 4"	A				213					
2	Hand exc TE mono	A						145			
3	Hand exc TS	A						2			
								147	5.4		

Row 1:

Count the entire SF of the slab for fine grading, don't subtract the bottom surface of the mono edge. If the sloped area of a large monolithic slab is several feet, this surface would be counted separately since it would have a slower labor factor than the adjoining grade (on the estimate).

Row 2:

From row 3 of the concrete takeoff.

Row 3:

From row 4 of the concrete takeoff.

Row 1: The entire SF of the slab is counted here for fine grading, including the bottom surface of the mono edge. See row 5 of the concrete takeoff.

Row 2: From row 3 of the concrete takeoff.

Row 3: From row 4 of the concrete takeoff.

Section 3 Brick ledges

Schoolhouse plans

See the online resources for diagrams 353.1, 353.2, 353.3, & 353.4

Plan interpretation

The drawing is named a "Slab Plan" although the footing pads are shown underneath. It is actually a slab and footing plan. The pads are no help in taking off the slab and should not be confused with the same type of dashed lines indicating the thickened slab.

The perimeter of the slab has a thickened edge. It is indented to create a ledge for the brick veneer, see Section A.

The interior of the slab is thickened at Sections B and C. The continuous TS at B is along the grid lines; the TS at C only occurs at the center.

Scope of the work

Include:

Form, reinforce, and pour the slab on grade.
Excavate earth required for concrete.
Additional fill is provided by the concrete trade.
Include perimeter edgeforms.
Include concrete labor and material and pumping.
Furnish and install curing compound and Visqueen.
Employ an independent testing service to take four cylinders per 50 cy of concrete.
Test one cylinder at 7 days, one at 14 days, and one at 28 days, hold one in reserve.

Exclude:

Footings.
Structural steel.
The Division 2 sub will provide the building pad at minus 18″.

Construction techniques

The perimeter plywood form will have "blocking" attached to it to provide the brick ledge; see Sketch 1.

Spread and level 1′ of fill dirt over the main part of the 6″ slab. Where the slab is a foot thick, at the grid lines at Section B and the center shown in Section C, spread and level 6″ of fill dirt.

Consider how this material would be deposited without a truck driving over the newly poured footings. The labor to handle the dirt operation will depend on where the earth is placed and whether too much or too little is put down. On the two long sides of the slab there is twelve feet and on the short sides there is ten feet. This is just enough room for a truck to enter and dump fill dirt without the truck driving on the footings.

See Sketch 1 for how to form a brick ledge.

See the online resources for diagram 353.5

Concrete takeoff

See Sketch 2 where the dimensions are converted to rectangle lengths. This sketch does not show the footing pads shown in Section A, which for this takeoff are assumed already poured.

See the online resources for diagram 353.6

CONCRETE TAKEOFF											353.7
job: Schoolhouse					number				date		
No.	**Description**	**Det**	**Qty**	**L**	**W**	**SF**	**Ht**	**CF**	**net CY**	**%**	**act. CY**
1	Main 6" slab on grade	A,B,C		43.67	49.67	2169	0.5	1085			
2	Add perimeter edge	A		178.0	3.17		1	564			
3	TS No/So 1,2,5,6	B	4	12.92	2.5		0.5	65	See Sketch 2		
4	TS E/W 3,4,7,8	B	4	10.92	2.5		0.5	55	See Sketch 2		
5	Add 12" slab	C		16.5	18.5		0.5	153	See Sketch 2		
6	Place and finish					**2169**					
7	Concrete material							1921	71.1	1.1	**78.2**
8	Curing compound		**10***	**gal**	at 250 sf per gal						
9	Visqueen		**3000**	**sf**							
			* This figures 9 gallons but curing compound is usually purchased in 5 gallon pails or in 55 gallon drums.								

Row 1: This is a count of the top 6″ slab; it excludes the brick ledge area.

Row 2: Convert the perimeter TE, Section A, into a rectangle 3′-2″ wide × 12″ high. The length of the TE is then twice the length of two equal sides counting full length plus the other two sides deducting the overlap at the corners, or 2 × 44′–8″ plus 2 × (50′–8″ less 6′–4″).

Rows 3, 4: The width of footing B, when converted to a rectangle, is 2′-6″ or 2.5.

Formwork takeoff

Count the contact area of the vertical plywood first. The best way to account for the brick ledge is by the linear foot, counting the 2 × 8s and the 1″ plywood. Note the plywood is counted as 8″ high but due to the waste inherent in cutting sheet goods into small strips, a 20% waste factor is used.

FORMWORK TAKEOFF									353.8
job: Schoolhouse		number					date		
No.	**Description**	**Det.**	**Each**	**Sides**	**L**	**Ht.**	**%**	**LF**	**SFSA**
1	Plywood perimeter 18"	A		1	191	1.5			**287**
2	Blockout for brick ledge:								
3	double 2 x 8		2		191		1.1	**420**	
4	1" plywood		1		191	0.67	1.2		**154**

Earthwork takeoff

The task here is to calculate how much fill dirt is needed underneath the concrete slab. The best way is not to directly count the contour of the earth shapes, which involves a number of rectangles and would resemble the concrete takeoff. A simpler method is possible. The easiest way is to first figure the entire 18″ depth from Section A (over the 2,263 sf of the slab) and then subtract the concrete from the already completed takeoff! See below. Instead of "adding up" as the concrete was figured, the quantity of concrete is subtracted from a beginning, easy-to-count volume.

Assume the footing pads below minus 18″ are completed (ignore them in this takeoff). Also assume there is no "overexcavation" beyond the perimeter of the slab. The amount of earth fill figured here only extends to the edge of the concrete, not beyond.

EARTHWORK AND GRADING TAKEOFF											353.9
job: Schoolhouse				number					date		
No.	**Description**	**Det.**	**Qty.**	**L**	**W**	**SF**	**Ht.**	**CF**	**net CY**	**%**	**act CY**
1	Total volume 18" deep			44.67	50.67	2263	1.5	3395			
2	Less brick ledge			188.7	0.5		0.5	47			
3	Less all concrete							1921			
4								1427			
5	amt of fill dirt							1427	52.85	1.3	**69**

See net concrete quantity before waste

30% is a compaction factor used in central Florida. Factors vary by region.

Section 4 Basketball courts and keyways

Keyways

This plan of four basketball courts features adjacent monolithic slabs. It is important that the edges of these slabs "join up" and remain at the same elevation; any movement (and there will be movement) in one slab needs to match the other one. One way to tie these slabs together is for them to share a keyway. A keyway is simply placing a notched recess in the edge of one slab and a key on the other. In the carpentry trade, it would be called a tongue and groove joint.

Basketball court plans

See the online resources for diagrams 354.1, 354.2, 354.3, & 354.4

Plan interpretation

There are four connected basketball half-courts. The main slabs are 6″ thick.

The perimeter is thickened to 18″ deep; see Section A. The angle of slope at the inside edge is not defined. By scale, it appears to be 45 degrees.

The joint between courts is shown in Section B. The slabs thicken to 12″ deep and are "keyed" together.

The main slab thickens against the goal post footing in Section C. The detail is interpreted to mean that the keyway exists on three sides of the goal post pad. A vertical "blockout" will provide a recess in the concrete for the goal post to be set and grouted later.

Scope of the work

Include:

Form, reinforce, and pour the basketball courts.
Excavate earth required for concrete.
Formwork design by trade contractor. Tolerance 1/4″ horizontal, 1/8″ vertical.
Goal post blockout installed by concrete trade, furnished by others.
Provide keyways.
Include concrete material, labor, and pumping.
Employ an independent testing service to take four cylinders per 50 cy of concrete.
Test one cylinder at 7 days, one at 14 days, and one at 28 days, hold one in reserve.
Furnish and install curing compound and Visqueen.

Exclude:

Division 2 subcontractor will provide the building earth pad at minus 6″.
Division 13 Goal posts.

Construction techniques

Pour the goal post footings first because the concrete is deepest here. By carefully placing the forms (having a good layout) for the goal posts at the proper elevation, these concrete pads give the rest of the pours an elevation to work towards. Make two more pours. The easiest way to make them, with edgeforms and keyways all around, is to pour the basketball courts in diagonal pairs.

The earth pad provided by Division 2 is at the bottom of the main slab, which means that the goal posts and thickened slab areas will be hand excavated. The main part of the slab is ready for fine grading.

Use a 2 × 12 to form the top one foot of the goal post concrete on all four sides; place a keyway on three sides. See Sketch 5. Earthform the remainder. Remove a small amount of earth (overexcavate) adjacent to the top of the pour to make room for the 2 × 12 form.

Pouring the goal post footings first means the keyway will be a reverse image of the key shown in Section C. This is a "means and methods" decision and is up to the contractor; the architect/engineer won't care which side the key is on; a key is a key.

Use 18″ plywood edgeforms at the perimeter of the slabs; see Sketch 4.

See the online resources for diagram 354.5

Concrete takeoff

First count the goal post pads, which are 3′ deep and 4′ square.

The two slab pours, being poured in diagonals, have the same quantity. The takeoff could show the entire slab pour (which is done here) or determine one pour and multiply by two.

The 6″ main slab has three thickened areas added to it, see Sections A, B, and C, and rows 5, 6, and 7 of the concrete takeoff. To determine the place and finish quantity, the area of the goal posts is subtracted from the total area of the court. Count the goal pads then count the entire slab.

CONCRETE TAKEOFF											354.6
job: Basketball Court				number				date			
No.	**Description**	**Det**	**Qty**	**L**	**W**	**SF**	**Ht**	**CF**	**net CY**	**%**	**act. CY**
	1st pour										
1	Concrete goal pads	C	4	4	4		3	192	7.1	1.1	7.82
2	Entire slab	plan		94	100	9400					
3	less goal pads		-4	4	4	-64					
4						9336	0.5	4668			
5	TE against goal pad	C	4	9.5	1.25		0.5	24			
6	TE Perimeter	A		362.0	2.5		1	905			
7	TE Interior	B		179.5	4.5		0.5	404			
8	Place and finish					9336					
9	Pour all court slabs							6001	222.2	1.1	244.5
10	Curing compound		38	gal		9400				divide by 250	
11	Visqueen		10270	sf		9336				1.1	
			c			a				d	

See the online resources for diagram 354.7

Row 5:

Length of TE at goal post, see diagonal lines on sketch 2:

4.0
2.75
<u>2.75</u>
9.5 see L column

Row 6:

Count the perimeter rectangles, exclude or deduct the 4 goal posts:

100
100
89 (47+47-2.5-2.5), the 2.5s are the equivalent rectangles at
<u>89</u> the perimeter.
378
−<u>16</u> less four goal post
362 see L column

Row 7:

The equivalent rectangle is 4′-6″ wide counting both slabs either side of a keyway. The lengths are shown below:

95 N/S (100 less two perimeter rectangles of 2.5 each)

See the online resources for diagram 354.8

Formwork takeoff

The goal pads are formed first and stripped after they are poured.

The slabs are formed and poured in checkerboard pattern two quadrants at a time. Per Sketches 1 and 3, the perimeter forms are used an additional 3′ at four locations. The exterior 18″ high form is turned 90 degrees towards the interior for 3′ then can reduce to 12″ high.

FORMWORK TAKEOFF									354.9
job: Basketball Court		number					date		
No.	**Description**	**Det.**	**Each**	**Sides**	**L**	**Ht.**	**%**	**LF**	**SFSA**
	1st pour :								
1	Form top 12" of goal pads	C	4	1	4	4			**16**
2	Form template for goal embed	C	4						
	2nd pour :								
3	Form 18" perimeter	A			198	1.5	see sketch 1,2		**297**
4	Form 12" keyway	B			182		1.1	**200**	
	3rd pour :								
5	Form 18" perimeter				186	1.5			**279**
6	Form 12" keyway				0				

See the online resources for diagrams 354.10 & 354.11

Row 1:
See Sketch 5.

Row 3:
The perimeter length of forms is different for pours 2 and 3. For pour 2:

$2 \times 50 = 100$ see Sketch 3, diagram 354.8.
$4 \times 21.5 = 86$
$4 \times 3 = \underline{12}$
198

Row 4:
The length of the keyway for pour 2:

$2 \times 47 = 94$ see Sketch 3, diagram 354.8.
$2 \times 44 = \underline{88}$
182

Row 5:
The perimeter length of the forms for the 3rd pour:

$2 \times 50 = 100$ see Sketch 3, diagram 354.8.
$4 \times 21.5 = \underline{86}$
186

Row 6:
The 12″ keyway for pour 3, none. The slab from pour 2 is already in place.

Earthwork Takeoff

The earth pad is at −6″ so the depth of excavation required for the 3′ deep goal posts is 2.5′. On the interior side of the goal posts (at the location of the TE against it), a small amount of earth is going to be removed in order to place the 2 × 12 form. See Section C. The cubic yardage for this is very small, and in row 2 the unit of measure shown is each, not cubic yardage. When a standard unit of measure is very small (but there is a labor operation nonetheless), estimators often change to "each" or "item" as a unit of measure. Some estimators would not show an earthwork operation for this because it is a small item.

There will be some overexcavation at the perimeter; see Sketch 4.

EARTHWORK AND GRADING TAKEOFF											354.12
job: Basketball Court				number					date		
No.	**Description**	**Det.**	**Qty.**	**L**	**W**	**SF**	**Ht.**	**CF**	**net CY**	**%**	**act CY**
1	Hand exc. post pads	C	4	4	4		2.5	160	**5.93**		
2	Exc. Area for forms		**4 ea**								
3	Hand exc TE @ goal	same as conc						24			
4	Hand exc. perim.	A		362	4		1	1448			
5	Hand exc. interior	same as conc						404			
								1876	**69**		
6	Fine grade 4 slabs					**9336**					

Row 4: The same length as concrete row 6. For the width, see Sketch 1.

This is the conclusion of Section 4 on monolithic slabs, *the last concrete section that includes earthwork.* The "earthwork required for concrete" is not needed for the upcoming chapters on walls, columns, beams, and suspended slabs.

6

CONCRETE WALLS

Section 1 Photos and drawing(s)

Photo 1 Rebar in walls
Drawing 1 Wall forms – walers, studs, and ties
Photo 2 Foundation sideforms
Photo 3 Elevator shaft wall forms and concrete
Photo 4 Concrete wall left rough
Photo 5 Wall forms and aligner braces
Photo 6 Forms on one side

Section 2 Wall formwork design
Section 3 Blockouts

Garage plans
Plan interpretation
Scope of the work
Construction techniques
Concrete takeoff
Formwork takeoff

Section 4 Retaining walls and waterstop

Waterstop
Retaining wall plans
Plan interpretation
Scope of the work
Construction techniques
Concrete takeoff
Formwork takeoff

Section 5 Concrete walls

Material bin plans
Plan interpretation
Scope of the work
Construction techniques
Concrete takeoff
Formwork takeoff

Section 1 Photos and drawing(s)

Photo 1 Rebar In Walls

One side of the wall is formed and rebar is being installed. After rebar inspection, the wall forms will be completed and the pour made.

Drawing 1 Wall Forms – Walers, Studs, And Ties

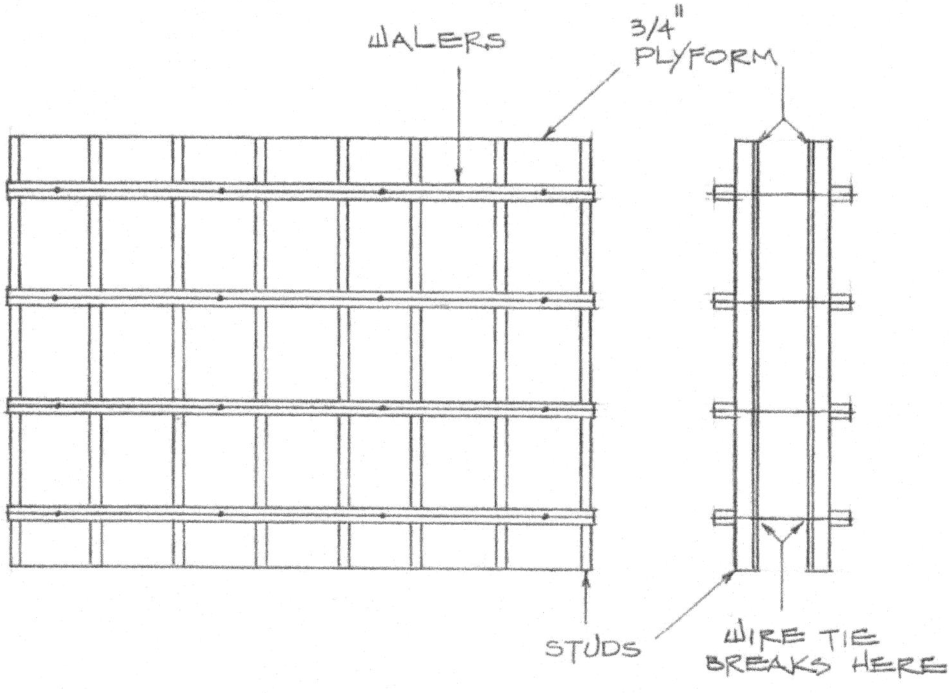

Photo 2 Foundation Sideforms

Plywood and 2 x 4 sections 8' long can be reused.

Forms were used here.

Photo 3 Elevator Shaft Wall Forms And Concrete

The four walls of an elevator shaft built on top of an isolated concrete pad.

The grain direction of the plywood is orizontal and the studs are vertical.

The horizontal double walers have steel ties sandwiched between them, and the "snap brackets" act as a cinch to pull the brace tight.

Foundation

Ties are small diameter steel rods perpendicular to the wall. The pressure of the concrete pushes outward and the ties, held by the visible snap brackets, hold the forms to a uniform width.

Freshly poured concrete

Photo 4 Concrete Wall Left Rough

The finish of this wall was purposely left rough.

No attempt was made to remove the ridges by grouting and rubbing the form edges.

"Pits" in a concrete surface can be caused by poor consolidation of concrete, lack of vibration, and a dry mix of concrete.

Photo 5 Wall Forms And Aligner Braces

The wall pour has been made (note concrete spillage on forms) and some of the vertical studding has already been removed.

A formwork accessory, called an "aligner brace", used to adjust the forms vertically.

Photo 6 Forms On One Side

The superimposed dark lines show what the outline will be of the wall after the next pour. The horizontal "belt" is probably precast.

Section 2 Wall formwork design

The height of fluid (wet concrete), plus the weight of concrete and the rate of pour, are the units of measure needed for determining *pressure* against the forms, which is what formwork design is all about. The resulting × pounds of pressure, in pounds per square foot, is the determining factor. This pressure is used to design a form to hold the form near motionless while concrete sets.

There is an important point here that may sound counterintuitive; *the width of a concrete wall does not factor into the lateral sideways pressure, nor does the length of the wall!* The most important considerations in wall formwork design are two things – the height of the wall and the rate of pour!

With regular concrete, weight is usually figured at 150 pounds per cubic foot. Therefore, concrete poured quickly to a height of 10′ exerts a pressure of 1,500 pounds per square foot laterally against the bottom of the form (10 x 150). That's a lot of pressure, and exceeds the design of most forms, which are often in the 600–1,000 pounds per square foot range.

The height of fluid concrete is commonly referred to as the *liquid head*. Note that a 4′ high pour already exerts a sideways pressure of 600 psf. If that is all that the forms are designed to withstand, the pour must stop at 4′. Otherwise, forms will do what they do, an event called a blowout, which is different from a ball game and a costly mistake that must be avoided.

As concrete starts to set (harden) and becomes more solid and less liquid, the side pressure against forms decreases quickly. Therefore, the limiting *rate of pour* (feet per hour) is a main factor in formwork design, but others include the temperature of the concrete. Once these factors are used to determine pounds per square foot, the resulting formwork "design" refers to all the form pieces, from the thickness of the plywood to the spacing of studs, walers, and ties.

Once the calculations have been made and the forms constructed, it is important that the rate of pour used in the design of formwork be conveyed to the concrete pour crew so that it is not exceeded!

Section 3 Blockouts

Blockouts are important components of forms. They are a method to create voids in the concrete, or openings. Sometimes a void is created in the middle of a wall, sometimes at the top or bottom. When wood or other material is placed between the forms, it "blocks out", or displaces, the concrete. The wall form continues uninterrupted on both sides of the wall.

It is simpler and faster to build the two sideforms for the full length and height of the wall, and then insert blockouts, than to only form the surface area of the concrete. This is how openings in concrete walls are made up to the size of doors and windows.

Garage plans

See the online resources for diagrams 363.1, 363.2, & 363.3

Plan interpretation

A stem wall is a short concrete wall, often a foundation wall as shown in Section A. The section is cut through the west side of the garage, and the stem wall is 8″ wide and 2′ tall. Section B, at the garage door, indicates the wall does not occur here. The garage slab thickens at the door where it abuts the driveway.

That's it; not much is shown except a rear opening, which is surely for a back door. The stem wall is interrupted here and the slab extends beyond the wall by 2″.

From plan interpretation, the stem wall extends all around the garage on three sides plus part of the east side, only interrupted by the garage door. Also by interpretation (there are no dashed lines to confirm it), the wall runs under the slab at the rear door; the stem wall is shorter at the rear door than elsewhere.

Scope of the work

Include:

Form, reinforce, and pour the concrete stem wall. Formwork design by trade contractor.
Tolerance 1/4″ horizontal, 1/8″ vertical.
Include concrete material and labor.

Exclude:

Concrete footing, slab.
No earthwork.

Construction techniques

Excavation was made for the footing pour, so none is anticipated for the stem wall. The earth is out of the way and there is no earthwork.

There will be plywood forms on both sides of the stem wall. The pour will be by ready mix truck.

Concrete takeoff

The unit of measure for wall pours is cubic yardage of concrete.

CONCRETE TAKEOFF											363.4
job: Garage					number				date		
No.	**Description**	**Det**	**Qty**	**L**	**W**	**SF**	**Ht**	**CF**	**net CY**	**%**	**act. CY**
1	Concrete stem wall	A		70.00	0.67		2	93.8			
2	Wall at door opening			3.33	0.67		1.33	3.0			
				73.33				96.8	3.6	1.1	**3.9**

Row 1:

16.67 west full length
23.33 north (one corner already counted, the NW one)

7.33 east (the north corner is already counted)
22.67 south (both corners are already counted)
70.00

Row 2:

The stem wall height is reduced, the top of it is beneath the slab at a height of 16″; see the Ht column above.

Formwork takeoff

The first method of figuring the forms assumes the 2′ high forms would stop at the edge of the rear door and lower forms would be used at the doorway. See "Formwork Takeoff Method 1" for walls of two different wall heights. The ends of walls are shown separately because the labor for them (per square foot) is higher than the sides and needs to be shown separately on the estimate.

FORMWORK TAKEOFF Method 1										363.5
job: Garage		number					date			
No.	**Description**	**Det.**	**Each**	**Sides**	**L**	**Ht.**	**%**	**LF**	**SFSA**
1	Stem wall sideforms	A		2	70	2			280
2	Sideforms at door			2	3.33	1.33			8.9
3	Total sideforms								**289**
4	Add ends		2		0.67	2			**3**

Rows 1, 2:

See the concrete takeoff for the L and Ht.

Another method of figuring the forms at the rear door is to build the 2′ high forms all the way around and through the rear door opening. After this is completed, a "blockout" would be placed in the top 8″ of the forms for the width of the door. This would be done with wood framing, or blocking.

Depending on the size of the blockout, the unit of measure can either be "each" or SF. For a blockout as small as the one in the residential garage, only 2 sf, the best unit of measure is simply one each. See "Formwork Takeoff Method 2″ for this second method.

FORMWORK TAKEOFF Method 2									363.6
job: Garage		number					date		
No.	**Description**	**Det.**	**Each**	**Sides**	**L**	**Ht.**	**%**	**LF**	**SFSA**
1	Stem wall sideforms			2	73.33	2			**293**
2	(a) Blockout for back door, or	B	**1**						
3	(b) Blockout for back door				3.33	0.67			**2**
4	Add ends		2		0.67	2			**3**

Section 4 Retaining walls and waterstop

Waterstop

Because of expansion and contraction, concrete needs to have joints every 20′ or so. Footings don't usually have joints, but slabs and walls do. Water must be prevented from passing through these joints. In concrete wall construction, "waterstop" is often used for this purpose. Waterstop material is designed to stop water *by its shape, often bulbous, and is embedded in the concrete.*

Waterstop is usually purchased and installed by the concrete trade and found in Division 3 of the specifications. But since it is a waterproofing product, it could be found within Division 7 of the specifications and furnished by the waterproofing trade, but it is usually installed by form carpenters.

Retaining wall plans

See the online resources for diagrams 364.1, 364.2, 364.3, & 364.4

Plan interpretation

The retaining wall has three different heights. The width of the two B walls are 1′ wide. Wall A is 1′ wide at the top and 2′ wide at the bottom, and sloped on one side.

The walls are separated by waterstop. The plan note points to only two locations, but notes that waterstop is located at all wall ends.

A guardrail is on top of the walls.

Scope of the work

Include:

Form, reinforce, and pour the concrete walls.
Formwork design by trade contractor. Tolerance 1/4″ horizontal, 1/8″ vertical.
Provide shop drawings reviewed and stamped by a structural engineer.
All staging and formwork designed by concrete trade.
Include concrete material, labor, and pumping.
MOT (maintenance of traffic).
Employ an independent testing service to take four cylinders per 50 cy of concrete.
Test one cylinder at 7 days, one at 14 days, and one at 28 days, hold one in reserve.

Exclude:

Division 2 All earthwork by others. The earth embankment will be 8′ horizontally away from the wall, and slope away at a minimum 45-degree angle of repose.
Division 5 Misc. steel

Construction techniques

Form and pour the tallest wall A first, then the two B walls, and the two C walls last. Erect scaffolding as needed on each side of the wall as the forms are installed.

Provide 8′ high barricades on the street side of the scaffolding. Access the scaffolding from the ends.

Provide a maintenance of traffic (MOT) plan to the city for approval prior to the work. Provide warning lights, lane change signage, and barricades for approaching traffic within two blocks from the intersection on all four streets.

The holes for the guardrail posts are to be core bored later, and there is no guardrail work for the concrete trade.

Concrete takeoff

Three concrete quantities are counted, walls A, B1, and B2. All three walls are poured separately because they are different heights and they are separated by waterstop.

CONCRETE TAKEOFF											364.5
job:Retaining Wall				number				date			
No.	**Description**	**Det**	**Qty**	**L**	**W**	**SF**	**Ht**	**CF**	**net CY**	**%**	**act. CY**
1	Wall A	A		20.67	1.5		11	341	12.6	1.05	**13.3**
4	Walls B1	B		15.17	1		8	121	4.5	1.05	**4.7**
5	Walls B2	B		13.00	1		4	52	1.9	1.05	**2.0**

Row 1:

The length of the two walls, best counted by the centerline method, is 9.33′ + 11.33′ = 20.67′.

Formwork takeoff

The wall ends are shown separately because they take more labor per square foot than the sides and are "different". Rows 3, 5, and 7 could be combined on the estimate. The inclined surface on row 2 would be kept separate, as it would take the most labor per square foot.

Waterstop is counted by the linear foot.

FORMWORK TAKEOFF									364.6
job: Retaining Wall		number					date		
No.	**Description**	**Det.**	**Each**	**Sides**	**L**	**Ht.**	**%**	**LF**	**SFSA**
1	Form vertical side of wall A	A		1	18.67	11			**205**
2	Form incline	A		1	21.67	11			**238**
3	Form wall ends		2		1.5	11			**33**
4	Form wall B1 north/south	B		2	15.17	8			**243**
5	Form 1 end each wall			2	1	8			**16**
6	Form wall B2 north/south			2	13	4			**104**
7	Form 1 end each wall			2	1	4			**8**
8	Place waterstop		2			4	1.1	9	
9	Place waterstop		2			8	1.1	18	
10								**26**	

See the online resources for diagram 364.7

Section 5 Concrete walls

Material bin plans

See the online resources for diagrams 365.1 & 365.2

Plan interpretation

Concrete walls of various widths and heights are used to make a "materials bin" such as would be found at a concrete plant. Structures of this type are used to keep piles of gravel and stone separated.

The biggest wall is at the center, 14′ tall and 2′ wide. The floor plan does not show the "ends" of the walls; all are shown connected. However, the plans have no requirement for a continuous pour. "Means and methods" are up to the contractor (including which walls are built first), and the requirement on the plans is to provide a keyway if there is a cold joint.

There are ten total bins including the one at each end of the structure.

Scope of the work

Include:

Form, reinforce, and pour the concrete walls.
Formwork design by trade contractor. Provide shop drawings reviewed and stamped by a structural engineer. Tolerance 1/4″ horizontal, 1/8″ vertical.
All staging by concrete trade.
Provide alignment bracing for concrete walls.
Include concrete material and labor and pumping. Pump all concrete.
Employ an independent testing service to take four cylinders per 50 cy of concrete.
Test one cylinder at 7 days, one at 14 days, and one at 28 days, hold one in reserve.

Exclude:

Division 2 Sitework.
Concrete slab.

Construction techniques

Wall forms can be constructed with plywood and 2 × lumber, or "hand-set panels" (prefabricated sections), which can be rented.

Scaffolding will be erected as the forms are built.

Form and pour the tall center wall first, providing good alignment for the remainder. Then build the walls running E/W, and then the short 4′ end walls parallel to the center wall.

Concrete takeoff

If the lengths of the walls are carefully noted, without double counting of intersections, the concrete quantities are easy to obtain. Give each wall a description.

CONCRETE TAKEOFF												365.3
job: Material Bin					number				date			
No.	**Description**	**Det**	**Qty**	**L**	**W**	**SF**	**Ht**	**CF**	**net CY**	**%**	**act. CY**	
1	Pour 1	A, Plan		43.50	2		14	1218				
2	2' center wall											
3	Pour 2	A, Plan	2	26.00	0.667		8	277				
4	8" walls	A, Plan	4	12.00	1		9	432				
5	1' walls	A, Plan	2	20.00	1.5		11.3	680				
6	18" walls							1389	51.45	1.05	**54**	
7												
8	Pour 3	A, Plan	4	4.00	0.67		4	43	1.588	1.05	**2**	
9	Short end walls											

See Sketch 1 for an illustration of the three pours.

See the online resources for diagram 365.4

Formwork takeoff

How does your company want the information presented on the estimate? The following formwork breakdown shows all of the walls on rows 1, 7, and 9, separating the walls in each of three pours. It may be that the estimator figures all of these walls take the same labor per square foot to install and could combine them, showing all of the surface area of the three pours together on one row. *But*, if the estimate shows the labor for wall forms separately for each of the three pours, better and more timely information is given to the project manager and superintendent. They have a labor budget for each pour, instead of grouped together.

FORMWORK TAKEOFF									365.5
job: Material Bins		number					date		
No.	**Description**	**Det.**	**Each**	**Sides**	**L**	**Ht.**	**%**	**LF**	**SFSA**
	1st pour								
1	2' center wall	A, Plan		2	43.5	14			**1,218**
2	2nd pour								
3	8" walls interior side	A, Plan	2	1	24	8			384
4	8" walls outside		2	1	26	8			416
5	1' walls	A, Plan	4	2	12	9			864
6	18" walls	A, Plan	2	2	20	11.33			906
7									**2,570**
8	3rd pour								
9	4' high walls form sides		4	2	4	4			**128**
10	Install keyways first pour:								
11	keyway for 8' high walls		2			8		16	
12	keyway for 9' high walls		4			9		36	
13	keyway for 11'-4" walls		2			11.33		22.66	
14	S/T keyways first pour							**75**	
15	Install keyways second pour:								
16	keyway @ 8" wall for 4' walls		4			4		**16**	
17	Wall ends:								
18	Form wall ends first pour		2		2	14			56
19	Form wall ends second pour:								
20	At 8" walls		2	2	0.67	8			21
21	At 12" walls		4	1	1	9			36
22	At 18" walls		2	1	1.5	11.33			34
23	S/T form wall ends 2nd pour								**147**
24	Form ends 3rd pour 8" walls 4' high		4		0.67	4			**11**

7

CONCRETE COLUMNS

Section 1 Photos and drawing(s)

Photo 1 Concrete column with chamfer strip
Photo 2 Column rebar cage
Photo 3 Round column forms
Photo 4 Rough concrete column

Section 2 Foundation piers

Shed on piers foundation plan
Plan interpretation
Scope of the work
Construction techniques
Concrete takeoff
Formwork takeoff

Section 3 Columns, chamfer strips, and recesses

Schoolhouse second floor column and beam plans
Plan interpretation
Scope of the work
Construction techniques
Concrete takeoff
Formwork takeoff

Section 1 Photos and drawing(s)

Photo 1 Concrete Column With Chamfer Strip

A concrete beam will connect to this column.

Chamfered vertical edges.

Photo 2 Column Rebar Cage

Typical column, or beam, reinforcing. The 4-sided smaller sized rebar, called "stirrups", holds this rebar "cage" together. Notice all the tie wire used. Large heavy cages are installed by crane.

Photo 3 Round Column Forms

The relatively thin round column form by Sonotube, a trademarked name, is strong enough to contain the concrete because it is strong in tension. "Platform staging" is at the top of the column.

Cast in place concrete with dovetail slots

Block wall with brick ties

"Tubes" used as concrete column forms

Photo 4 Rough Concrete Column

Section 2 Foundation piers

Shed on piers foundation plan

See the online resources for diagrams 372.1 & 372.2

Plan interpretation

All of the columns are 12″ square. The four end columns are at an elevation of 9′-4″, the four inner columns are 8′-4″ high. A W12X72 beam is 12″ high and weighs 72 lbs per LF. Steel weld plates are embedded in the concrete. The bottom of all columns are 12″ below grade.

Scope of the work

Include:

Form, reinforce, and pour the concrete columns.
Formwork design by trade contractor.
Embed steel weld plates (furnished by others) in the concrete.
Include concrete pumping.
Employ an independent testing service to take four cylinders per 50 cy of concrete.
Test one cylinder at 7 days, one at 14 days, and one at 28 days, hold one in reserve.

Exclude:

Division 2 Sitework.
Division 5 Structural and Misc. Steel.

Construction techniques

There is no earthwork; the top of the concrete footing is visible and exposed because of the footing excavation.

The column sides will be formed with plywood. The weld plates will be a part of the formwork of the four inside columns. They will be hand placed into the top of fresh concrete. The steel angles at the outside columns will be installed after the pour by the steel trade.

Concrete takeoff

There are two different column heights; the cubic yardage of both is added together before they are sent to the estimate.

CONCRETE TAKEOFF											372.3
job: Shed On Piers					number				date		
No.	**Description**	**Det**	**Qty**	**L**	**W**	**SF**	**Ht**	**CF**	**net CY**	**%**	**act. CY**
1	East and west cols		4	1.00	1		10.33	41.3			
2	Interior cols		4	1.00	1		9.33	37.3			
								78.6	2.913	1.05	**3.1**

Formwork takeoff

The column forms are added together before being sent to the estimate.

FORMWORK TAKEOFF									372.4
job: Shed On Piers		number					date		
No.	**Description**	**Det.**	**Each**	**Sides**	**L**	**Ht.**	**%**	**LF**	**SFSA**
1	East/west cols		4	4	1	10.67			171
2	Interior cols		4	4	1	9.67			155
3									**325**
4	Set embeds top of col		4						

Section 3 Columns, chamfer strips, and recesses

Schoolhouse second floor column and beam plans

See the online resources for diagrams 373.1, 373.2, 373.3, 373.4, & 373.5

Plan interpretation

The perimeter columns shown in Sections A (north, west, and south sides) and C (east side) are 2′ square. The A columns are built to an elevation of 12′-4″ and the steel beams connect to them from the side. The steel beams on the east side of the structure sit on top of the 11′-4″ high C columns, which are 1′ shorter than the A columns because the tube steel shown in Section C is 1′ high.

The six interior columns are 1′ square. The drafter has chosen to point arrows at only three of these six columns. If designers were to "point" to all of the elements that a note covers, the plan would be too busy to read. It is only incumbent on a designer to provide enough information to allow proper interpretation, and the correct interpretation is that six columns are 1′ square. All six columns are to be exposed and are 15′-8″ high. They have chamfered edges and decorative recesses; see Sections B and B1.

Scope of the work

Include:

Form, reinforce, and pour the concrete columns.

Exclude:

Division 2 Sitework.
Division 5 Structural steel.

Formwork Scope:

Design by trade contractor.
Formwork Scope:
Include concrete material and labor and pumping.

Construction techniques

Provide scaffolding and build all of the forms at one time. At the B columns, provide continuous chamfer strips at the four vertical edges plus the horizontal chamfer at the top of the column. Use 1 × 8 lumber (3/4″ × 7-1/4″), cut to a 6″ width to make the recesses, and install them to the inside face of the forms.

Chamfer strips and recesses in concrete are usually easy to form. Chamfer strips are often simply triangular pieces of continuous "trim" that are attached to a concrete corner or edge (see Section D). They can be horizontal or vertical. A formed recess can be many shapes, but often it is just a board attached to the face of a form that concrete will be poured against, with the board providing a recess in the concrete. See Sections B, C, and D.

Concrete takeoff

CONCRETE TAKEOFF												373.6
job: Schoolhouse					number				date			
No.	**Description**	**Det**	**Qty**	**L**	**W**	**SF**	**Ht**	**CF**	**net CY**	**%**	**act. CY**	
1	Type A columns	A	8	2.00	2		12.33	395				
2	Type B and B1 columns	B	6	1.00	1		15.67	94				
3	Type C columns	C	4	2.00	2		11.33	181				
								670	24.81	1.05	**26.1**	

Formwork takeoff

The formwork square footage of columns A, B, and C, are collected on row 10.

FORMWORK TAKEOFF									373.7
job: Schoolhouse		number					date		
No.	**Description**	**Det.**	**Each**	**Sides**	**L**	**Ht.**	**%**	**LF**	**SFSA**
1	Plywood form A col's	A	8	4	2	12.33			**789**
2	Plywood form B col's	B	6	4	1	15.67			**376**
3	B chamfer strips col sides	D	6	4	15.67		1.1	414	
4	B chamfer col top horiz.		6	4	1		1.1	26	
5								**440**	
6	B add 3/4" x 6" recesses top		6	4		1.83	1.15	51	
7	B add 3/4" recesses bottom		6	4		9	1.15	248	
8								**299**	
9	Plywood form C col's	C	4	4	2	11.33			**363**

Note the 15% waste factor on rows 6 and 7. As explained at the beginning of this book, contractors use various waste factors, and a textbook could ignore them and simply present exact answers. However, they are shown in these case studies to illustrate not only where waste factors need to be evaluated, but also their range. The recesses in the concrete are only 6″ wide; when a much larger piece, such as a 4′ × 8′ sheet of plywood, is cut into much smaller sections, a lot more waste will occur than if whole sheets are used.

8

CONCRETE BEAMS

Section 1 Photos and drawing(s)

Photo 1 Poured Tie Beam With Some Forms Remaining

Concrete Tie Beam On Block

Beam spans an opening, post shoring still underneath

"Dove-tail" slots at the rectangular columns are for placement of brick ties

Form removal at several stages. Some walers and studs remain at the right.

Photo 2 Poured Concrete Beam Supported On Platform And Post Shores

A concrete beam is being supported by a beam bottom "platform". The sideforms have been removed. To the right, the platform has been removed but a vertical shoring brace is still in place.

Dovetail slot

Post shore

The platforfm is big enough for carpenters to stand on and is also the form for the bottom of the beam.

Photo 3 Formed Beam With Platform Beam Bottom

The platform consists of aluminum beams on vertical scaffolding frames. The horizontal plywood on top of the aluminum is 3/4" plyform, same as the vertical plywood, where some of the vertical studs have been removed.

Photo 4 Classic Beam Forms With Vertical Post Shores

Note the number of vertical post shores supporting the beam bottom. At the side of the beam form there are two rows of double walers with wall ties spaced about 16" o.c. horizontally.

Section 2 Tie beams and beam bottoms

Beam bottoms

Beam bottoms are especially important to the estimator for several reasons. Beams are heavy, and their bottoms are typically supported with post shoring. The labor for vertical support is captured in the labor factors for building beam bottoms. Also, the formwork for beam bottoms has the potential for staging a crew and material. If construction is to continue above the beam, the beam bottom may be widened and a workforce platform built.

In either case, post support or platform, the labor to build a beam bottom form is greater per square foot than beam sides. For this reason, beam bottoms are shown separately on the takeoff.

Although a beam bottom may range far wider than its actual surface area in contact with concrete, the unit of measure is still the square feet of surface area, although platform construction can be shown separately if the estimator chooses to.

Church plans

See the online resources for diagrams 382.1, 382.2, & 382.3

Plan interpretation

Tie beam A, 30″ deep, continues around the perimeter except at the four gable end walls. Section A1 occurs where the tie beam has an "open bottom" and is unsupported for the length of the stained glass openings. Section B is cut at the doorways, where there is an unsupported 16″ wide, 10′ long beam.

The walls up to the concrete beams are built of 12″ block; above the concrete beams there is wood construction.

Scope of the work

Include:

Form, reinforce, and pour the concrete beams.
Formwork and staging design by trade contractor. Provide shop drawings reviewed and stamped by a structural engineer. Tolerance 1/4″ horizontal, 1/8″ vertical.
Provide shoring for beam bottoms and alignment bracing for concrete walls.
Include concrete material and labor and pumping.
Employ an independent testing service to take four cylinders per 50 cy of concrete.
Test one cylinder at 7 days, one at 14 days, and one at 28 days, hold one in reserve.

Exclude:

Division 4 masonry.

Construction techniques

Provide scaffolding and forms to pour the lower door beams at Section B first, and then make a continuous pour of tie beams A and A1. Tie beam A will be poured 2′-8″ deep. The plans indicate a 2′-6″ deep beam, but since this doesn't match blockwork (13′-6″ is not blockwork, but 13′-4″ is), the contractor elects to pour 2″ of extra concrete instead of building the block wall to a 13′-6″ height. Above the windows, at Section A1, the beam is 2′-6″ high, and of course the concrete beam is built to this size at the window openings.

Tie beam A is formed on two sides; beams A1 and B have beam bottoms and three sides are formed. The formwork takeoff only counts the contact area, but more than any other type of pour, beams are "overformed"; the actual sideform may be double the contact area. This is because beam forms often lap down the sides of block or concrete walls. The sideforms used for Section A might be 4′ high; there is no need to cut the sheet of plywood, the bottom 18″ of it can be used to cinch it tight against the wall.

At the A1 window openings, use a horizontal 2 × 12 for a beam bottom form at the window head. Support the 2 × 12 with 4 × 4 posts on the block wall below. For the formwork at Section B, see Sketch 1.

Sometimes it is not clear how far past an opening the beam is to extend; for estimating this length can be assumed to be the same as the depth of the beam. For the project manager and superintendent, this assumption will not be made, as an exact bearing length will have to be shown on the rebar shop drawings prior to construction.

See the online resources for diagrams 382.4 & 382.5

Concrete takeoff

CONCRETE TAKEOFF											382.6
job: Church					number				date		
No.	**Description**	**Det**	**Qty**	**L**	**W**	**SF**	**Ht**	**CF**	**net CY**	**%**	**act. CY**
1	Continuous tie beam	A		137.33	1		2.67	367			
2	Continuous tie beam	A1		204	1		2.5	510			
3	Beam above doors	B	2	13.33	1.33		1.67	59			
								936	34.66	1.05	**36.4**

After the decision is made about having two depths for beams A and A1, the accuracy of the takeoff rests with getting the length of the beam correct. The combined length of beams A and A1 is illustrated in Sketch 3 and totaled below:

Total A plus A1	$(34.67 \times 4) + (2 \times 35.67) + (2 \times 65.67) = 341.33$
Length of window tie beams at A1	$(20 \times 8) + (22 \times 2) = \underline{204}$
Length of tie beams at A	137.33

See the online resources for diagram 382.7

Formwork takeoff

Show beam bottoms separate.

FORMWORK TAKEOFF									382.8
job: Church		number		Add W ↓			date		
No.	**Description**	**Det.**	**Each**	**Sides**	**L**	**Ht./W**	**%**	**LF**	**SFSA**
1	Form sides tie beam A, A1	A		2	341.33	2.5			**1707**
2	Form beam bottom A1	A1	2		22	1			44
			8		20	1			160
									204
3	Form sides door beams	B	2	2	13.33	1.67			**89**
4	Form beam bottoms doors	B	2		10	1.33			**27**

Section 3 Concrete beams

Schoolhouse and second floor beam plans

See the online resources for diagrams 383.1, 383.2, 383.3, & 383.4

Plan interpretation

Section A applies to three sides of the structure; Section C applies to one side only. All of the beams are continuous and run across the top of columns. The columns in Sections B and B1 are first poured to 13′-6″ and then continue to a height of 19′-4″ by being formed and poured on top of the beam.

Scope of the work

Include:

Form, reinforce, and pour the concrete second floor beams.
Formwork and staging design by trade contractor. Provide shop drawings reviewed and sealed by a structural engineer.
Tolerance 1/4″ horizontal, 1/8″ vertical.
Provide shoring for beam bottoms and alignment bracing for concrete walls.
Include concrete material and labor and pumping.
Employ an independent testing service to take four cylinders per 50 cy of concrete.
Test one cylinder at 7 days, one at 14 days, and one at 28 days, hold one in reserve.

Exclude:

Columns.

Construction techniques

Provide scaffolding and pour all of the beams at the same time. For the 2′ wide beam bottoms, use form plywood placed on top of adjustable scaffolding for support. For the 1′ wide beam bottoms, support with 4 × 4 wood posts or adjustable steel post shoring.

Concrete takeoff

Only one quantity is needed, the cubic yardage of all the beams.

CONCRETE TAKEOFF												383.5
job: Schoolhouse					number				date			
No.	**Description**	**Det**	**Qty**	**L**	**W**	**SF**	**Ht**	**CF**	**net CY**	**%**	**act. CY**	
1	Pour beam A	A		130.00	2		2	520				
2												
3	Pour beam B, B1			182.00	1		1.5	273				
4												
5	Pour beam C			46.00	2		2.5	230				
6								1023	38	1.05	**39.8**	

Formwork takeoff

With the bottoms and ends shown separately, the estimator can combine them in any way on the estimate.

FORMWORK TAKEOFF										383.6
job: Schoolhouse		number				date				
No.	**Description**	**Det.**	**Each**	**Sides**	**L**	**Ht.**	**%**	**LF**	**SFSA**	
1	Beam A inside				126	2			252	
2	Beam A outside				138	2			276	
3	Beam B, B1				211	1.5			316.5	
4	Beam C inside				44	2.5			110	

FORMWORK TAKEOFF									383.6
job: Schoolhouse		number					date		
No.	**Description**	**Det.**	**Each**	**Sides**	**L**	**Ht.**	**%**	**LF**	**SFSA**
5	Beam C outside				50	2.5			125
6									**1080**
7									
8	Beam bottoms :					W			
9	A				114	2			228
10	B, B1				176	1			176
11	C				43	2			86
									490

Section 4 Rake beams

When counting the concrete of rake beams, which are *beams on an incline*, they usually need to be "sized upwards" from the plan details when supported on block construction. The up and down pattern of blockwork creates the need for extra yardage in the concrete pour beyond the minimum depth shown on plan details.

The plywood (with 2 × bracing) sideforms of a rake beam (or a horizontal tie beam) extends down the face of the block below, beyond the contact area, usually at least 8″ but sometimes a foot or two.

Despite knowing that a 4′ piece of plywood (for a 16″ or 24″ or 32″ beam) is going to be used for a sideform, most estimators still use "contact area" as a unit of measure.

The plans here are for the horizontal tie beam at 24′ and the rake beam above and excludes a tie beam at 16′.

Church plans

See the online resources for diagrams 384.1 & 384.2

Plan interpretation

There are two gable end walls 36′ long and two more are 42′ long. These walls are all 16″ wide. They are 16′ high at their ends and rise at a 45-degree angle. There is a horizontal beam with a bottom at 24′, plus a 2′ deep rake beam that follows a 45-degree angle at the top of the gable walls. The configuration is shown in Sketch 1.

Scope of the work

Include:

Form, reinforce, and pour the horizontal beam at 24′ and the rake beams.

Construction techniques

The block walls and tie beam below 16′ are already completed.

Sequencing:

Construct the gable block walls to a height of 24′, the elevation of the horizontal tie beam.
Form and pour the rake beams from 16′to 24′, first pour.
Form and pour the horizontal rake beams at 24′, second pour.
Lay block from elevation 25′-4″ to the top.
Form and pour the rake beams from 25′-4″ to the ridge, third pour.
See Sketch 1.
Note the first and second pours could possibly be combined but are separate pours here.

See the online resources for diagram 384.3

Concrete takeoff

CONCRETE TAKEOFF											384.4
job: Church					number				date		
No.	**Description**	**Det**	**Qty**	**L**	**W**	**SF**	**Ht**	**CF**	**net CY**	**%**	**act. CY**
1	1st pour rake beams :	B									
2	36lf gable walls		4	11.00	1.33		2.5	146			
3	42lf gable walls		4	11.00	1.33		2.5	146			
4			8 legs*					293	10.84	1.05	**11.38**
5	2nd pour, horiz. beams :										
6	36lf gable walls		2	19.00	1.33		1.33	67			
7	42lf gable walls		2	25.00	1.33		1.33	88			
8			4 legs*					156	5.77	1.05	**6.05**
9	3rd pour rake beams :										
10	36lf gable walls		4	10.00	1.33		2.5	133			
11	42lf gable walls		4	14.50	1.33		2.5	193			
12			8 legs*					326	12.07	1.05	**12.67**
*	1 Leg is 1 side of a gable, one straight run of concrete.										

Rows 2, 3:

There are eight "legs" all the same length, from elevation 16′ to 24′. The rake beam lengths are best measured from elevation sheets; here, use Sketch 1 and scale measure, or figure exactly by geometry.

Rows 6, 7:

There are two horizontal beams on the 36′ gable walls and two more on the 42′ long walls.

Rows 10, 11:

Scale measure. There are eight "legs".

Formwork takeoff

FORMWORK TAKEOFF									384.5
job: Church		number					date		
No.	**Description**	**Det.**	**Each**	**Sides**	**L**	**Ht.**	**%**	**LF**	**SFSA**
1	1st pour rake beams :								
2	36lf gable walls		4	2	13	2			208
3	42lf gable walls		4	2	13	2			208
4									**416**
5	2nd pour horiz beams :								
6	36lf gable walls		2	2	20	1.33			106
7	42lf gable walls		2	2	26	1.33			138
8									**245**
9	3rd pour rake beams :								
10	36lf gable walls		4	2	12.5	2			200
11	42lf gable walls		4	2	17	2			272
12									**472**

Rows 2, 3:

The lengths of the forms are a bit longer than the concrete length. When beams don't have perpendicular ends, and are pointed, such as these, an exception is made to figuring net square feet of contact area. The form is figured "tip to tip".

9

ELEVATED CONCRETE SLABS

Section 1 Photos and drawing(s)

Photo 1 - Upper Slab Pour

Thickened Slab What is this wood doing here? Flexible conduit

Leveling "screeds" are being pulled by the men bending over.

Photo 2 Elevated Slab Forms

A beam is at the top of the photo. A concrete floor will be poured on top of the plywood, which is seen toward the interior supported by wood 4 x 4's.

Aluminum beams support the 4 x 4's, which in turn are supported by vertical post jacks. Itakes a lot of forms and labor to build a platform strong enough to support the weight of an elevated slab pour.

Photo 3 A Busy Slab Ready To Pour

Two "mats" of rebar plus electrical distribution

Photo 4 Telegraphed Forms On Bottom Of Finished Slab

This slab was cast in place, the supporting forms high in the air

Section 2 Slab on deck

Schoolhouse second floor slab and joist plan

See the online resources for diagrams 392.1, 392.2, 392.3, & 392.4

Plan interpretation

Top-bearing bar joists are supported by concrete beams. A steel deck is on top of the bar joists and supports a 4″ slab. At the edge of the pour is a steel angle. The center portion of the floor is open.

Scope of the work

Include:

Form, reinforce, and pour the slab on deck.
Include concrete material and labor and pumping.
Employ an independent testing service to take four cylinders per 50 cy of concrete.
Test one cylinder at 7 days, one at 14 days, and one at 28 days, hold one in reserve.

Exclude:

Division 5 structural steel.

Construction techniques

Provide scaffolding and formwork for a continuous slab pour. For the trade contractor, this is an easy pour to prepare for – the edgeforms are made of steel and by others, and there is no earthwork!

The steel angles are bolted before the pour, there are no embeds and no formwork except for screeds that the concrete finisher uses.

Concrete takeoff

There is an important "conversion factor" that is used for the depth of the concrete. The steel deck will not be flat, and corrugations will vary by manufacturer. *Each manufacturer provides product data sheets that show how to calculate concrete when using their deck.*

The primary plan dimensions are to the centerline of columns; as is often the case, a bit of arithmetic is required to determine the length of the slab sides.

Assume the slab will average 4.25″ deep (determined by studying manufacturer's data). The areas of the slab are:

40.67 × 15.18 = 617 sf north slab
13.18 × 16.33 = 215 sf center
40.67 × 15.18 = 617 sf south slab
1,449 sf

There is no reason for the above arithmetic to show up on the estimator's takeoff – only the summation of 1,449 sf. With computers, the slab outline can be traced to get the overall square footage. When figuring by hand, the estimator's best way to leave behind something that can be checked later is to handwrite on the plans.

CONCRETE TAKEOFF											392.5
job: Schoolhouse					number				date		
No.	**Description**	**Det**	**Qty**	**L**	**W**	**SF**	**Ht**	**CF**	**net CY**	**%**	**act. CY**
1	2nd fl slab place and finish					1449					
2	Concrete material						0.354	513	19.00	1.05	**19.9**

Section 3 Second floor porch and stairs

Like beam bottoms, the bottom of slabs, especially stairs, involve a great deal of labor. Sometimes a formwork platform can extend well beyond the footprint of the slab, as sometimes occurs with beams. Nevertheless, although the contact area might be quite different from the work area, the square feet of contact area is what most estimators use when determining labor and material on the estimate. So, that is the number needed on the takeoff.

The width and length of stairs are shown "in plan". Note that the plan length is a horizontal distance and the stairs are at an incline. Measure the length of stairs from an elevation sheet or a section *using the sloped distance, the incline, for the length*. For the average thickness of the slab, measure perpendicularly to the incline instead of vertically.

Condos porch and stair plan

See the online resources for diagrams 393.1, 393.2, 393.3, 393.4, & 393.5

Plan interpretation

This is a cast-in-place elevated concrete slab with a slope of 2″ from north to south. There is a beam underneath on three sides that is part of the pour. On the north side, the porch slab will sit on top of the exterior block wall at 9′-4″.

There are two identical sets of concrete stairs with 6″ thick landings at an elevation of 7′-6″. The treads will have metal nosings. The minimum thickness of the stairs is 6″, the risers are 7″ high, and the treads are 11″ wide. At the bottom, the stairs are 8″ below grade; see Section B.

Scope of the work

Include:

Form, reinforce, and pour the porch slab and two sets of stairs.
All staging and formwork design by concrete trade.
Provide shoring for beam bottoms and support for the slab.
Include concrete material and labor and pumping.

Exclude:

Division 4 masonry.

Construction techniques

There are many ways to support an elevated slab. One is shown below in Sketch 1. Protect the first floor slab with 3/4″ plywood first. Then erect scaffolding to the elevation of the beam bottoms. The scaffolding, like a table, provides the lateral stability that vertical posts do not. On top of this, the tabletop is made with 3/4″ regular plywood, say CDX, lying flat. Two by ten joists sit on top of this, on which a layer of 3/4″ plyform becomes the surface for the bottom of the main porch slab.

We're thankful for the shorthand the industry uses here, which is figuring the labor and material cost of a slab form from just the contact area. Formwork such as this is too complicated for the estimator to stop and figure out at the estimating stage.

The stairs and landings are a second pour.

See the online resources for diagram 393.6

Concrete takeoff

CONCRETE TAKEOFF											393.7
job: Condos					number				date		
No.	**Description**	**Det**	**Qty**	**L**	**W**	**SF**	**Ht**	**CF**	**net CY**	**%**	**act. CY**
1	Pour 1										
2	Main slab			30.67	6.67	205	0.58	119			
3	add doorways	A	1	4	3.33	13	0.67	9			
4	add doorways		1	8.67	3.33	29	0.67	19			
5	drop beam lower	D		40	0.67		0.58	16			
6	drop beam upper	D		40	0.833		0.25	8			
7	Place and finish										
8	Concrete material							172	6.4	1.05	**6.7**
9	Pour 2										
10	concrete steps upper	D	2	4	3.33	13	0.75	20			
11	lower steps		2	16	3.33	53	0.75	80			
12	landings	plan	2	7.33	3.33	24	0.5	24			
13	Place and finish					91					
14	Concrete material							124	4.60	1.05	**4.8**
15	Curing compound		**2**	**gal**							

Row 2: The slab is 8″ thick against the building and 6″ thick at the south edge for an average thickness of 7″ (Ht column).

Rows 3, 4: The separate rectangles of slabs are usually added together and a total square footage would be used on row 2. However, this is for where slab thickness remains the same across the slab. This is an example of various slab thicknesses; although the door areas here are small enough for the difference to be negligible, on a larger slab it would matter.

Rows 5, 6: From Section D the beam is centered 1′ from the edge of the slab; use the plan dimensions to figure the length of the beam. Use the centerline method to determine 40 LF. The beam has two widths, so it is divided into a lower beam 8″ wide and an upper beam 10″ wide.

Row 10: The 4′ length from the landing up to and across to the beam of rows 5 and 6 is scaled from Section D. The width of 3′-4″ is taken from the plan.

Row 11: To scale the length, there isn't an elevation or sketch to measure the lower run of stairs. However, the horizontal distance is shown on the plan to be 12′-10″. The landing elevations are 7′-6″ (see plan) and the bottom of both stairs is at minus 15″ (see Section B). This vertical distance is 8′-9″; dividing by the riser height of 7″, there are 15 risers and 14 treads (there is always one more riser than treads). The treads are 11″ wide (see Section D) for a total horizontal distance of 12′-10″.

The hypotenuse of this triangle (8′-9″ and 12′-10″) is 16′, which is used for the length of the stairs.
The average depth (Ht column) of the stairs is 9″, as measured from Section D.

Formwork takeoff

This is a complicated little formwork job. Defining it with several quantities will help determine labor hours accurately. Some formwork of this slab is figured by the contact area, some by linear feet. The slab bottom, the biggest job, is calculated by contact area. However, the edges of the slab and the sides of the beam are less than 12″ and they are shown in linear feet. Common 2 × lumber can be used to form them.

The stairs are measured by the square feet of their bottoms and the linear feet of their sides. The stair slab lengths are measured on the incline, not horizontally!

FORMWORK TAKEOFF									393.8
job: Condos		number					date		
No.	**Description**	**Det.**	**Each**	**Sides**	**L**	**Ht./W**	**%**	**LF**	**SFSA**
1	Slab bottom (net)				30.67	6.67			205
2	add doorway		1		4	3.33			13
3	add doorway		1		8.67	3.33			29
4	less area of beam				40	0.83			-33
5	Net area of slab bottom								**214**
6	Beam sides 10" high			2	40		1.1	**88**	
7	Beam bottom 8" wide				40		1.1	**44**	
8	Slab edge say 2 x 12 cut				44		1.1	**48**	
9	All risers 2 x 8 edgeforms		2	18	4		1.1	**158**	
10	Embed metal nosings		**36**						
11	Stair bottoms		2		20	3.33			133
12	Landing bottoms		2		7.33	3.33			49
									182
13	Stair sides say 12" edge		2	2	20		1.1	**88**	
14	Landing sides 6"		2		14.67		1.1	**32**	

Row 5: This quantity sent to the estimate.

Rows 6, 7: The length of 40′ has already been determined; see rows 5 and 6 of the concrete takeoff.

Rows 9, 10: From row 11 of the concrete takeoff, it was determined that there were 15 risers in the first flight of the stairs. From Section D, there are 3 more for a total of 18 risers. The top riser in Section D is the edge of the slab; see row 8.

There are two separate stairs, so the number of nosings is 2 × 18 risers = 36 total.

PART 4

Masonry

1

PRODUCTS AND METRICS

Section 1 Introduction

The modular units of block and brick are the subjects of this text. Stone, marble, and granite are not covered. A course of block or brick is one horizontal row. Two block or brick courses laid side by side have two "wythes". Block and brick have joints of mortar that bind the courses together. In years past, brick walls several courses thick were used for bearing walls, but now brick is mostly used as a veneer.

Concrete masonry units, or CMUs, combined with concrete and reinforcing steel (rebar), are the choice for many buildings and load-bearing situations. When the voids of block, which are called cells, are filled in horizontal coursing, the resulting beams are called bond beams. The blocks that are used for bond beam coursing are called lintel blocks in this text, which allow the horizontal passage or rebar. The tops of single-story block walls are made with bond beams, and they occur every 8 to 12 feet in height in multi-story block construction.

Bond beams have continuous rebar running their length. Another piece of block reinforcement is a ladder-style type made with thick wire and installed in horizontal block mortar joints. This reinforcement is often placed 16″ o.c. (on center) vertically, or every other course in 8″ block masonry. The vertical cells of block are also often filled with rebar and concrete and extend for the full height of the wall. In building construction, vertical cells often occur every 4′ o.c. horizontally, but this can vary. The spacing of concrete and rebar within block construction, both horizontally and vertically, is designed by structural engineers.

Masonry, having a high density, retains its temperature for many hours. Its energy benefit is that a lot of heat energy is required to change the temperature, giving masonry a high "thermal mass". When exterior block walls are heated during the day by the sun, some of the heat is retained through a cold night. However, the insulating R-value of masonry materials is not high by itself, even when empty cells are filled with insulation. Block wall construction is often combined with wood furring or frame walls that can contain additional insulation.

Laying concrete blocks involves at least three trades – steel reinforcement, concrete (pouring of block cells), and block laying. These chapters cover the last two and reinforcement is neglected, because a supplier will often quote rebar, saving the estimator a lot of trouble. Counting rebar accurately is an arduous task requiring a good knowledge of techniques and lap and waste factors. The accessory items of mortar mix and sand are included in this text, as well as precast beams – known here as standard "precast U lintels". There is no coverage of specialty work such as fireplaces, stonework, or horizontal pavers. Equipment, scaffolding, and material handling are discussed but not included in these takeoffs.

Section 2 Masonry contractors and products

Contractors who count masonry

Consider three kinds of contractors who count blocks and bricks – a masonry subcontractor, a general contractor, and a construction manager. A masonry subcontractor purchases material, lays block and brick in-house, and needs a takeoff and estimate reflecting this detail. The estimator for a masonry sub only takes off masonry, and provides quotes to general contractors and construction managers (GCs and CMs). These prime contractors, preparing a bid for the entire project, may take a quote from a masonry contractor and be done with Division 4. However, there are times when they must take off block and brick.

The GC, in an effort to be more competitive, will sometimes buy all of the masonry products in-house and hire a subcontractor for labor only (perhaps a small group of a few masons that work for a set unit price per block). In this case, the GC will have to complete a takeoff for the material and be able to provide the management for purchasing and handling masonry products if a subcontractor's markup is going to be saved.

The CM's involvement in Division 4 estimating stems from having to provide several iterations of budgeting. Perhaps the final price for this work is delivered by subcontractors, but the CM is going to be responsible during the design phase for budgeting blocks and bricks from incomplete plans, perhaps multiple times. However, the CM may only complete a square foot count.

It is the estimators for these contractors who count masonry products, and they do so in various levels of detail.

Mortar and types of blocks

Type N is a mortar of moderate strength in common usage for exterior above-grade locations and interior bearing walls. Type S is for exterior use at grade and below. Other types are "M" (the strongest) and "O". Mortar types are named in the project specifications, and mortar types are not further discussed in this textbook.

Regular concrete blocks (yes, they are called regulars!) and concrete lintel blocks are the two most common block types and the main focus of this textbook. Lintel blocks are made for the purpose of containing concrete and rebar, and after they are filled they are called bond beams.

Slightly different wall thicknesses of standard block are made all around the United States. There is no uniform interior shape of a filled concrete block cell, although they have the same exterior dimensions.

The masonry products included in these takeoffs are:

1 Regular block.
2 Lintel block.
3 Header block.
4 Brick.
5 Straight precast "U" lintels.
6 Precast door headers.
7 Precast window sills.
8 Concrete (inside block cells).
9 Masonry sand.
10 Mortar mix.
11 Steel shelf angles.
12 Truss anchors.
13 Rigid insulation.

Rigid insulation

Often it is in the mason's scope to install rigid insulation against block walls. Insulation is a Division 7 item, but its location sometimes makes it awkward for insulators to install it. Sometimes it is sandwiched between masonry materials. If the mason is onsite building block walls, readying for brick veneer, it doesn't make sense to interrupt this work and bring an insulator jobsite for a simple installation. Sometimes the purchase of rigid insulation remains in Division 7, but the insulator furnishes it to the mason for installation.

Section 3 Block openings

Precast U lintels

The typical precast unit is made to span a door, window, or other opening and support the wall above. They are 8″ wide and 8″ high, and *typically* bear 8″ at each end, making them 16″ longer than the opening. They are called *8″ straight U lintels.* In this text, precast lintel lengths can be determined by simply adding 16″ to the opening width. However, a precast plant may not manufacture a U lintel the exact length of an "opening plus 16″. In real practice, estimators typically get a list of the local product lengths and use it when doing takeoffs.

There are two units of measure for precast U lintels, one is "each" and the other is "LF". Preferably, the lengths of each precast are known and can be listed on the takeoff as descriptions, such as "4′ precast straight U lintels", and a quantity given for them, such as "10 each". If the plans are not clear enough to determine specific lengths, the count of precasts is made by the linear foot.

As a unit of measure, if the takeoff leaves precasts counted in linear feet, someone later will have to figure out the exact lengths. However, this is better than the estimator guessing at some of the precast lengths, because a project manager or superintendent may later use the lengths in a purchase order without verifying and get the wrong material on the job. *Either take them off exact or leave them to a round number in linear feet. Imprecise lengths should not be on the takeoff.*

The unit of measure for precasts is sometimes shown in both linear feet and each; both get counted because the estimate uses one to price the labor (LF) and one to count the material (each). Each length of a precast has a specific material price, so counting each specific length is the best way to quantify the material. However, if the LF of precasts is used to price material, a blended (average) price for precasts will have to be used on the estimate.

If all of this is confusing, that's because it is. That's one reason why masons, when figuring block labor, sometimes keep things simple and figure block counts straight through openings (and forget about precasts and openings, at least as far as block labor goes), and price a wall as if it were solid block.

Precast door headers

A door header is a particular type of precast U lintel.

Doors have all kinds of opening heights. Consider a residential door 6′-8″ high with a 2″ frame head, for a total height of 7′, which varies 2″ from blockwork. Six foot eight inches is blockwork, not seven feet. How can a masonry opening be made 6′-10″ high? The use of a "precast door header" solves the arithmetic here. Precast plants make precast U lintels with 2″ legs at either end. The precast lintel is 8″ high at each end but only 6″ high across the opening. Precast door headers come in handy above lots of doors.

Buildings can have hundreds of 7′ hollow metal doors in masonry openings, which is the height of most commercial doors. The heads for these frames are usually 4″; the door plus frame height equals 7′-4″, which is blockwork.

Steel angles

Steel angles (sometimes called shelf angles) are the typical method of carrying brick across an opening such as a door or window. A typical size would be 4″ × 4″ × 3/8″, and it would extend 6″ or so past the opening (i.e. angles bear on brick support for 6″ at both ends of the angle). It is the task of the structural engineer to specify the size and thicknesses of steel angles.

Section 4 Counting block

Counting wall lengths

Blocks and bricks are sold by the piece and "each" is the unit of measure for block *material* on the takeoff. This count may be converted to the whole number of pallets required, but "each" is the primary unit of measure.

The key to correctly counting masonry quantities is to simply get the length of the wall correct. A wall type means a like kind wall, such as a 10′ high wall with two courses of lintel block at the top. It might be for 10′ long or 1,000 LF.

With walls all over the place at different bottom elevations and various heights, with twists and turns in their layout and openings for this and that both high and low, determining the lengths of walls can be quite confusing. Plan reading, such as the interpretation of where one wall section applies and another one begins, takes some experience. Finding the wall height is usually easy (although this can quickly get complicated with multiple wall heights and stories). The hard part is defining the types of walls and getting their lengths correct.

Blocks are modular

Blocks are modular, meaning their dimensions allow them to be stacked and turn corners without having odd dimensions left over. Including mortar joints, blocks are 8″ high and 16″ long. The mortar is 3/8″ thick, and the actual size of these blocks are 7-5/8″ × 7-5/8″ × 15-5/8″ long.

Wall *heights* of 2′, 4′, 6′, etc. all match blockwork. Blockwork means that regular block sizes can be used to build to the required height. A wall 8′ high has 12 courses of regular block and 8′ matches blockwork. With the use of 4″ high block, a readily available size, block walls can be built in 4″ incremental heights.

Multiples of 16″ block *lengths* create even-numbered wall lengths of 4′, 8′, 12′; all match blockwork. Adding or subtracting 8″ or 16″ to these lengths are all blockwork (with the use of regular and half blocks). With the use of halves (8″ L) and three-quarter block (12″ L), horizontal dimensions 4″ apart can be achieved.

The nominal face area of a block is 128 square inches (16 × 8). Since one square foot equals 144 square inches, there are 1.125 blocks per square foot of wall surface (remember, one block is *less* than one square foot). Blocks are purchased in "cubes" of 90 blocks on a 4′ square wood pallet stacked five blocks high.

The conversion factor of 1.125 used in this book, to convert SF into each, is for "two cell blocks" with a modular size of 8″ W × 8″ H × 16″ L. The same square foot methodology used in these takeoffs can be used for any size of stone or brick, only the conversion factor changes. This conversion factor of *1.125* is the all-important number in counting block.

Mortar mix is purchased by the bag. Assume that each 80 lb. bag is enough to lay 55 regular blocks. Masonry sand is purchased by the cubic yard (old way) or now more usually by the ton. Assume that one ton of sand is enough to lay 290 regular block. Often, sand is taken off by the cubic yard and converted to tons. A cubic yard of sand weighs (in Florida) 1.25 tons.

Some block metrics are:

1 sf = 1.125 blk (8″ × 16″ face)
1 bag mortar mix will lay 55 regular block

1 cubic yard sand = 1.25 tons
1 ton sand needed to lay 290 regular block
1 sf block or precasts requires 0.35 cf of concrete fill
Typical block factor on breakage and waste is 5%

Block takeoffs

For the estimator, each division of work has its own characteristics. Masonry takeoffs are not burdened, like concrete, with accessory trades like formwork and earthwork that aren't on the plans. The estimator does not require a thorough knowledge of block construction techniques to count blocks. However, taking off block is a geometric exercise that requires an ability to see conceptually. And taking off masonry accessories (precasts, steel lintels, flashing, etc.) takes a good knowledge of techniques to interpret locations. Brick takeoffs can quickly get complicated by the use of multiple brick patterns.

Before the takeoff begins, the "bid form" should be reviewed to see if there are alternates or unit prices. If so, they require separate takeoffs and estimates. Masonry takeoffs then proceed with these steps:

1 Review the structural plans, note block and brick at the lowest elevation.
2 Scan the architectural, civil, and MEP plans for more blockwork.
3 Review the specifications.
4 The takeoff begins, the estimator using the S plans first, determining SF of walls beginning at the bottom of the structure. Use a block takeoff sheet.
5 Deduct for true outs (clear openings).
6 Count the precasts.
7 Takeoff the horizontal lintel blocks, then the vertical lintel blocks.
8 Determine concrete-filled cells last. Use a concrete takeoff sheet.

The first task is to identify a few wall types before the counting begins. Like kind walls have the same height, but not all walls with the same height are like kind (because they may be different in other ways). Like kind walls share a sectional view on the plans. If a wall has different components than others that might cause it to have a different labor price, show it separately with a name (description) and quantity. Separate block and brick not only by wall height and type but by area (floors or elevation level).

While the ending unit of measure for blocks is "each", *it is easiest to start by counting square feet and make the conversion to "each" a last step*. The first step on the takeoff sheet, after identifying a few wall types, is to describe one of them on row 1 of the takeoff. Then, the length is multiplied times the height and the gross square footage of the wall is shown. Note that the masonry takeoff has started with the *total square feet of wall area for one wall type*. Step 1, length times height for wall type 1.

Doors and *windows* are called "outs" on the takeoff. The quantities of doors and windows are taken off from the floor plan, and their size from door and window schedules. Outs of less than 10 sf are omitted. The square foot quantity of outs are subtracted from the gross wall area.

If the wall on this first row of the takeoff consists only of regular block, and there are no outs, no precasts, no other products, then the counting is done and the SF of wall area is converted to "each" by the conversion factor of 1.125 (see metrics). The quantity of regular block has been determined in this wall that is made totally of solid regular blocks.

Taking this example further, if the wall has one clear opening in it, the opening is shown on row 2 of the takeoff and is subtracted from the total wall area on row 1. The counting is done and the remaining SF of the wall is converted to the number of regular block.

Using this same wall, if it also has, beyond one clear opening, two courses of lintel blocks at the top of the wall, then row 3 of the takeoff sheet is used to show the area of the lintel blocks, which is subtracted from the beginning SF already lessened by the clear opening. SF is then converted to regular block.

The process continues with other masonry components such as precasts. The count of the regular block is the most important part of the block takeoff but it is never directly measured. Note that if an out or product is missed and not deducted from the total sf, at least the regular block count will be high.

This text does not include takeoffs for half or three quarter block.

This way of taking off block is the "total area method". From a beginning square footage of the total wall area, *all products except regular blocks are treated like an "out"* and subtracted. What remains is regular block. Precasts and lintel blocks are outs just like clear openings. Clear openings are the only "true outs", but the total area method treats everything as a "negative out" to the square footage and the positive number left consists of regular blocks.

The following format is used in this text to count blocks. Work from left to right. Precasts and other masonry products can be added next to the lintel blocks.

MASONRY TAKEOFF											414.1
job:			number						date		
No.	**Description**	**Det.**	**Qty.**	**L**	**Ht.**	**SF**	**Reg blk SF**	**Lintel blk SF**	**Precast LF**		

Block *labor* is also priced by "each" but sometimes a mason's count for *labor* is different from the *material* count made by the GC or CM who hires him or her. The difference is that masons (when they are figuring labor) often count straight through door and window openings, not accounting for outs (avoiding the time and clutter), figuring that the extra labor to lay out and build an opening is the same as laying blocks there. The mason's count figures the wall solid with blocks.

So estimators may have to figure the block two ways, one for the labor and one for the material. Phew!

Section 5 Counting concrete in blocks

Concrete-filled cells and precasts

Concrete is typically poured into horizontal and vertical filled cells and precasts; that's three different products (regular block, lintel block, and precasts). Concrete should be taken off last, after all of the regular blocks, lintel blocks and precasts are counted. The unit of measure for *concrete-filled cells* is the cubic yard, but concrete-filled cells, like blocks, is counted by the square foot first. To count the cubic yardage of concrete-filled cells, the surface area of a wall that contains concrete is figured. This square footage is then multiplied times a factor to obtain cubic yardage.

Continuous horizontal filled cells are usually poured into courses of lintel blocks which become a "bond beam". Typically, horizontal filled cells are at the top of the wall (sometimes multiple courses). If the wall is over 8′ or 10′ high, horizontal bond beams may also be located at mid heights. Horizontal poured cells are also found underneath windows and other openings, as seen by sectional views on the architectural drawings, perhaps not shown on the structural plans.

The concrete mix

The concrete mix for filled cells is quite different from that used for slabs because the concrete plant uses smaller aggregate. The aggregate used in concrete to fill block cells is a "small pea gravel"-sized aggregate, small enough to be pushed through a 2″ hose.

Concrete conversion factor

The interior configuration of a block varies depending on manufacturers. One region of the country may favor blocks of a particular interior contour that are different from another region. What this means for the estimator is that it may take a different quantity of concrete to fill 100 blocks in different states. A universal "factor" cannot be given in this text to fit all circumstances. To make this topic more confusing for the estimator, filling concrete in a single course of blocks 100′ long and 8″ high (67 sf of block surface area), does not take, exactly, the same amount of concrete to fill a vertical cell 8″ wide and 100′ high (67 sf of block surface area).

Then there are precast U lintels to be filled, located above doors and windows, with the open part of their "U" shape always filled with concrete. The amount of concrete that it takes to fill precasts is slightly different than filling either horizontal or vertical block cells. Phew!

The author has averaged three different "rules of thumb" to arrive at an easy conversion factor for use in this text. This factor is used to determine the amount of concrete it takes to fill the voids of lintel block, whether horizontal or vertical, straight precast U lintels, and precast door headers.

The factor is 0.35 cubic feet of concrete per sf of wall area. The wall area includes horizontal and vertical lintel blocks plus the area of precasts. Area is figured in nominal dimensions of 4″ or 8″ increments. To count one horizontal course 100′ long, the height is 8″ or 0.67′ and the total is 67 sf. To count wall area for one vertical filled cell 100′ high, the width is 8″ or 0.67′ and the total is 67 sf. Multiply the sf times 0.35 and then divide by 27 to get cubic yardage.

Horizontal filled courses of block are easier to count than vertical cells and should be figured first. Typical locations of vertical concrete-filled cells are at corners, wall intersections, and door and window jambs. Additionally, structural engineers often stipulate that one vertical cell be filled every few feet (say 4′ o.c.). Note this vertical concrete may pass through regular blocks courses, and lintel blocks and precasts, that have already been counted. Vertical cells are poured in "lifts" of 4′–8′ at a time.

There are other ways to calculate the quantity of concrete in filled cells besides the one presented here. What the estimator should remember is that regional differences in the size of the block prevent the use of a universal standard means of measurement. Factors should be obtained from local masons and suppliers.

Section 6 Bricks

Brick metrics

There are many sizes of brick and only one will be used in the takeoffs of this book. This is a modular brick with a dimension of 3-5/8″ W × 2-1/4″ H × 7-5/8″ L. Whatever the size in the specifications, it is referred to as the *specified size*. The *actual size* is the exact measurement of any given clay brick after it comes out of the oven, which will be ever so slightly different than specified because of the manufacturing process. A third type of description is *nominal size*, which rounds off a 7-5/8″ length to be 8″. The modular relationship of this brick is that two brick widths plus a mortar joint equals one brick length, and three brick heights plus two mortar joints equal one brick length.

To figure masonry sand, it takes one ton of sand to lay 1,000 brick. For mortar, it takes one bag to lay 143 brick.

The following conversion factors, for the various brick orientations, include a 3/8″ mortar joint.

See the online resources for diagram 416.1

Row 1: Use the conversion factors on this row when brick are one course high and separated by a mortar joint. Multiply LF times the factor to determine the quantity of brick. Use for rowlocks at window sills, intermediate header bricks, or sailor or soldier coursing.

Row 2: The nominal area of a brick face including mortar on two sides.

Row 3: Use the conversion factors on this row when brick are multiple courses high. Divide the factor by 100 (685 becomes 6.85) and multiply times SF of wall. Or, divide the SF of wall by 100 and multiply times the factor in the table.

Five percent is added to all of the factors for waste and breakage.

Brick orientations

Bricks have a face, an end, and a side. Each one can be positioned horizontally or vertically to create six "orientations", as shown below.

See the online resources for diagram 416.2

Brick patterns and mortar joints

Of the six patterns shown, only the first three consist completely of stretchers. Without knowing the brick pattern, wall brick cannot be quantified.

See the online resources for diagram 416.3, 416.4, 416.5, 416.6, 416.7, 416.8, & 416.9

2

FOUNDATION BLOCKS

Section 1 Photos and drawing(s)

Photo 1 Header Block

At the center of this photo is a header block, sometimes called a shoe block. They are often used in foundations at slab level. The elevation of the top of the block and the slab are the same. An imaginary slab edge is shown by the dark line.

Photo 2 Lintel Block

Lintel blocks are used to make bond beams, which are horizontal continuous cells filled with concrete and reinforced with rebar, often one #5.

The lintel block shown are nominal size 8" high x 16" long x 8" wide, and actually sized 7-5/8"H x 15-5/8"L x 7-5/8"W.

Rebar placement shown by the dark line.

Photo 3 Masonry Sand

The unit of measure for masonry sand is cubic yards. These small white/yellow grains of sand are measured with a sieve size described in specifications.

The sand piles at a concrete or brick plant are tested periodically by an independent testing lab (the same companies that test earth compaction and concrete cylinder testing).

They furnish the sand characteristics to the plant, and the plants provide these testing reports to contractors in need of submittal information for a project.

I am on the other side with Alice.

Photo 4 Split-Face Header Block

Continuous concrete footing with formed sides	Regular block on top of the footing, split face block above. The split-face block is more expensive and will be visible when the work is completed. The regular block will be out of sight below finish grade.	Split-face header block at perimeter of slab pour

Photo 5 Regular, Half, And Three Quarter Block

Photo 6 Vertical Rebar Dowels And Lintel Blocks

Section 2 Header blocks

Garage foundation plans

Refer to the photo of header block in the last chapter. These block do not have the top half of one side. The end view of a header block is "L" shaped. This allows concrete to pour into and down the cell of the block. Header block are used to tie a slab and a foundation together.

See the online resources for diagram 422.1

The concrete within header block cells is usually poured with the slab; it is included in the slab takeoff and is not a part of the masonry-filled cells.

See the online resources for diagrams 422.2, 422.3, 422.4, & 422.5

Plan interpretation

In plan view, a garage slab is shown surrounded by block walls. The main house slab is on two sides of the garage, a yard is to the north and a garage door and driveway are on the east side. There are three walk doors. The high point of the garage slab is on the east side – note the minus 7″ elevation on the plan. From Section B, the elevation of the slab at the garage door is minus 12″, so the overall drop in the slab is 5″. The main slab of the house is at 0″.

Section A details a foundation wall made of four courses of regular block plus a header block for the top course.

There are two courses of regular block underneath the garage door slab, see Section B. The footing height remains the same on all four walls. At the top of the section "regular block" are pointed out to be beyond, not header block.

There are three courses of block under the walk door on the north side, see Section C. The slab has a thickened edge and there is an opening in the foundation wall. The foundation wall resumes beyond the opening with a top course of regular block.

Scope of the work

Include:

Regular block and header block.
Mortar mix and sand.

Exclude:

Concrete-filled cells.

Construction techniques

The concrete footings are in place and rebar dowels turned up out of the top of the footings. The foundation walls are ready to be built, and batter boards surround the perimeter of the structure. The earth building pad will be spread and compacted after the foundation walls are built.

All of the blocks will be installed in one continuous operation.

The north wall of the garage may need bracing when the fill dirt is compacted; see Section C. The grade appears to be almost 3′ lower than the slab on the exterior side. On the west and south sides of the garage, fill dirt can be placed on both sides of the wall at the same time and bracing will not be needed.

After the building pad is compacted and fine graded, the soil will be sprayed for the prevention of termites. Then a vapor barrier will be placed on the ground and wire mesh laid on top of it. The slab would then be poured.

Masonry takeoff

The foundation wall can be described as "one wall" on the takeoff. After interpreting the details, most of the wall is five courses high (3′-4″) with a header block at the top, interrupted by the doors.

MASONRY TAKEOFF											422.6
job: Residential Garage			number						date		
No.	Description	Det.	Qty.	L	Ht.	SF	Reg blk SF	Lintel blk SF	Precast LF	8" reg block SF	8" header block SF
1	Foundation wall total area			85.33	3.33	284					
2	Less garage door out	B	-1	12	2	-24					
3	Less doorway	C	-1	4.67	1.33	-6					
5	Header block	A	-1	36.67	0.67	-25					25
6	Total block SF					229				229	25
7	Conversion factor sf to block								x	1.125	1.125
8	Waste and breakage								x	1.05	1.05
9	Quantity of block sent to the Estimate									**271**	**29**
10	Mortar mix	**5.5**	say		6	bags					
11	Masory sand	1	sy		**1.25**	ton					

Row 1: The total block wall area is determined on the first row of the takeoff. Length multiplied by height, including the openings. The length of the wall, starting at the bottom left hand corner, is 20′ + 23′-4″ + 19′-4″+22′-8″ = 85′-4″.

Rows 2, 3:

The SF of the outs are deducted on rows 2 and 3. Note the use of a negative sign in the "Qty" column. The quantity on row 2 is one garage door; since the purpose of row 2 is to subtract this opening, or "out", from the total area, designate the quantity as a negative number. This is how to get row 1 to "subtract" the other rows beneath it (when an Excel spreadsheet is being used).

The height of the garage door out is the distance from the top of wall, zero elevation, to the top of the bottom two foundation blocks, or 2′.

On row 3 the door out is 16″ high.

Rows 5–9:

The length of header block is west (20 − 3.33 = 16.67) plus south (23.33 − 3.33 = 20). The header blocks are subtracted down *and* moved across to a column designated for header blocks. The net area of header block is shown on row 6 (25 sf).

Rows 7, 8, 9:

The area of regulars and headers are converted to each on row 7, by multiplying SF times 1.125, and a waste factor is added on row 8, for a final quantity shown on row 9.

Row 10: Uses the factor of 55 blocks per bag, round up to six bags.

Row 11: Use 290 blocks per ton, round up to 1.25 tons of sand. Sand is sold in quarter-ton increments.

Section 3 Elevator shaft

Escher's elevator shaft

See the online resources for diagrams 423.1 & 423.2

Plan interpretation

There is something odd about these plans, no doubt the influence of Mr. Escher.

The base slab of this unusual elevator shaft has four different elevations. The interpretation of the plan is that the two walls shown in Section A are on all four sides. The walls have the same top elevation of minus 12″ but have different bottom elevations.

All of the blocks are regulars. There are no lintel blocks or precasts.

The dotted line on the plan is not the outline of the wall; rather, it appears to be the transition line at the edge of the thickened slab. It is curious that this dashed line is shown in plan instead of the outside edge of the block wall. Ask Mr. Escher why.

The slab plan is going to be used to measure the wall lengths; unless the estimator correctly interprets their *exact* locations, the lengths of the walls will not be measured accurately.

Scope of the work

Include:

Regular and lintel blocks.
Sand and mortar mix.

Exclude:

Concrete slab and filled cells.
Earthwork.

Assume:

Access limited. Earth will slope up and away from the slab. Staging limited to top of slab or beyond the pit area. Division 4 contractor responsible for all material handling of masonry products.

Construction techniques

The base slab will be in place when the mason arrives onsite. The earth surrounding the slab will be up and down and not usable for staging. There will only be enough room on the slab for one mason and some block and mortar. The remainder of the material will be staged away from the "pit".

Starting at the corners, all of the block would be laid at one time.

Masonry takeoff

Row 1: All of the walls are similar and are built with regular blocks. They have different heights so each one has to be counted separately. The wall on the lowest slab is 11′-4″ long and it is 8′ high. The surface area of the wall is determined.

Rows 2–4: Length multiplied by height.

Row 5: The length of all walls is 44 LF and the total area is 291 sf. The area is copied to the right in a column designated for 8″ regular block.

Row 6: Conversion factor square feet to each.

Row 7: 5% waste. This is a common waste factor but companies figure waste differently.

Row 8: Regular block count is completed.

Row 9: Figure 55 blocks per bag of mortar.

Row 10: Figure 290 blocks per ton of sand.

MASONRY TAKEOFF											423.3
job: Escher'sElevator Slab			number						date	8" reg	8" lintel
No.	Description	Det.	Qty.	L	Ht.	SF	Reg blk SF	Lintel blk SF	Precast LF	block SF	block SF
1	Wall 1 at -9'			11.33	8	91					
2	Wall 2 at -8'-4"			9.33	7.33	68					
3	Wall 3 at -7			11.33	6	68					
4	Wall 4 at -6'-4			12	5.33	64					
5	Regular block SF			44.0		291				291	
6	Convert SF to each								x	1.125	

MASONRY TAKEOFF											423.3
job: Escher'sElevator Slab			number						date		
No.	**Description**	**Det.**	**Qty.**	**L**	**Ht.**	**SF**	**Reg blk SF**	**Lintel blk SF**	**Precast LF**	**8" reg block SF**	**8" lintel block SF**
7	Waste factor								x	1.05	
8	Total regular block									**344**	ea
9	Mortar mix		344 Blocks/55 blocks per bag of mortar							**6 bags**	
10	Masonry sand		344 Blocks/290 blocks per ton of sand							**1.1 tons**	

Section 4 Lintel blocks

Condos foundation

See the chapter on foundation blocks and refer to the photos of lintel blocks. They are made with a longitudinal passageway for receiving continuous horizontal rebar and concrete. In block wall construction, a course of lintel blocks may be laid horizontally around the perimeter of a building every 8′ or so in height. These solid (filled with concrete) block courses are called "bond beams".

A lintel block is shown from an end view. The sides are full height but the webs are half high.

See the online resources for diagrams 424.1, 424.2, 424.3, 424.4, & 424.5

Plan interpretation

Block foundation walls are shown on the foundation plan in continuous unbroken lines. The dashed lines indicate the footing. The drafter has not shown all of the footing lines and figures enough is drawn; the rest can be inferred.

The 6′ high exterior wall shown in Section A has lintel block for a top course and is poured solid. Section C is also at the exterior wall but cuts through the three exterior doors. The lintel blocks under the doors are also poured solid.

Section B is at the interior wall, and is made with 6″ block. The 6″ wide horizontal lintel blocks at the top of wall B are solid (poured with concrete). However, there are no vertical filled cells. The footing depth is not shown. In "drawing speak" the drafter is saying, "I don't have to finish drawing the bottom half of this detail because I have already shown it twice, in Sections A and C!" The interpretation is that the footing is minus 6′ like the other sections. The plans could *not* be interpreted to mean that the contractor has the discretion to place the footing for Section B at a higher elevation than minus 6′ and save money with fewer foundation block.

An important fill dirt note is adjacent to the sections. The entire site is to be excavated to minus 4′. New fill dirt will be imported.

Poured vertical cells are shown for corners and jambs, plus 4′ o.c. as noted on the plan and Section C.

Scope of the work

Include:

Regular and lintel block.
Mortar mix and sand.

Exclude:

Concrete footing and slab.
Earthwork.
Concrete-filled cells.

Construction techniques

When the mason arrives onsite, the building pad would be level at minus 4′. Footing trenches would be filled with concrete. The top of footing (TOF) would be 2′ below the earth pad.

The block walls would not be completed at one time because access is needed for dirt trucks to import earth within the foundation walls. One end, either the north or south, would not be built so that trucks could enter.

With foundation walls 6′ high, bracing against the exterior walls would be accomplished by backfilling the outside of the walls at the same time that the interior fill is spread and compacted.

To ensure proper compaction, fill dirt is deposited in "lifts" of approximately a foot, compacted, and then tested for compaction before another lift is deposited. After most of the fill dirt has been deposited, the block walls are "buttoned up" at the open end. Fill and compaction occurs adjacent to this last wall construction.

Masonry takeoff

The block takeoff is divided into exterior and interior walls because the block width is different. Both walls are figured 6′ high to get a beginning wall surface area.

Row 1: Starting at the lower left, the length of the walls is 60.67 + 58.67 + 53.33 + 3.33 + 6.67 + 54.67 = 237.33.

Row 2: The surface area of the concrete TE is deducted.

Row 3: There are two purposes for this row. One is for the area of the lintel block to be deducted from the beginning square footage of the wall. The other purpose is to get a count of the lintel block. First, use a negative 1 for the quantity so that the area subtracts from row 1. Then, copy the SF of lintel blocks to the right in the column designated for lintel blocks. Reverse the negative number!

Row 4: This is the area left after making the deductions from the total SF from row 1. This is the *net* SF of regular block, and it is copied to the right in the column designated for 8″ regular block.

MASONRY TAKEOFF											424.6
job: Condos			number						date	8" reg block SF	8" lintel block SF
No.	Description	Det.	Qty.	L	Ht.	SF	Reg blk SF	Lintel blk SF	Precast LF		
1	Exterior walls	A,C		237.33	6	1424					
2	Less T.E.	C	-1	18	0.67	-12					
3	Lintel block top course	A	-1	237.33	0.667	-158					158
4	Regular block SF					1254				1254	
5	Interior walls	B	2	68.67	6	824					
6	Lintel block top course	B	-1	137.33	0.667	-92					92
7						732				732	
8	Total sf of block									1986	250
9	Convert SF to Each								x	1.125	1.125
10	Add waste and breakage								x	1.05	1.05
11	Total block each									**2346**	**295**
12	Mortar mix	36.33	say	**36**	bags						
13	Masonry sand	6.89	say	**7**	tons						

Row 3: The lintel blocks at the perimeter of the slab are interrupted by the front doors; see Section A. However, lintel blocks are placed on the next course below slab level as shown on Section C. Therefore, the perimeter of the exterior walls is used for this count.

3

SINGLE-STORY BLOCK WALLS

Section 1 Photos and drawing(s)

Photo 1 Precast U Lintels

Straight precast U lintels are used to span across openings. They typically bear on 8" of block at each end. The slits at the ends are for vertical rebar placement extending to the 8" jamb below.

Photo 2 Precast Door Headers

Precast door headers (shown upside down). The concrete is 8" high and 8" wide (actual 7-5/8" x 7-5/8") at the ends, which matches blockwork, but is only 6" thick above the door opening. The holes at the ends are for vertical rebar.

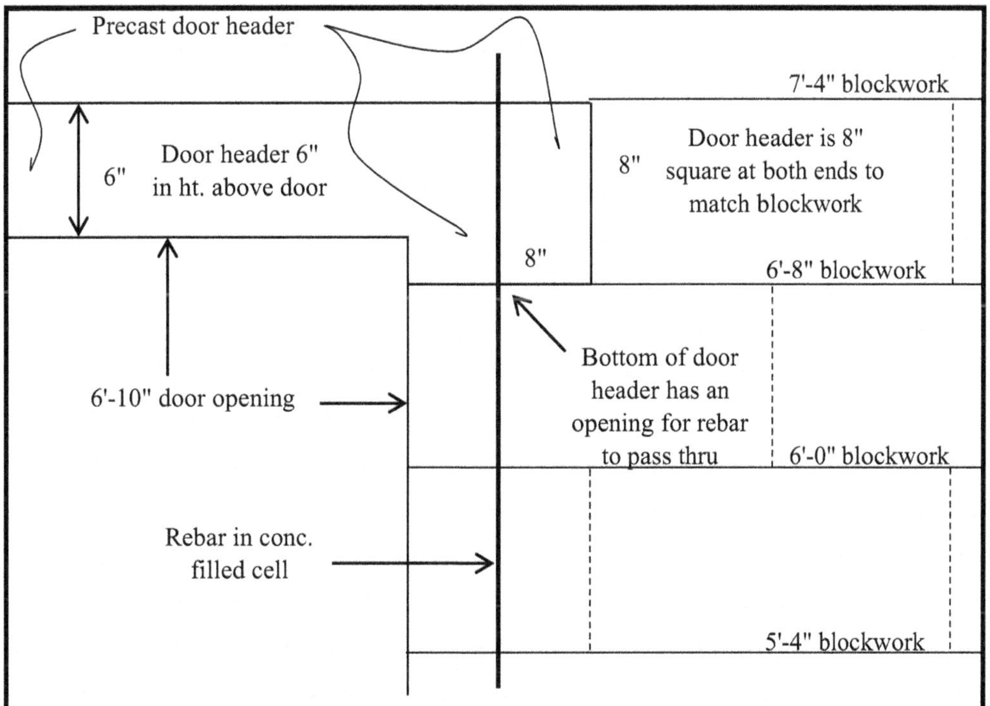

Drawing 1 PRECAST DOOR HEADER 1-1/2"=1'-0"

Photo 3 Precast Window Sills

Precast Window Sills, thin version. The turned down edge is to the exterior, which extends down a block wall, and the ridge near the center is for window placement.

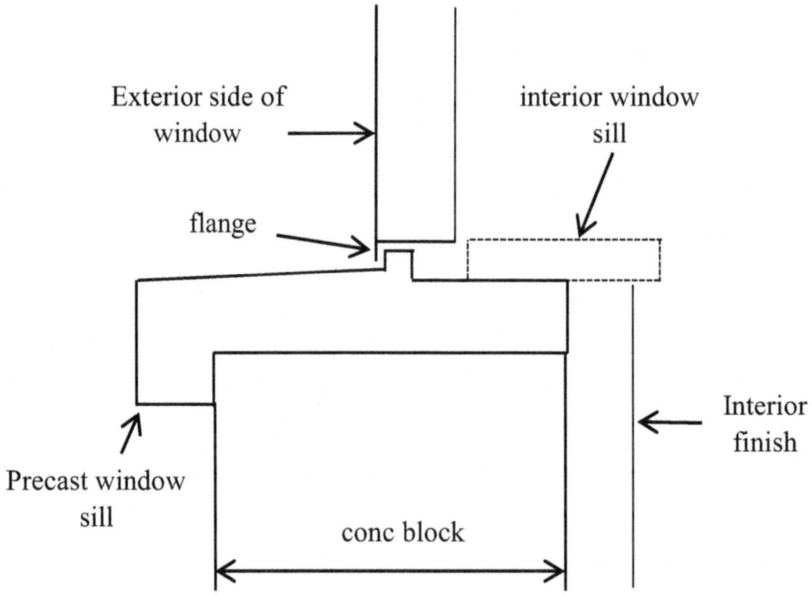

Drawing 2
WINDOW ON PRECAST SILL NTS

Photo 4 Precast Window Sills, Thick Version

Photo 5 Block Walls

Section 2 Block wall case study

Counting the linear feet of wall is no problem here! With a given length of 100 feet, focus can be made on *how to count regular block, lintel block, and other wall components that will repeat in the following chapters.* The key to counting block by the square foot method is to start with the total surface area of the wall. To this square footage, individual items are subtracted out and what is left over is regular block. Along the way, lintel block and precasts are counted.

100′ Wall

The 100-foot wall here is presented to allow the 1, 2, 3 takeoff instruction of a generic wall.

See the online resources for diagrams 432.1 & 432.2

Plan interpretation

The elevation and Section A are all that is given. There are three doors and six windows in the 100′ wall.

At the 7′-4″ elevation there are nine each precast straight U lintels, three for doors and six for windows. Regular block are laid at 7′-4″ between the precasts. Above the precasts, from 8′ to 10′, are lintel block. The top two horizontal courses of lintel block continue for the length of the wall and are filled with rebar and concrete. Concrete fills the entire vertical space between the precasts and the top of the wall.

The vertical cells beside the openings are poured solid.

Scope of the work

Include:

Regular block and lintel block.
Precast straight U lintels.
Precast concrete sills.

Concrete-filled cells.
Mortar mix and sand.

Exclude:

Doors and windows.
Rebar.

Construction techniques

The concrete slab is in place and rebar dowels extend vertically 3′ or so above the slab at the location of vertically filled cells.

The blocks at the ends of the wall and the door jambs are built first. When the block coursing reaches 7′-4″, the nine precasts (above the doors and windows) would be set. The sills are usually installed last to keep them from being damaged.

Building codes and/or structural specifications will dictate whether the filled cells will be poured at one time (one lift) or two. A concrete pump and operator will be employed (or this will be done in-house) by the mason, and a hose will transport the concrete (and people will carry the hose) to the point of the pour.

Masonry takeoff

In this text, the total wall area is found first. The takeoff then proceeds in this order:

2 True outs.
3 Lintel blocks.
4 Straight precast U lintels.

The use of a "negative one" in block takeoffs allows for some efficiency by using the same number for two purposes across the page. This occurs in the case of counting *wall components other than regular block*. Consider lintel blocks and precasts. First, their square footage and the door and window outs are subtracted from the 1,000 sf block wall to arrive at net 680 sf of regular block.

Note the negative numbers in the "Qty" column above. The minus sign is used so that the multiplication steps to arrive at the "SF" column will produce a negative *area*. That way, it can be used as an "out". And that is the task, to list all of the outs so that what is left is the area of regular block.

The door and window outs are "true outs"; a masonry wall component will not be placed there, and regular block will not be laid in the opening. List true outs first; they are the easiest to count and it's possible that while roaming the plans some facts will be learned about the remaining outs, the lintel blocks and precasts.

The "lintel blocks" are treated as an out, a deduction, of the area of regular block. So are the precasts. Their negative areas are used in the SF column. Lintel blocks and precasts now need to be counted in their own units of measure. Their descriptions and data are already there. Using some columns to the right, the SF of lintel blocks and the LF of precasts are counted up and summed below. The second use of a negative one changes these quantities to positive numbers.

Use of the minus sign in block takeoffs saves time in the duplication of descriptions and units of measure.

The unit of measure for precasts can vary depending on how the cost of labor and material is shown on the estimate. Often two counts are needed, each and LF.

Each length of precast has a unique price; they do not cost the same per linear foot. The quantity of each length must be shown on the estimate (and therefore the takeoff), to count the material exactly. The takeoff is used to count how many there are of each length, and on the estimate the *unit cost of the material* is shown and multiplied by the quantity to provide material pricing. That takes care of the material, but what about the labor?

The labor for setting precasts, on an estimate, depends on who is doing the estimating – a mason doing the work in-house, or perhaps a general contractor buying the material and paying labor only for a sub, or a CM completing a budget. Whoever is figuring it, the labor is often calculated from the total linear footage of precasts.

In this text, two units of measure are determined for precasts – *each* and total *LF*.

MASONRY TAKEOFF									432.3
job: 100' Wall		number						date	Precast
No.	**Description**	**Qty.**	**L**	**Ht.**	**SF**	**Reg blk SF**	**Lintel blk SF**	**Precast LF**	**window sills**
1	block wall entire area		100	10	1000				
2	less door outs	-3	3.33	7.33	-73				
3	less window outs	-6	3	3	-54				
4	lintel block top 2 courses	-1	100	1.33	-133	——→	133		
5	lintel blk abv windows	-6	4.67	0.67	-19	——→	19		
6	lintel blk abv doors	-3	4.67	0.67	-9	——→	9		
7	Precast door lintels 4'-8"	-3	4.67	0.67	-9			14.01	
8	Precast wdw lintels 4'-8"	-6	4.67	0.67	-19			28.02	
9	Precast window sills	-6	3	0.17	-3				18
10	SF of block				680	→680	161	**42**	**18**
11	convert sf to each				x	1.125	1.125		
12	waste factor				x	1.05	1.05		
13	Total of reg and lintel blocks					**804**	**190**		
						each	each		
		Mortar Mix	Sand						
14	Total all blocks	994	994						
15	Block per bag	55	290	Blocks per ton			Bolded items sent to the estimate		
16	Bags	18.1	3.4	Tons					
	say	**20**	**3.5**						

Row 1: Begin with the total wall area of 1,000 sf and subtract everything within this area except regular block. Regular block is the remainder, 680 sf; see row 10.

Rows 2, 3: Deduct the "true outs" first, the clear openings of the doors and windows. This area does not include the precast above.

There is one "arithmetic trick" that must be used. These outs must be a negative number, so introduce a negative one into the quantities column. This is so that *the area is a negative number and will be subtracted from the SF from Row 1.*

Rows 4, 5, 6: These three rows count the SF of lintel block. Row 4 counts the top two continuous courses (height 1.33′) and rows 5 and 6 count the lintel block at the elevation of 8′. The lintel blocks at 8′, directly above the precast lintels, are not continuous but only are placed above the precasts. The SF quantity of lintel block has a dual purpose and shortens the steps taken on the takeoff. First, the lintel block area is used as an "out" to the beginning square footage of 1,000 sf. Second, this same SF is sent to the right in a designated column for 8″ lintel block. The negative from above has to be reversed and *changed to a positive.*

Rows 7, 8: Continue with counting the surface areas of the precasts with the same dual purpose as above – both as a negative out and as a positive quantity in a designated column. The length of a precast is 16″ longer than the opening to allow 8″ bearing on each side.

Row 9: Two inches is used here for the height of the precast sills, and there are six each. Their SF is subtracted from the wall SF and the LF of precast sills is sent to a column on the right. Many of the masonry takeoffs in this textbook ignore the areas of precast sills because they do not add up to much.

Row 10: The remainder of 680 sf is the quantity of regular block and is copied to the right in the designated column for regular block. All columns are totaled on row 10.

Row 11: The factor of 1.125 is used to convert SF to each.

Row 12: Waste factor 5% is an average waste factor for waste and breakage.

Row 13: These totals are sent to the estimate!

Concrete takeoff

See the online resources for diagram 432.4

There are no negative numbers to account for when counting concrete-filled cells! It does help, though, to count the horizontal cells first, vertical cells second. The filled cells at one of the counts have to be lessened by their intersection with the other.

The SF of surface area is used to get a total area of filled cells. In this text, the amount of concrete that it takes to fill a precast lintel (per sf) is the same as the concrete that it takes to fill horizontal cells, and this is equal to the amount of concrete that it takes (per sf) to fill vertical cells. When the total area for all of these is found, it is converted to cubic yardage.

Nominal dimensions are used. A block cell, whether vertical or horizontal, is measured 8″ × 8″. A precast lintel measures 8″ high.

CONCRETE TAKEOFF 432.5

job: 100' Wall | number | date

No.	Description	Det	Qty	L	W	SF	Ht	CF	net CY	%	act. CY
	To count concrete in filled cells, an exception is made here in not working left to right so that the takeoff form remains the same - multiply columns 1, 2, and 3 together to get column 4.										
						↓					
1	Filled cells		*1*	*2*		*4*	*3*				
2	lintel block top 2 courses			100.00		133	1.33				
3	Pc and lintel blk abv drs		3	4.67		19	1.33				
4	Pc and lintel blk abv wdws		6	4.67		37	1.33				
5	vert cells dr jambs		6	0.67		29	7.33				
6	vert cells wdw jamb		12	0.67		59	7.33				
7	vert cells ends of wall		2	0.67		12	8.67				
8						289					
9	Convert SF to CF		0.35			289		101.1	3.7	1.05	**3.9**
	20 total vertical filled cells		1	2		4	3				

Row 2: Count the surface area of the bond beam.

Rows 3 and 4: These rows count the area above the doors and windows, including the precast and one lintel block above.

Rows 5 and 6: The door and window jambs are counted 7′-4″ high; don't count through the precasts because they are already accounted for.

Row 7: There is one vertical cell at each end of the wall, 8″ wide and counted here up to the bottom of the bond beam.

Row 9: To convert SF to cubic yards, use 0.35 cubic yards per SF of block area.

Section 3 Wall length quiz

Crazy wall length plan

See the online resources for diagram 433.1

Plan interpretation

You're on your own!

Count the LF of the wall as if all the blocks were all laid in a straight line at the same height of 10′. Why aren't more dimensions given? Practice is needed in being able to think in modular dimensions, turning corners, and counting columns. Estimators spend time chasing around block walls determining their LF. Whether the count is made with a scale or a mouse, it takes judgment.

What is the number that, when multiplied by the given height of 10′, provides the correct square footage of the wall?

Block wall lengths are often difficult to determine because of their configuration. Estimators have to accurately count the "ins and outs" of the wall length.

See if your count is the same as others'.

Section 4 Bond beams and precast U lintels

Ticket booth plans

See the online resources for diagrams 434.1, 434.2, 434.3, & 434.4

Plan interpretation

The ticket booth is constructed of 8″ block walls with a 16″ bond beam of lintel block at the exterior perimeter. The interior wall is frame.

There are two walk doors, both 3070. The plan gives their width as 3′-4″, and the head height is 7′-4″ from Section C. The 16″ bond beam is interrupted at the doors; see Section C, where the bond beam is only one course high.

There are two windows shown on the plan, both with openings 2′-8″ wide. Section D gives the height of the openings, and there are lintel block under the sills.

Vertical cells are poured at the jambs and corners, see Sections A and C.

Scope of the work

Include:

Regular block and lintel block.
Precasts.
Concrete-filled cells.
Mortar mix and sand.

Exclude:

Doors and windows.
Concrete slab.
Wood trusses.

Construction techniques

The concrete slab is in place when the mason arrives onsite. Rebar dowels extend vertically out of the slab at the filled cells. Staging is no problem; there is plenty of room. The block are delivered and placed in the middle of the slab, away from the perimeter.

The corners and door jambs would be built first with regular block. The precasts above the windows are one course lower than at the doors, where the precast is at the top course of the wall.

Masonry takeoff

MASONRY TAKEOFF										434.5
job: Ticket Booth			number						date	
No.	**Description**	**Det.**	**Qty.**	**L**	**Ht.**	**SF**	**Reg blk SF**	**Lintel blk SF**	**Precast LF**	
	a	b	c	d	e	f	g	h	i	
1	Exterior walls	A-D		64.00	8.0	512				
2	Less door outs	C	-2	3.33	7.33	-49				
3	Less window outs	D	-2	2.67	4.0	-21				
4*	P.C. U lintels abv drs 4'-8"	C	**-2**	4.67	0.67	-6			**9.34**	
5	P.C. U lintels above window	D	-2	4.00	0.67	-5			8	
6	Lintel block top 12th course	A	-1	54.66	0.67	-37		37		
7	Lintel block 11th course	A,D	-1	49.33	0.67	-33		33		
8	Lintel block under windows	D	-2	2.67	0.67	-4		4		
9	Subtotals					357	357	73	17	
10	Convert SF to Each					x	1.125	1.125		
11	Add breakage and waste					x	1.05	1.05		
12	Block each						**422**	**87**		
13	Mortar mix	9.24	say	**11**	bags					
14	Masonry sand	1.752	say	**2**	tons					
*	Because the 4'-8" length of the precast is known, the L is in the description and the qty is 2 ea.									
	If the length of the precasts are being approxmated, then a LF quantity is used from column i.									

Row 1: The length of the wall is the total perimeter, counting the corners once. The total surface area of the wall, including outs, is 512 sf.

Rows 2, 3: The doors and windows, true outs, are subtracted here. There are two doors of the same size and two windows of the same size.

Rows 4, 5: The precast lintels of the doors and windows are 16″ longer than the opening. The quantities and lengths are shown in bold type because this is sent to the estimate; the LF can be collected to the right as another unit of measure if the estimate needs it.

Row 6: The lintel block at the top 12th course have a length equal to the wall perimeter (64′) less the two precasts (9′-4″) shown in Section C, or 54′-8″.

Row 7: The lintel blocks at the 11th course have a length of the perimeter less the width of two door openings (6′-8″) and the precasts (8′) above the two windows shown in Section D.

Row 8: This counts the lintel block for the width of the windows for the course at 2′.

Row 12: After the multiplication of factors, the number of blocks and lintel blocks going to the estimate is found.

Concrete takeoff

CONCRETE TAKEOFF											434.6
job: Ticket Booth					number				date		
No.	**Description**	**Det**	**Qty**	**L**	**W**	**SF**	**Ht**	**CF**	**net CY**	**%**	**act. CY**
1	Horizontal conc. filled cells FROM THE MASONRY TAKEOFF:										
2	Precast U lintels above doors	C	2	4.67		6	0.67				
3	P.C. U lintels above window	D	2	4.00		5	0.67				
4	Lintel block top 12th course	A	1	54.67		37	0.67				
5	Lintel block 11th course	A,D	1	49.33		33	0.67				

CONCRETE TAKEOFF											434.6
job: Ticket Booth					number			date			
No.	**Description**	**Det**	**Qty**	**L**	**W**	**SF**	**Ht**	**CF**	**net CY**	**%**	**act. CY**
6	Lintel block under windows	D	2	2.67		4	0.67				
7	All horizontal precasts and lintel block filled cells					**85**					
8	Vertical conc. filled cells:										
9	Corners, intersections and 4' o.c.		15	0.67		67	6.67				
10	Door jambs		4	0.67		20	7.33				
11	Window jambs		4	0.67		16	6				
12	All vertical filled cells		23			103					
13	All filled cells (row 7 + 12)					188					
14	convert SF to CF with factor of .35					0.35		65.67	2.43	1.05	**2.6**
	TOTAL FILLED CELLS										**2.6**

Rows 2–6: Concrete will be poured into the products on rows 2–5, which are copied from the masonry takeoff rows 4–7. Only the total SF is needed, see row 7. When counting concrete that is poured inside *blocks* and *precasts*, the square feet of each are added together. This textbook assumes the concrete volume for these two products are the same.

Row 7: The concrete takeoff can itemize, as on rows 2–6, all of the horizontally filled cells made up of lintel blocks and precasts, or they can be left off the concrete takeoff and their total SF simply carried forward from the masonry takeoff. Either way, the rest of the filled cells, the vertical poured cells, are shown on rows 9–11.

Row 9: First, there are cells that go all the way up to the tie beam (the bottom of the tie beam is 6′-8″). They are located at corners, wall intersections, and 4′ o.c. There are 15 of these. Counting vertical cells can be tedious!

Row 10: The four vertical cells at door jambs extend from 0″ to 7′-4″.

Row 11: There are four vertical cells at the window jambs. The course at 2′ is already counted, as well as everything above 6′-8″. This leaves 6′ of uncounted filled cells.

Row 12: The total quantity of vertical filled cells is 103 sf.

Row 13: This is the total of the horizontal filled cells from row 7 and the vertical cells from row 12.

Row 14: The conversion factor of 0.35 cubic yards per SF is multiplied by row 13 to get a total of 2.6 cubic yards of concrete, which is forwarded to the estimate.

Section 5 Block columns and outs

Mad Hatter plans

See the online resources for diagrams 435.1, 435.2, 435.3, 435.4, 435.5, & 435.6

Plan interpretation

The Mad Hatter's exterior walls are built of 8″ block walls 12′ high. There are six block columns 24″ square built to the same height. There is a 16″ high bond beam at the perimeter.

All three sections have different window heights. Section A is for five windows, Section B for one, and Section C for three. They have 8″ precast U lintels above them except for the 16″ high precast lintel in Section B, at the same bottom and top elevation of the bond beam.

All of the window sills are cast-in-place concrete; there are no precast sills.

Note that the sections state that windows 1, 5, and 6 are "shown". This means, in drawing language, that the correct height for these windows are drawn to scale in the drawing. It also means that there may be other windows that share these sections but, if so, the drawing might not be to scale.

The masonry openings, noted "M.O." on the door and window schedules, give the rough openings of the block.

There is no section through the doors. To determine the wall construction above the doors, refer to the door schedule. While the door size is shown to be 3070, the height of door as noted in the M.O. is 12′, which is the height of the wall.

Door 1 is a storefront door, and apparently there are no block above it; it has a transom above (see remarks), which is a fixed piece of glass. Door 1 occupies the full height of the wall; no masonry is there.

However, door 2 is only 7′-4″ high and there is no transom. Without a specific note about what is above door 2, it can still be inferred that there is a precast lintel and block above. The architect would say that there are precasts and block drawn above all those other openings, so a drawing does not have to be shown above this door.

Vertical cells are at the corners, wall ends, and jambs. They are shaded at some generic locations but are not shown at others. The vertical cells of the block columns are noted to be poured solid. Note that the center 8″ square of the columns is "open" and concrete is not placed there.

Scope of the work

Include:

Regular and lintel blocks.
Precasts.
Concrete-filled cells.
Mortar mix and sand.

Exclude:

Doors and windows.
Aluminum storefront.

Construction techniques

The block columns, being at the corners of the structure and providing the layout, would be started first. The walls between them would then be laid, butting into the columns.

There would be intermediate concrete pours. A logical pour would include a first lift of vertical cells and some or all of the horizontal window sills, which would be formed. The precast lintels above the windows shown in Sections A and C could be included in this first pour, and all of the vertical cells up to the elevation of the precasts.

A second lift would be poured that would include the bond beam, the remainder of column cells, and the vertical cells above the first lift.

The frame wall would be built after the block walls are up.

Masonry takeoff

MASONRY TAKEOFF										435.7
job: Mad Hatter			number						date	
No.	**Description**	**Det.**	**Qty.**	**L**	**Ht.**	**SF**	**Reg blk SF**	**Lintel blk SF**	**8" p.c. LF**	16" h p.c.
1	Ext. 8" walls 12' high	ABCD		140.00	12	1680				
2	Add all columns		6	5.33	12	384				
3	Total sf of block walls					2064				
4	Less concrete surface area					-196				
5	Less window outs # 1	A	-1	2.67	4	-11				
6	# 2	A	-1	3.33	4.67	-16				
7	# 3	A	-1	4.00	2.33	-9				
8	# 4	C	-1	6.67	4.67	-31				
9	# 5	B	-1	14.00	6	-84				
10	# 6	C	-1	6.67	3.33	-22				
11	# 7	C	-1	12.00	4	-48				
12	# 8	A	-1	4.00	2.67	-11				
13	# 9	A	-1	3.33	4.67	-16			*List everything that is not a regular block!*	
14	Less door outs # 1		-1	4.67	12	-56				
15	# 2		-1	3.33	7.33	-24				
16										
17	Precast lintels at windows # 1		-1	4.00	0.67	-3			4	
18	# 2		-1	4.67	0.67	-3			4.67	
19	# 3		-1	5.33	0.67	-4			5.33	
20	# 4		-1	8.00	0.67	-5			8	
21	this precast is 16" high # 5		-1	15.33	1.33	-20			15.33	15.33
22	# 6		-1	8.00	0.67	-5			8	
23	# 7		-1	13.67	0.67	-9			13.67	
24	# 8		-1	5.33	0.67	-4			5.33	
25	# 9		-1	4.67	0.67	-3			4.67	
26	Precast at door 2		-1	4.67	0.67	-3			4.67	
27										
28	Lintel block top course	ACD	-1	124.67	0.67	-83		83		
29	Lintel block 17th course	AC	-1	95.00	0.67	-63		63		
30	The remainder is the sf count for regular block				→	1334	1334	147	**73.67**	**15.33**
31	Convert SF to Each					x	1.125	1.125		
32	Add waste and breakage					x	1.05	1.05		
33	Total regulars and lintel block needed						**1576**	**173**	ea	
34	Mortar mix	31.0	say	**32**	**bags**					
35	Sand	5.9	say	**6**	**tons**					

Row 1: Because all of the exterior walls have the same height of 12′, they can be treated as a like kind wall. There is one total length, one height, and one total wall square footage shown on the first row of the takeoff for all of the walls on this plan. Since door 1 is 12′ high, the same as block wall height, *its length is not counted as wall length.* If it were, extra arithmetic would have to deduct the opening.

Row 2: The length of the columns is 5′-4″!

Row 3: This is the sum of the continuous 12′ walls and the 12′ columns. This is the total SF surface of block area.

Rows 4–12: This is the deduction of true window outs, simple arithmetic taken from the window schedule and counting quantities from the floor plan.

Rows 13, 14: Deducts the door outs.

Rows 16–25: A list of the precasts; add 16″ to the door/window opening width. The numbers 1 through 9 in the description rows are window numbers. On row 25, the precast is shown that is assumed to be above door 2. There is no precast above door 1.

Row 26: The length of the top course of lintel block (see row 26) is found by subtracting the precast lintel shown in Section B from the perimeter. This is 140′ less 15.33′ or 124′-8″.

Row 27: The length of the lintel block course at 11′-4″ is found by subtracting the precast lintels shown in Sections B and C (windows 4–7) from the perimeter. This is 140′ less (8′ + 15.33′ + 8′ + 13.67′) or 95′.

The quantities in bold type are sent to the estimate.

Concrete takeoff

	CONCRETE TAKEOFF										435.8
job: Mad Hatter					number				date		
No.	**Description**	**Det**	**Qty**	**L**	**W**	**SF**	**Ht**	**CF**	**net CY**	**%**	**act. CY**
1	Horizontal filled cells:										
2	All horizontal precasts and lintel blocks					196					
3	Columns	plan	6			384	12				
4	Vertical filled cells exterior walls:										
5	Wall ends		12	0.67		86	10.67				
6	Wall corners		1	0.67		7	10.67				
7	Wdw jambs section A		10	0.67		67	10.00				
8	Wdw jambs section B		2	0.67		14	10.67				
9	Wdw jambs section C		6	0.67		40	10.00				
10	Dr 2 jamb		1	0.67		7	10.00				
11			32			801					
12	Total SF times conversion factor of .35					0.35		280	10.4	1.05	**10.9**
13	Cast in place concrete sills			L	W	HT	SF	CF	CY	%	net CY
14	Window 1	A		2.67	0.67	0.67	1.7889	1			
15	Window 2	A		3.33	0.67	0.67	2.2311	1			
16	Window 3	A		4.00	0.67	0.67	2.68	2			
17	Window 4	C		6.67	0.67	0.67	4.4689	3			
18	add sill projection	C		6.67	0.33	0.33	2.2011	1			
19	Window 5	B		14.00	0.67	0.67	9.38	6			
20	add sill projection	B		14.00	0.33	0.33	4.62	2			
21	Window 6	C		6.67	0.67	0.67	4.4689	3			
22	add sill projection	C		6.67	0.33	0.33	2.2011	1			
23	Window 7	C		12.00	0.67	0.67	8.04	5			
24	add sill projection	C		12.00	0.33	0.33	3.96	1			
25	Window 8	A		4.00	0.67	0.67	2.68	2			
26	Window 9	A		3.33	0.67	0.67	2.2311	1			
27							50.9511	30	1.1	1.05	**1.2**
	The SF column (L x HT, row 27) is used on the masonry takeoff as an "OUT"										

Row 1: The horizontal filled cells can be determined by using arithmetic already calculated on the masonry takeoff; see rows 17–29.

Row 3: The SF of the columns, which are completely filled with concrete, is copied from row 2 of the masonry takeoff.

Rows 5–10: The vertical filled cells cannot be taken from the masonry takeoff. Each "condition" of vertical cells is listed and counted. The maximum height of a vertical cell is 10′-8″ to the bottom of the tie beam. This height is reduced by 8″ at the precast U lintels for the doors and windows except for window 5 at Section B.

Row 10: Door 2 has two jambs but one has already been counted as a "wall end".

4

MULTIPLE BLOCK WALL HEIGHTS

Section 1 Photos and drawing(s)

Photo 1 Block elevator shaft with concrete tie beams

Photo 1 Block Elevator Shaft With Concrete Tie Beams

This is an elevator shaft with cast in place concrete beams at the two floor levels shown. Another story is going to be built - observe the scaffolding and rebar.

Photo 2 Multi-story block wall with concrete tie beams

Photo 2 Multi-story Block Wall With Concrete Tie Beams

A concrete tie beam is formed at the top of the wall

Section 2 Material bin block walls

Material bin plans

See the online resources for diagrams 442.1 & 442.2

Plan interpretation

The material bins are a maze (at least it is symmetrical!) of short and tall walls all bearing at the same slab elevation. But once the turns in each wall are determined and the *lengths* of each wall are found, the components of this structure become easy to quantify.

The key to reading this plan is interpreting the extent of each wall section. First, note the floor plan has four types of wall sections, J–M, all with a sectional view. Second, notice the "wall ends" on the plan, which are designated by the small but important perpendicular (to the length of the walls) lines the drafter has used to designate the end of a wall. Look at any wall and observe where the wall starts and turns and stops.

Note that the two walls at the ends of the structure are described on the plan, not in section.

There is an E/W "J" wall at the middle of the plan. At the N/S ends there are two more "J" walls, each shaped in plan like a capital "I". The J walls have four each horizontal lintel block courses.

There are four each "K" walls, all 8′ long and 2′ high. All courses are poured solid.

The 10′ high L walls are all straight sections, four each. There is a 16″ bond beam at the top of the wall plus a mid-height course of lintel block. Note the mid-height lintel course at 4′ is at the same location for walls sections J, L, and M.

There are two "M" walls that are 12′ high. They have three lintel courses.

The vertical filled cells are shown in plan view. They are very handy for counting individual cells along wall K, until the estimator realizes they are diagrammatic; a 9′-4″ long wall has 14 cells, not the 12 shown. There are no precasts.

The dashed line at the perimeter is noted to be the "extent" of the slab, meaning the outer edge.

Scope of the work

Include:

8″ regular blocks and 8″ lintel blocks.
Concrete-filled cells.
Sand and mortar mix.

Exclude:

Concrete slab.
Rebar.

Construction techniques

There is plenty of room for staging on the slab. The wall ends would be built first. Since there is a horizontal course filled solid at 4′ everywhere (except for wall K, which stops at 2′ high), this defines the first lift. A concrete pour would occur when the walls are 4′-8″ high.

The J walls are going to require three pours; walls L and M could be managed with two pours. As each horizontal bond beam is poured, concrete drops down to fill vertical cells to the elevation of the prior pour.

Masonry takeoff

MASONRY TAKEOFF											442.3
job: Material Bins			number						date		*Total wall length*
No.	**Description**	**Det.**	**Qty.**	**L**	**Ht.**	**SF**	**Reg blk SF**	**Lintel blk SF**	**Precast LF**		
1	14' walls total area	J	2	37.33	14	1045					*74.66*
2	14' Center wall total area	J	1	14	14	196					*14.00*
3	Total 14' walls					1241					
4	All four lintel block crs.	J	-1	88.66	2.67	-237		237			
5	2' walls total area (all lintel blk)	K	-1	37.33	2	-75		75			*37.33*
6	10' walls total area	L	4	14	10	560					*56.00*
7	All three lintel blk. crs.		-1	56	2	-112		112			
8	12' walls total area	M	2	14	12	336					28
9	All three lintel blk. crs.		-1	28	2	-56		56			
10	8' walls total area	plan	4	19.33	8	619					*77.32*
11	Both lintel blk. crs		-1	77.32	1.33	-103		103			
12	Total block SF					2174	2174	582			**287.3**
13	Convert SF to Each					x	1.125	1.125			
14	Add breakage and waste					x	1.05	1.05			
15	Total block each						**2568**	**688**			
16	Mortar mix		59.2	say	**52**	**bags**					
17	Masonry sand		11.2	say	**10**	**tons**					

There are five different walls because there are five different heights. The different walls are simply listed on the takeoff (rows 1, 5, 6, 8, and 10). The wall heights are taken from the sections or the floor plan notes about the end walls. *Getting the correct length of the walls is the key to everything else.*

With a repeating pattern like this, the estimator should check the individual wall lengths against the total. Whether the work is being done by scale or mouse, human error can cause arithmetic mistakes. For this structure, after the individual wall lengths are counted, they can be combined differently by adding all the N/S walls to the E/W walls.

Each of the four wall sections, plus the fifth wall type on the ends of the structure, has a given number of horizontal lintel block. For the takeoff, it doesn't matter where they occur, just how many courses there are. See rows 4, 5, 7, 9, and 11.

Compare the parts to the whole by counting the wall lengths another way. Count the walls in both directions.

Compare the parts to the whole by counting the wall lengths another way. Count the walls in both directions.

All walls running N/S 10 ea × 9.3 = 93.3
End walls running E/W 2 ea × 34 = 68
Center walls running E/W 3 ea × 42 = <u>126</u>
287.3 Checks. Compare to takeoff.

Concrete takeoff

CONCRETE TAKEOFF											442.4
job: Material Bins					number				date		
No.	**Description**	**Det**	**Qty**	**L**	**W**	**SF**	**Ht**	**CF**	**net CY**	**%**	**act. CY**
1	Fill all horizontal lintel blocks (get SF from masonry takeo					582					
2	Vertical cells 14' walls	J	18	0.67		137	11.3				
3	Vertical cells 2' walls	K	0								
4	Vertical cells 10' walls	L	8	0.67		43	8				
5	Vertical cells 12' walls	M	4	0.67		27	10				
6	Vertical cells 8' walls	plan	20	0.67		89	6.67				
			50			878					
	Convert SF to CF .35					0.35		307	11.4	1.05	**11.9**

Row 1: Take the total square feet of lintel blocks from the masonry takeoff.

Rows 2–6: Rows 2–6 are the vertical cells counted for each wall type. The heights of the cell pours are the total wall heights (from the wall sections) less the 8″ courses of lintel block already counted in the horizontal count of lintel block.

Section 3 Block walls and door headers

T plans

The plan presented in this chapter is an example of architectural information conveyed in a different way than usual. Instead of wall sections, the plan and schedules below contain notes that explain wall construction. No sections are given, only a plan and a couple of schedules, but all the information needed for the takeoff is shown.

See the online resources for diagrams 443.1 & 443.2

Plan interpretation

The exterior walls are 10′ and 12′ high, the interior wall is 8′ high.

There are three walk doors (one type 1 and two type 2) and one garage door type 3.

There are four window openings shown on the plan, see A–D. They have various widths and heights as shown on the window schedule. A precast is above each opening. Precast U lintels (horizontal filled cells) are above all four windows on the window schedule.

The location of vertical concrete-filled cells is shown in plan view.

The 12′ wall has two horizontal courses of lintel block at the top of the wall filled with concrete.

The 8′ and 10′ walls have one top horizontal course filled with concrete.

A precast door header is used above door 1 (not a precast straight U lintel). Note the use of the word "precast" on the door schedule, which does not define the specific type of precast. Door 1 and 2 are 6′-8″ tall, the height of many residential doors.

Using a 2″ head frame on top of a 6′-8″ door, a masonry opening of 6′-10″ is needed. This is not blockwork, and to accomplish a masonry opening of 6′-10″ the precast industry makes a product called a "door header", which is a precast U lintel with legs.

Scope of the work

Include:

- 8″ regular block and lintel block.
- Concrete-filled cells in 4′ lifts.
- Precasts.
- Sand and mortar mix.

Exclude:

- Doors and windows.
- Garage door.

Masonry takeoff

View how the one-page takeoff of the three walls is organized. All three walls are listed separately; see rows 2, 7, and 18. On these rows the total square footage of the walls is determined. From these totals, the true outs are first subtracted. Next, the precasts are accounted for, being subtracted from the area of the wall and also counted in linear feet in a column to the right.

MASONRY TAKEOFF										443.3
job: T Plan			number						date	
No.	**Description**	**Det.**	**Qty.**	**L**	**Ht.**	**SF**	**Reg blk SF**	**Lintel blk SF**	**Door hdrs**	**U lintels**
1	All block 8"	plan								
2	8' high wall			14.67	8	117				
3	Less door 1 out		-1	3.33	6.83	-23				
4	Precast door header above door 1		-1	3.33	0.67	-2			3.33	
5	Lintel block top course		-1	14.67	0.67	-10		10		
6						83	83			
7	10' high walls	plan		53.33	10	533				
8	Less window A out		-1	3.33	2.33	-8				
9	Less window B out		-1	4.33	2.33	-10				
10	Less window C out		-1	8.33	4.33	-36				
11	Less door 2 out		-1	3.33	6.83	-23				
12	Precast door header above dr 2		-1	3.33	0.67	-2			3.33	
13	Precast U lintel above wdw A		-1	4.67	0.67	-3				4.67
14	Precast U lintel above wdw B		-1	5.67	0.67	-4				5.67
15	Precast U lintel above wdw C		-1	9.67	0.67	-6				9.67
16	Lintel block top course		-1	53.33	0.67	-36		36		
17						405	405			
18	12' high walls	plan		101.3	12	1216				
19	Less window D out		-1	12.33	4.33	-53				
20	Less door 2 out		-1	3.33	6.83	-23				
21	Less door 3 out		-1	10	8	-80				
22	Precast U lintel above window D		-1	13.67	0.67	-9				14
23	Precast door header above door 2		-1	4.67	0.67	-3			4.67	
24	Precast U lintel above door 3		-1	11.33	0.67	-8				11
25	Lintel blk top two courses		-1	101.3	1.33	-135		135		
26	Lintel blk continuous at 6'-8"		-1	71.66	0.67	-48		48		
27						857	857			
28	Subtotals						1345	228	**11.3**	**45.0**
29	Convert SF to Each					x	1.125	1.125		
30	Add waste					x	1.05	1.05		
31	Total regular block and lintel blk						**1589**	**270**		
32	Mortar mix		33.8	say	**35**	**bags**				
33	Sand		6.4	say	**6.5**	**tons**	(buy sand in half ton increments)			

The sizes of the door and window outs are taken from their schedules. Note door 1 (row 3) and door 2 (row 20) *have jambs and head 2″ thick*. The masonry opening is then 3′-4″ × 6′-10″.

The masonry openings of the windows are given on the schedule, giving the information for the outs on rows 8–10 and 19.

The area of the precasts is deducted on rows 4, 12–15, and 22–24. The LF of these precasts is shown in a column to the right. The lintel blocks are shown on rows 5, 16, and 25–26. Their square footage is shown in a designated column to the right.

Row 28 contains the square footage of all of the regular block, which is a total of the areas of the three walls shown in the column for 8″ regular block. This row also shows the total area in SF of the lintel blocks, and the total LF of the precasts.

Row 29 is the conversion for changing SF into each. Row 30 adds an allowance of 5% to the quantity of regular and lintel blocks. The totals in bold type from row 31 are forwarded to the estimate.

Concrete takeoff

CONCRETE TAKEOFF											443.4
job: T Plan					number				date		
No.	**Description**	**Det**	**Qty**	**L**	**W**	**SF**	**Ht**	**CF**	**net CY**	**%**	**act. CY**
1	Concrete fill horizontal cells:										
2	SF of all lintel blocks					228					
3	SF of all precasts			56.34		38	0.67				
4	Concrete fill vertical cells:										
5	8' wall, jambs at wall ends		2	0.67		10	7.33				
6	8' wall, door jambs		2	0.67		9	6.67				
7	10' wall, cells at wall ends		6	0.67		38	9.33				
8	10' wall, window jambs		6	0.67		35	8.67				
9	12' wall at corners, ends		6	0.67		43	10.7				
10	12' wall at wdw D jambs		2	0.67		13	10				
11	12' wall at door 2 jambs		2	0.67		13	10				
12	12' wall at door 3 jambs		2	0.67		13	10				
14	Total SF of filled area					440					
15	Multiply SF times conversion factor .35 to get CF					0.35		154	5.70	1.05	**5.99**
16	Amount of concrete in filled cells and precasts										**6.0**

Row 2: The area of the lintel blocks is taken from row 28 of the masonry takeoff.

Row 3: The linear feet of precasts, including door headers and U lintels, is taken from row 28 of the masonry takeoff.

Rows 5–6: The vertical cells in the 8′ wall have two conditions. The two cells at the wall ends extend up to the single course tie beam with a bottom at 7′-4″. The two cells beside the door also tops out at 7′-4″ but the cell at the door precast is already poured; the vertical dimension of the pour at the door jambs is 6′-8″.

Rows 7–8: The vertical cells in the 10′ wall have two conditions. At the corners and wall ends there are six cells that extend uninterrupted to the bottom of the tie beam, 9′-4″. The vertical cells at the window jambs have one less cell to fill, the precasts, so their height is 8′-8″.

Rows 9–12: The vertical cells in the 12′ wall have two conditions. At the corners and wall ends there are six cells that extend uninterrupted to the bottom of the tie beam, 10′-8″. The remaining six cells pass through a precast, so their takeoff height is 10′.

Section 4 Block walls

Two rectangle plans

See the online resources for diagrams 444.1 & 444.2

Plan interpretation

The walls of 8″ block are built to various heights shown in three wall sections. There is a fourth wall type, 4′ high, described on the plan without an accompanying section or detail.

There is a front door D shown on the plan, an overhead (the "oh" is shorthand for overhead door) door 8′ wide, and a 4074 opening (note the dashed line indicates there is something overhead). Straight north of the dashed line, there is another break in the wall but without a dashed line. This indicates that the wall actually starts and ends – nothing continues overhead.

The walls shown in Section P are 12′ high, have a 16″ high bond beam at the top of the wall and a precast above the windows. Section P is cut through window 6.

Although Section P is only shown to cut through one window, *by interpretation the wall is the same at the second window 6 to the north.* Continuing further along the wall, there is a wall section labeled "P similar". The wall where the cut is made has no window. *By interpretation, "Section P sim" is the same as Section P except for the window.*

There are two type 6 windows, one type 7, and one type 8. The width and height of window 6 is shown on the plan and in section P; the widths of windows 7 and 8 are shown in plan, but their height is given in a note above the sections. This note about window heights is a bit unusual, but wherever the drafter puts information, it is the estimators' job to find it.

The S walls are on the right side of the plan. They are 10′ high and have a 16″ high bond beam. There are two window openings, 7 and 8, and one overhead door. Precast lintels are used above the windows; see Note 1. There is a precast U lintel above the overhead door (note on plan) and door D (note on plan).

Section R is the wall running N/S and has a 4074 opening.

The location of vertical filled cells are shown in plan.

Scope of the work

Include:

Regular blocks and lintel blocks.
Precasts.
Filled cells.
Mortar mix and sand.

Exclude:

Doors and windows.
Overhead door.

Construction techniques

A lintel course is to be poured horizontally at 3′-4″ (under the number 6 windows) and at 4′ for the remainder of the walls.

The two window precasts in Section P would be poured next, including the jambs down to the lift at 4′. The bond beams at Section S and R could be poured at this time or with the next poor.

The block above the windows in wall P would then be completed. The tie beam at the top of the wall could then be poured.

Masonry takeoff

MASONRY TAKEOFF - (first half)									444.3
job: Two Rectangle			number						date
No.	Description	Det.	Qty.	L	Ht.	SF	Reg blk SF	Lintel blk SF	Precast LF
1	4' wall entire area			10.67	4	43			
2	Lintel block top course		-1	10.67	0.67	-7		7	
3						**36**	36		
4	8' walls entire area	R		13.33	8	107			
5	Less 4074 opening	R	-1	4	7.33	-29			
6	Precast U lintel above opening	R	-1	5.33	0.67	-4			5.33
7	Lintel block at 4'	R	-1	9.33	0.67	-6		9	
8	Lintel block top course	R	-1	8	0.67	-5		5	
9	Regular block SF	R				**62**	62		
10	10' walls entire area	S		41.67	10	417			
11	Less window 8 out	S	-1	2	3.33	-7			
12	Less window 7 out	S	-1	3	3.33	-10			
13	Less overhead door out	S	-1	8	8	-64			
14	Precast U lintel wdw 8	S	-1	3.33	0.67	-2			3.33
15	Precast U lintel wdw 7	S	-1	4.33	0.67	-3			4.33
16	Precast U lintel overhead door	S	-1	9.33	0.67	-6			9.33
17	Lintel blk top two courses	S	-1	41.67	1.33	-55		55	
18	Lintel blk horizontal at 4'	S	-1	33.67	0.67	-23		23	
20	Regular block SF	S				**247**	247		

MASONRY TAKEOFF (second half)											444.4
job: Two Rectangle			number						date		
No.	Description	Det.	Qty.	L	Ht.	SF	Reg blk SF	Lintel blk SF	Precast LF		
21	12' walls entire area	P		53.69	12	644					
22	Less window 6 out	P	-2	4	4	-32					
23	Less door D out	P	-1	2.67	7.33	-20					
24	Precast U lintel wdw 6	P	-2	**5.33**	0.67	-7			11		
25	Precast U lintel door D	P	-1	**4**	0.67	-3			4		
26	Lintel block top two courses	P	-1	53.69	1.33	-71		71			
27	Lintel block at 4'	P	-1	43.02	0.67	-29		29			
28	Lintel block at 7'-4"	P		39.03							
29	Lintel block under windows	P	-1	8	0.67	-5		5			
30	Regular block SF					477	477				
31	s/t						822	204	**37**		
32	Convert SF to Each					x	1.125	1.125			
33	Add breakage and waste					x	1.05	1.05			
34	Total blocks each						**971**	**241**			
35	Mortar mix		say	22	bags						
36	Masonry sand		say	4.25	tons						

The areas of each wall are shown on rows 1, 4, 10, and 21.

There are lintel blocks at the top of the 4′ wall; see row 2.

The note on the floor plan at door D shows the masonry opening; this information goes on row 5. The precast U lintel above it is 16″ longer and shown on row 6.

Rows 7–8:

There are two horizontal courses of lintel blocks shown in Section R, one at 4′ and the other at the top of the wall. They are shown separately on the takeoff because they have different lengths. Because the wall is only 8′ high, and a precast over the door opening is at that height, the length of the top course of lintel block is the length of the wall (13′-4″) less the precast (5′-4″) or 8′, see row 8. The lower lintel block at 4′ is for the full length of the wall less the 4′ opening, see row 7.

Rows 10–18:

The takeoff for wall S starts on row 10. The two window outs are shown first, rows 11 and 12, and then the overhead door out is deducted, see row 13. The next three rows, 14–16, are used to account for the precast lintels.

The two horizontal lintel courses, one at 4′ and the other at the top of the wall, are shown on rows 17 and 18. The course at 4′ is interrupted by the 8′ overhead door, so the length for this course is 41′-8″ less 8′, or 33′-8″, row 18. This course is not interrupted by windows 7 and 8.

All of the areas are deducted from the square footage (down) and their linear footage is sent across to designated columns.

Rows 21–29:

The last wall listed on the takeoff, wall P, has a length of 53′-8″ on row 21. The area of two windows are deducted on row 22. On row 23, the area of the front door is deducted.

The precast lintels of wall P are accounted for on rows 24 and 25.

The top two courses of lintel blocks for wall P are for the full wall perimeter of 53′-8″ and shown on row 26.

On row 27, the lintel blocks at 4′ are interrupted by the window and door outs. This is 53′-8″ total length less 8′ for window openings and 2′-8″ for door D, or 43 LF.

On row 28, see the note above window 6 – the "lintel blocks beyond", which is interpreted to mean that lintel block continue for the length of the wall at the height shown for the precast lintels, which is 7′-4″. This is the length of the wall, 53′-8″, less two precasts 5′-4″ long and one above the door 4′ long, or 39 LF.

The lintel block "under all openings this wall or *this section*" in Section P is accounted for by the width of two type 6 windows (2 × 4′). See row 29.

Concrete takeoff

CONCRETE TAKEOFF												444.5
job: Two Rectangle					number				date			
No.	**Description**	**Det**	**Qty**	**L**	**W**	**SF**	**Ht**	**CF**	**net CY**	**%**	**act. CY**
1	Fill all lintel block					204					
2	Fill all precasts			37		25	0.67				
3	Vertical cells 4' wall up to 3'-4"			10.67		35.5	3.33				
4	Vert cells wall R	R	6	0.67		26.8	6.67				
5	Vert cells wall ends S	S	6	0.67		32.2	8.00				
6	Vert cells wdw jambs	S	4	0.67		21.4	8.00				
7	Vert cells gar dr jambs	S	2	0.67		9.8	7.33				
8	Vert cells wall ends P	P	7	0.67		46.9	10.00				
9	Vert cells dr/wdw jambs	P	6	0.67		37.5	9.33				
						439					
10	Convert SF to CF use .35					0.35		154	5.7	1.05	**5.97**
11	Concrete in filled cells and precasts										**6.0**

Row 1: Copy the SF of lintel block from the masonry takeoff row 31.

Row 2: Copy the LF of precasts from row 31, and paste it on row 2 of the concrete takeoff. Multiply this length by 0.67, which 8″ high. Determine SF.

Row 3: The 4′ wall is poured solid; the top course of lintel block is already accounted for in row 1, so use 3′-4″ for the height. Its length is from row 1 of the masonry takeoff.

Row 4: There are six vertical cells in wall R. They all have the same height, being poured up to the bottom of the bond beam at 7′-4″, and passing through one cell at 4′ that is already poured, for a height of 6′-8″.

Row 5, 6: There are six cells at wall ends that are 8′ high. There are four cells at window jambs that are 8′ high. These cells extend to 8′-8″ but pass through one horizontal cell that gets deducted, so the vertical dimension needed here is 8′.

Row 7: The two cells at the garage door jambs extend to 8′ in height; when the 8″ course at 4′ is subtracted, this vertical measurement is 7′-4″.

Row 8: At wall P, there are seven cells at wall ends. They extend to the bottom of the tie beam at 10′-8″ but pass through the horizontal course at 4′, so the height needed here is 10′.

Row 9: There are six cells beside the door and window openings. They all extend to the bottom of the tie beam at 10′-8″, but pass through both the lintel course at 4′ and a precast at 7′-4″, which leaves a height of 9′-4″.

5

SLOPING BLOCK WALLS

Section 1 Photos and drawing(s)

Photo 1 Split-Face Block With Concrete Tie Beam

A three wall addition to an existing building.

Photo 2 Fluted Block Gable End

Section 2 12″ block

Zig-Zag addition plans

See the online resources for diagrams 452.1, 452.2, 452.3, 452.4, & 452.5

Plan interpretation

This is a three-wall addition to an existing building. A note on the floor plan and one on Section B signifies that the walls are built with 12″ wide block. The vertical filled cells are shown in plan view. There are two horizontal bond beams 8″ high.

There are two fans shown on the floor plan plus an unmarked doorway. An "Opening Schedule" is provided that gives the size of the masonry openings.

Although the north wall is longer than the south wall, the north wall is interpreted to be similar to the South Elevation provided.

Section B is typical for the east wall. Note the walls that run E/W are sloping. Some identifying letters have been given to the east wall that would not appear on a floor plan. They are 1E to 7E.

Scope of the work

Include:

Regular block and lintel block.
Precasts.
Concrete-filled cells.

Exclude:

Doors and windows.
Exhaust fans.
Concrete tie beam.

Construction techniques

The block wall construction would stop upon completion of the lintel block at 2′-8″. A concrete pour would be made, and regular block would be laid up to the next bond beam at 6′-8″, where another pour would occur.

On block structures with multiple bond beams and/or concrete beams, the sequencing of construction is like layers of a cake. Lay a few horizontal courses up to a certain elevation, pour some filled cells, lay a few courses, pour again, and so on.

This structure has a concrete tie beam at the top. For those walls running N/S, it is easy enough to lay block to a constant elevation and maintain 16″ of minimum tie beam concrete. For those walls running E/W, the elevation of the block wall rises in stair-step fashion. Careful layout is maintained in order to balance the considerations of (a) maintaining a minimum of 16″ of concrete, and (b) putting enough block in place so the tie beam doesn't become (expensively) too deep!

Masonry takeoff

Before the block can be taken off, the wall heights of the sloping walls must be determined using the 4/12 slope shown on the south elevation. This is "where the work is" for these kinds of walls – determining the heights. Many times an elevation or side view can be used to measure them from the plans. If not, a quick drawing similar to Sketch 1 is useful.

The south wall is 10′ high on the low end and 16′-8″ on the high end; see the South Elevation.

If the elevation is correctly shown to scale, it can be used to get approximate wall heights (method 1), or geometry (method 2) can be used.

Method 1: *Scale* the drawings or a sketch. See the dimensions in bold type below for walls 1E, 3E, and 5E. These heights are used on rows 4, 6, and 8.

See the online resources for diagram 452.6

Method 2: Geometry. Use rise and run. Solve for vertical leg and add to 8′-8″. (8′-8″ is the height of the tie beam low end.)

See the online resources for diagram 452.7

452.8

MASONRY TAKEOFF

job: Zig-Zag			number						date		
No.	**Description**	**Det.**	**Qty.**	**L**	**Ht.**	**SF**	**Reg blk SF**	**Linte l blk SF**	**Precast LF**		
	All block 12" wide										
1	South Elevation			18.67	12.25	229	average height of wall				
2	North Elevation			20.00	12.00	240	average height of wall				
3	East wall:										
4	1E			6.33	9.08	57					
5	2E			7.67	10.41	80					
6	3E			5.67	11.67	66					
7	4E			5.00	10.83	54					
8	5E			9.00	10.00	90					
9	6E			4.00	9.42	38					
10	7E			4.00	8.67	35					
11				80.34							
12	Less outs 2 exhaust fan		-2	4	4	-32					
13	Less out 4 door		-1	3.33	7.33	-24					
14	Precast above exhaust fan		-2	5.33	0.67	-7			7		
15	Precast above door		-1	4.67	0.67	-3			3		
16	Lintel block at 2'-8"		-1	76.00	0.67	-51		51			
17	Lintel block at 6'-8"		-1	76.00	0.67	-51		51			
18	Lintel block at 10'-8"		-2	14	0.67	-19		19			
19	S/T block, total precasts					701	701	121	10		
20	Convert SF to each					x	1.12 5	1.125			
21	Waste					x	1.05	1.05			
22	Total regular block and lintel blks						**829**	**142**			
23	Mortar mix	26		say	**26 bags**						
24	Masonry sand	5.0		say	**5 tons**						

With the heights of the N/S walls determined, the simplest method to determine the height of the E/S walls (2E, 4E, and 6E) is to average the height of the walls at their two ends.

Row 5: Wall 2E is 9′-1″ + 11′-8″ = 20′-9″ divide by 2 = 10′-5″.
Row 7: Wall 4E is 11′-8″ + 10′ = 21–8″ divide by 2 = 10′-10″.
Row 9: Wall 6E is 8′-8″ + 10′ = 18′-8″ divide by 2 = 9′-5″.

For the heights of the north and south walls:

Row 2: North wall (8′8″ +15′-4″) / 2 = 12′.
Row 1: South wall (9′-2″ + 15′-4″) / 2 = 12′-3″.

Rows 16, 17: For the length of the lintel block at 2′-8″ and 6′-8″, use the total wall length (79.33′) less the door out (3.33′).
Row 18: The lintel block at 10′-8″ only run partially across the N/S walls before they run into the concrete beam. The easiest way to determine the length of this course is to scale it, and it measures 14′.
Row 23: Figure 55 block per bag of mortar as usual, and then multiply by 1.5 to account for the block being 12″ wide.
Row 24: Figure 290 block per ton of sand as usual, and then multiply times 1.5 to account for the block being 12″ wide.

Concrete takeoff

The square feet of precasts and lintel block (rows 1 and 2) are copied from the masonry takeoff.

The vertical filled cells are counted in lifts, the first being from zero to 2′-8″, the distance to the first horizontal bond beam. There are 37 each filled cells as counted from the floor plan. Note that the 8″ wide cells are a reminder that this is an area count of the face of the block. Since these block are 12″ wide, the quantity of concrete first figured with the factor used for regular block is multiplied by 1.5.

CONCRETE TAKEOFF											452.9
job: Zig-Zag					number				date		
No.	Description	Det	Qty	L	W	Ht	SF	CF	net CY	%	act. CY
	Horizontal concrete:										
1	SF of all precasts						10				
2	SF of all lintel block						121				
3	Vertical concrete:										
4	1st lift 8" wide cells		37	0.67		2.67	66				
5	2nd lift 8" wide cells		37	0.67		3.33	83				
6	3rd lift 8" wide cells		12	0.67		3.33	27				
7							307				
8	convert SF to cubic feet .35						0.4	107.28	4.0		
9	Multiply x 1.5 for 12" wide block							x	1.5		
	(the .35 factor on row 8 is for								6.0	1.05	**6.3**
	8" block)										

Section 3 Special block

Schoolhouse plans

See the online resources for diagrams 453.1, 453.2, 453.3, & 453.4

Plan interpretation

The two room schoolhouse has walls of 8″ block and 24″ concrete columns.

Typical locations of vertical filled cells are shown on the plan.

The front elevation has a block gable wall with a slope of 12/12, defined by Section B. There is no door shown in the front doorway; it is an "opening" in the wall. Since Section B is cut through the mid-point of the gable wall, the 24′ height shown is to the peak of the concrete tie beam.

The N/S/W elevations are all similar to the front. The note under the Front East Elevation is "WEST, SOUTH AND NORTH ELEV'S SIMILAR". The rear wall is the same length as the front wall, and its gable would have the same area. The interpretation is that the rear wall is the same as the front except for the doors and windows and has the same wall dimensions.

Section B reveals that there is a fifth gable wall. The B section line on the plan runs all the way through the building. The left and right "outside" walls are the east and west sides. The middle wall of Section B is the inside entry wall.

Section A shows the typical exterior wall to be 16′ high with a 16″ concrete tie beam at the top. Precast lintels are at 9′-4″ above the 40″ high fixed window units. A precast sill 4″ high and 12″ wide is underneath the windows.

The windows shown in Section A have a precast concrete sill and a precast lintel above. There is *lintel block under the window, which is interpreted to be under all windows.* The exterior window openings are not labeled, but two separate notes describe them. One is that all units are 40″ high; the other is that the frames are 2″. The widths of the openings are obtained from plan dimensions.

The N/S walls are a shorter length than the front wall. The "similar" note is interpreted to mean that they "look" like the front wall although they have different dimensions. They have a concrete tie beam and the same components above and below the windows. Note that the N/S gable walls will be shorter than the front and rear walls.

Scope of the work

Include:

- All blocks, regular and lintel.
- Concrete-filled cells.
- Precasts.
- Mortar mix and sand.

Exclude:

- Doors and windows.
- Concrete tie beam.

Construction techniques

The columns are the key to a proper layout and they would be started first.

The elevation of 5′-8″ is important for two reasons. Refer to Section A. One, since windows are everywhere and lintel blocks are under the windows, block construction would halt at this elevation, the top of the lintel block, for a concrete pour.

The other reason that 5′-8″ is an important elevation is that it does not fit 8″ block coursing. A "special size" of block will need to be used *somewhere under the lintel block* to create this elevation. A 4″ high block will work nicely, and a full horizontal course of these will have to be used for one of the courses below 5′. This is *not* something that would be shown on the plans!

At elevation 9′-4″ there is either a lintel block or a precast, and it is poured solid.

The block construction stops at 14′-8″ for those walls that are not gables.

The concrete tie beam (and the block construction) is highest on the front and rear sides. The columns are farther apart here than on the sides. The instructions are for a 45-degree slope and that the "west, south and north elevations are similar to the front".

Masonry takeoff

			MASONRY TAKEOFF								453.5
job: Schoolhouse			number						date	4" high block	
No.	Description	Det.	Qty.	L	Ht.	SF	Reg blk SF	Lintel blk SF	Precast LF		
1	Perimeter walls to 14'-8"			170	14.67	2494					
2	Add gable walls 14' wide		3	14	3.5	147					
3	Add gable walls 12' wide		2	12	3.0	72					
4	Total wall area					2713					
5	Less window outs		-1	96.00	3.67	-352					
6	Less front opening		-1	6.67	9.33	-62					
7	P.C. above windows 14' L walls		-4	11.33	0.67	-30			45		
8	P.C. above windows rear wall		-2	5.33	0.67	-7			11		
9	P.C. above windows 12' L walls		-6	9.33	0.67	-38			56		
10	One course 4" high block 16" long		-1	170	0.667	-113				113	
11	Lintel block under windows		-1	96.00	0.67	-64		64			
12	s/t					2046	2046	64	112	113	
13	Convert SF to Each				X		1.125	1.125			
14	Add breakage and waste						1.05	1.05			
15							**2416**	**76**	**112**	**113**	
16											
17	Mortar mix	45.3		say	**46**	**bags**					
18	Masonry sand	8.6		say	**9**	**tons**					

The area of all walls are counted together (up to 14′-8″ in row 1. Rows 2 and 3 add the gable end triangles on five walls. There are three walls similar to the front elevation and two side elevations.

Row 1: The wall length is seven each at 14 LF and six each at 12 LF for a total of 170 LF.
Row 2: The gable wall width is 14 LF so the rise at 45 degrees is 7′; half of this is 3.5′, which is used as the height.
Row 3: The gable wall width is 12 LF so the rise at 45 degrees is 6′; half of this is 3′, which is used as the height.
Row 5: All of the windows are 40″ high so they can be deducted on one row. Their total length is 96 LF.
Row 6: The width of the opening is on the floor plan; the height is on Section A.
Row 7-9: These rows count the precasts above windows.

Row 10: The bottom of the stone window sills are at 5'-8", see Section A. This does not match 8" modular blockwork, and a 4" high block must be used. Since there are so many windows, almost the entire perimeter wall, a continuous course of 4" high block will occur somewhere between the floor and the stone sills. Also see this topic in Construction techniques.

Row 11: See Section A for the lintel course underneath stone sills. There are 96 LF of windows.

Concrete takeoff

	CONCRETE TAKEOFF										453.6
job: Schoolhouse					number				date		
No.	**Description**	**Det**	**Qty**	**L**	**W**	**Ht**	**SF**	**CF**	**net CY**	**%**	**act. CY**
1	Fill all precasts			112		0.667	75				
2	Fill all horizontal lintel block						64				
3	*Fill vert. cells in horizontal walls (non-gable) up to 14'-8":*										
4	Vert. cells at end of walls		16	0.67		14.67	157				
5	Vert. cells at wdw jambs		16	0.67		14.00	150				
6	*Fill vert. cells in N/S gable walls:*		16	0.67		18.00	193				
7	*Fill vert. Cells in E/W gable walls:*		8	0.67		17.00	91				
			56				730				
	Convert sf to cf		56			x	0.35	255.544	9.5	1.05	9.9
	Concrete in filled cells and precasts										**10.00**

The areas for the first two rows are taken from the masonry takeoff.

Each vertical height of filled cells is different on a gable wall. Also, there can be numerous conditions of horizontal courses that interrupt the vertical measurement of vertical cells. It is not worth the time to count each vertical location separately – an average height is used.

The count of the horizontal filled cells is fixed and is accurate if the lintel block and precast quantities are correct. For vertical cells at gable walls, and walls with lots of openings that create multiple heights of vertical cells, *use judgment and select a common height.*

An example of taking the time to differentiate between two conditions is shown on rows 4 and 5. Row 4 counts the vertical cells at the end of walls, which counts from the floor up to the bottom of the tie beam. Row 5 lessens this distance by 8″ because of the horizontal course already figured at the precast above windows. If rows 4 and 5 had been both counted at the same height, there would be very little difference in quantity.

An example of using an average height of vertical filled cells is shown on rows 6 and 7. The heights of 18′ and 17′ are approximations, using a scale to measure the elevation plan.

Counting vertical filled cells can take a lot of time without always adding value to the estimate.

6

BRICK

Section 1 Photos and drawing(s)

Photo 1 English Cross Bond

A cross bond pattern is made with alternating courses of stretchers or headers. The stretchers do not align vertically.

Photo 2 Rigid Insulation On Frame Wall

Galvanized metal wall ties nailed through insulation to wood studs, bent horizontally into brick mortar joints

Insulation board

Taped corners

Tape bottom first, sides second, top last

Moisture barrier on top of concrete brick ledge.

Section 2 Brick veneer

Ticket booth plans

A veneer wall is by definition non-load-bearing, and brick are laid on one side of block, wood stud, or steel studs. There are various methods to tie the two walls together, often done with galvanized metal "wall ties".

The space between the brick and the "backup" wall (which is often structural) is called a cavity. Moisture will collect in this air space because the brick are porous. This moisture has to be collected and sent back to the exterior. Several construction techniques are employed to guard against moisture penetration to the interior of the building. One is a vapor barrier, another is flashing, and there are a myriad of waterproofing products.

Vapor barriers are usually applied against the structural wall. It is often thin plastic sheeting similar to concrete slab vapor barriers or roofing felt. Flashing, often in a three-sided "Z" pattern, is made of thin galvanized metal or one of many prefabricated products. It is important for flashing to be placed at any "change" in the face of the brick veneer, e.g. a door, window, or anything that interrupts the regular repeating courses of brick.

When water moisture collects at the bottom of a cavity, it is allowed to travel through "weep holes" in the brick mortar at the base of the wall and pass to the exterior.

See the online resources for diagrams 462.1, 462.2, 462.3, & 462.4

Plan interpretation

The brick and block sit on the slab and both walls are 8′ high. The brick pattern is stack bond as noted on the floor plan and Section A. Sections A and C show the wood-trussed roof, but for some reason Section B does not. Two out of three is good enough, and the estimator knows that if a wood roof is over A and C then it has to be at Section B also.

The bricks are held up (supported) over the door and window openings with steel angles. The angle is described in Section B. Since Section C shows an angle but does not describe it, the drafter is inferring that it is identical to the other detail.

There is galvanized flashing shown at the base of the walls. It is 8″ high and let into the block wall. At the bottom, it extends completely under the brick veneer and turns down when it reaches the exterior wall.

The rigid insulation is conveniently in 16″ high sheets, because the brick ties described in Section A are at every other course in height.

A vapor barrier, needed for all exterior walls, is provided by the waterproofing shown in Section A. This product comes in gallon cans and is rolled or brushed on.

Scope of the work

Include:

Brick, rigid insulation, wall ties, steel angles, mortar.

Exclude:

Concrete, block, rebar, waterproofing, flashing.

Construction techniques

The concrete blocks are laid and the filled cells poured before bricklaying begins. Wall ties are left embedded in the mortar joints of the block wall and extend out so they can be embedded within the mortar joints of the brick wall. The flashing shown at the base of the wall is installed by the block mason. The brickwork is not ready to begin until the waterproofing is installed (by others) and the rigid insulation is in place.

The corners of a brick wall are built first. Care is taken to make the corner plumb and to begin each course with the desired brick pattern.

The vertical legs of some steel angles are bolted to backup walls; they are not fastened to the block wall of this project, they just sit, or bear, on the brick underneath the angle beyond the opening.

Brick takeoff

On the very first row of the takeoff, the area of the brick wall is calculated. For the length of each side of the building, subtract 4″ at the corners so they are not counted twice.

The "out" for the windows has a height of 4′-8″, which includes the space taken by the rowlock brick.

Review the headings of the takeoff format. It includes the units of measure "quantity, length, and height". Width is not usually needed, but the structure in this chapter has some horizontal brick at the window sills. When the estimator needs to use a different unit of measure, it is either added in a column or an existing column is changed.

Note how the rowlock brick are counted. When viewed "in plan", the two courses of rowlocks have the same count per square foot as brick in a running bond or stack bond.

TICKET BOOTH BRICK TAKEOFF						462.5
No.	**Description**	**Det**	**Qty**	**L**	**Ht.**	**SF**
1	Stack bond brick total area			69.33	8	555
a	Less door outs	C	-2	3.33	7.33	-49
b	Less window outs	B	-2	2.67	4.67	-25
c	Net square feet of vertical brick					481
					W	↓
2	Rowlock brick total area	B		2.67	1.33	4
a	Net square feet of brick					484
b	Factor for standard brick					7.2
	see Introduction to Masonry					**3488**
3	Sand (one ton per 1,000 brick)	**3.5 tons**				
4	Mortar mix (one bag per 143 brick)	**25 bags**				
5	Rigid insulation, same as 1c	481 x 1.1 = **529** *				
*	When this quantity is sent to the estimate it will be rounded to 536 sf to match whole sheets of 2' x 4' insulation.					

Section 3 Solid brick walls

One room schoolhouse plans

See the online resources for diagrams 463.1, 463.2, & 463.3

Plan interpretation

All of the walls are 8″ brick, as noted on the floor plan. Section A shows that the wall is composed of two wythes. The concrete slab has a thickened edge and is recessed 3/4″ where the brick wall bears on it. This will keep water from entering the interior of the building.

The front elevation shows the brick to be standard size in a running bond pattern. Wall Section A shows there to be three other patterns – header, sailor, and soldier.

Both doors are 36″ wide, and the frames must have 2″ legs because the brick opening is shown to be 3′-4″. The front elevation shows that the frame heads are 4″ high so that a masonry dimension of 7′-4″ can be met.

The window sizes are shown on the floor plan, as well as the brick opening widths. It appears there are two header courses that extend all around the building, per the note on the front elevation. Note the rear west windows are also 2′ high, so the rear elevation is going to be similar to the front. However, the windows on the north and south ends of the building are 4′ high.

There is a problem with these north/south openings being 4′ high. These windows will interrupt one of the brick courses (the higher one) that are supposed to extend around the building.

This kind of conflict found on plans is common. While this is not of great significance here, when it happens on a large building the consequences can be great, and the contractor has the duty, per contract language, to report "inconsistencies".

The proper action for the contractor is not to second-guess the designer but to write an RFI (request for information) pointing out that the north/south windows interrupt the lower header course. What happens at the bottom of the north and south windows? The RFI below is how the contractor asks the architect questions.

See the online resources for diagram 463.4

In this example, for takeoff purposes, *assume that a subsequent addendum issued by the architect changes the size of the north/south windows from 4040 to 4050, and that the bottom of these windows occur on top of the header course at 2′.*

Scope of the work

Include:

Brick, steel angles, mortar fill at top of wall, and truss anchors. Install door frames furnished by others.

Exclude:

Concrete, windows, doors.

Construction techniques

The steel shape is an upside-down T. It is made of two pieces of flat plate, one 7″ wide and one 3″ wide. When these two pieces are welded together, they make a strong bridge across the opening, supporting both wythes of brick.

The "truss anchor" at the top of the wall is exactly that, a metal strap embedded in the wall. The strap, about the width of a thin belt, has holes in it for nails to be fastened to wood trusses.

This is a solid wall: two brick wythes and solid mortar.

Brick takeoff

The length of the wall, shown on the first row of the takeoff, is found either by the centerline or rectangle methods.

The length of the header courses is the outside perimeter less the width of two doors.

To count the truss anchors at 2′ o.c., the perimeter can be divided by two, but the number of corners should be added. This should be considered an "exact" count, to which a small waste factor should be added.

463.5	**SCHOOLHOUSE BRICK TAKEOFF**								
No .	**Description**	**Det**	**Qt y**	**L**	**Ht.**	**SF**	standar d brick	heade r	sailo r
1	Brick perimeter total square feet	pla n		81.3 2	10.06 3	818			
2	Less door openings		-2	3.33	7.33	-49			
3	Less window openings e/w		-5	4.00	2.00	-40			
4	Less window openings n/s		-4	4	5.00	-80			
5	Less two header courses		-1	74.6 6	0.438	-33		33	

463.5	SCHOOLHOUSE BRICK TAKEOFF								
No.	Description	Det	Qty	L	Ht.	SF	standard brick	header	sailor
6	Less one sailor course		-1	74.66	0.67	-50			50
						567	567	33	50
7	Factor for standard brick						7.2		
8	Total of standard brick incl waste						4081		
9	Factor for header brick							14.4	
10	Total of header brick incl waste							470	
11	Factor for sailor brick								4.73
12	Total of sailor brick incl waste								237
13	Total of all brick	**4788**							
14	Masonry sand (one ton per 1,000 brick)	**5 tons**							
15	Mortar mix (143 brick per bag)	**34 bags**							
16	Truss anchors 1' oc	81 divided by 2'o.c. = 40 plus 4 corners plus say 2 extra = **46 ea**							

PART 5
Steel

1

STRUCTURAL STEEL

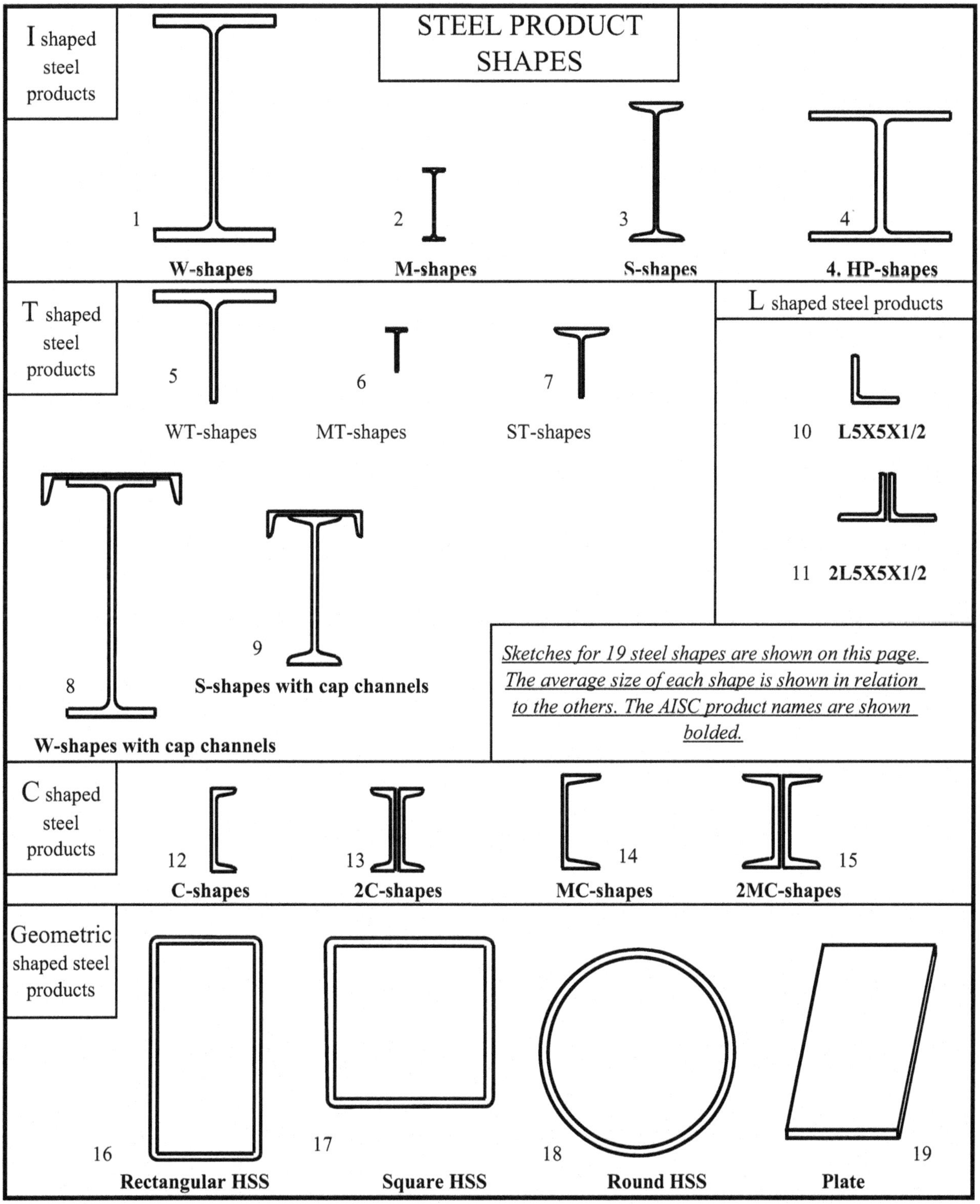

Section 1 Introduction

Steel is durable, with the highest strength-to-weight ratio of any building material. The unit of measure for steel strength is "ksi", one kip being 1,000 pounds per square inch. The weight of steel, according to the AISC, is 490 pounds per cubic foot.

Steel is an alloy of approximately 98% iron and 2% carbon. For much of the century after approximately 1860, steel was produced by the "Bessemer process", named after its English inventor Henry Bessemer. Steel had been made before this, but was expensive and limited to things like swords and knives. The manufacturing process in the U.S. and Europe improved quality and reduced price to where steel was used in buildings and bridge construction beginning in the late 1800s.

Steel has been recycled since the 1800s, when folks went door to door asking for old knives and utensils made of steel. In 2012, the steel recycling rate was 88%, the most recycled material on the planet. Steel can be recycled continually without degradation in performance from one product to another. The LEED rating system for buildings, the U.S. Green Building Council's Leadership in Energy and Environmental Design, promotes the use of recycled materials, and many LEED points for new buildings are gained by using steel.

Even while two out of every three tons of new steel are produced from old steel, it is still necessary to continue to use some quantities of virgin materials. This is true because many steel products remain in service as durable goods for decades at a time, and demand for steel around the world continues to grow.

This first chapter about steel construction concerns "rough iron", the structural steel of Division 5, composed of columns and beams and girders. Together with the smaller pieces of steel that connects or joins them, a building's skeleton is built.

Using steel is a speedy way to build onsite, because the planning and fabrication process can occur in a shop offsite. This preparation is significant and usually occurs while the building foundation is being built. The shop drawings and fabrication process must proceed in lockstep fashion to keep up with the timely progress onsite. Once started, onsite erection is rapid and can (mostly) proceed in all seasons.

The steel contractor faces huge logistical issues when building a structural steel building. Close tolerances are required for large and heavy steel products that are spaced far apart, detailed sequencing of both on- and offsite work is required, and multiple deliveries of big loads of steel often travel far distances. The steel trade is a complex division of work sometimes operating on a gigantic scale.

A substantial amount of preplanning is involved in building a multi-story structural steel building. The shop drawing process requires close coordination between detailer, the shop, contractor, and engineer. There is a recurring process of ordering material, fabrication (the industry term for shop assembly), planning the onsite work, delivery, and erection (the industry term for installation) of steel members (a member is the industry term for individual pieces of structural steel). Taken together, estimating, shop drawings, fabrication, planning and scheduling, and erection of a structural steel building is a huge operation. Close coordination of all parties is required to achieve the small dimensional tolerances required for field erection. It must fit.

Weight is the unit of measure needed for a "budget". A steel subcontractor will factor in many more line items to determine a precise bid than a GC or CM will.

This introduction to the steel industry, from the viewpoint of the GC or CM, is to acquaint the student with the common steel shapes, some language, and typical "connections" (the industry term that describes the attachment of two or more steel members). A familiarity with the name, or designation, of the products is essential to the estimator, as well as how to determine their length and weight.

Section 2 The AISC and the steel manual

The *American Institute of Steel Construction (AISC)*, founded in 1921, is the nonprofit technical standards developer and trade organization for the fabricated structural steel industry in the United States.

The *American Welding Society* specifies how welds are made.

OSHA (Occupational Safety and Health Administration) publishes *Safety and Health Regulations for Construction*, with steel erection addressed in 1926 Subpart R, which must be addressed in the design, detailing, fabrication, and erection of steel structures.

The AISC publishes the *Steel Construction Manual*, hereafter referred to as the "manual" or "steel manual". The manual has 17 parts. Part 1 names the structural steel products commonly used in the steel industry. Several of these are called "shapes," such as "W-shapes and S-shapes", that manufacturers produce. Alas, this nomenclature does not help the student remember the shape of the steel product, because products do not match, for the most part, their letter names.

There are not very many structural steel products. Several of them are simply back to back (side by side) versions of single pieces doubled up, and a couple of products are repurposed as connected top and bottom pieces, making a single product from two individual ones. A few products are even made by simply cutting others in half. The steel industry is very efficient!

After a few pages of naming and defining these products, the remainder of Part 1 of the manual is made up of tables showing the dimensions and properties (depth, thickness of web and flanges, weight per foot, etc.) of these products. These tables are primarily used by structural engineers, but estimators use these tables to find the weight per foot of any given product because *every single size of every single shape is listed in Part 1 of the manual.*

Much of the rest of the manual, Parts 2–17, contains tables that structural engineers use in the design of these structural shapes. They are about designing for compression, tension, flexure, and the connection of structural shapes with welds or bolts.

A manufacturer does not make a structural product that is not listed in the manual unless an engineer has created a custom design. There would be no data for it, no testing, no loading tables; *all the structural shapes that can be used in the United States to carry loads in a building are listed in Part 1 of the manual, as well as every size that each shape is made.* When an estimator reviews the structural bid plans, they usually contain these and only these products.

Steel products can be used for various "elements" (AISC language). A W-shape can be a *girder* used to support multiple joists in one building and a *beam* in another. An element defines how a product is used.

Part 16, near the end of the manual, contains two important documents often referenced by bid plans and specifications. They are:

AISC Specification for Structural Steel Buildings.
AISC Code of Standard Practice for Structural Steel Buildings and Bridges.

The AISC also publishes *Detailing for Steel Construction*, not a part of the manual, which covers the standard practices and recommendations for steel detailing, including the preparation of shop and erection drawings.

Section 3 Drawings

Structural plans

The structural steel portion of drawings include floor plans and details. The layout of columns and beams are shown in plan view, and details and sections define the connections, although sometimes in a general nature. Although structural drawings may appear to be very specific, they are often schematic when it comes to connections, just as architectural plans are diagrammatic.

The structural engineer, although ultimately responsible, often leaves connection work up to detailers. The AISC provides standard details and practices (see *Detailing for Steel Construction*). However, *the engineer of record (the one hired by the architect or owner) is always responsible for the adequacy of design.*

Steel shop drawings

Detailed shop drawings are prepared and approved before parts can be fabricated in a shop assembly. Shop drawings are prepared by "detailers", who are the drafters of the steel industry. Their drawings are sent to the prime contractor and on to the engineer of record for approval. Shop drawings on large projects are completed in stages, medium sized buildings taking perhaps two or three months of sequenced shop drawings. For example, when drawings for a few floors are ready, the batch is sent to the structural engineer, who usually returns them in a couple of weeks.

Many detailers are independent contractors and work on their own or in small firms. Sometimes they are employed in-house by steel companies. On certain jobs, the contract documents may require that the shop drawings be signed and sealed by a separate engineer *before* being submitted to the engineer of record.

The detailer provides connection details not shown on the structural plans, and this interpretation of details can help customize the work for the shop they are being produced for. The detailer can influence shop verses field work. The detailed part of connections can be customized somewhat to the best practices of the steel fabrication shop and can result in the most efficient labor. Beam-to-beam and beam-to-column details are often left up to the detailer (and the fabricator, who the detailer is usually working for). This can provide some economies and small cost savings for the shop fabricator.

Steel shop drawings are also called fabrication drawings. They illustrate details and dimensions, locate bolt holes, and describe welds. They contain instructions to the shop about how to assemble connections. Bolting requires very exacting shop drawings and precise measuring in the shop, where the punching and drilling takes place.

An "erection" drawing is a type of shop drawing that is helpful for all parties to use on larger projects. It depicts "configuration" in a geometric way that allows the viewer to pinpoint where a member is within the building. Each structural member is given a "piece mark". A set of erection drawings helps for visualizing the geometry of the structure. These drawings are called "E sheets" and can be used to define the sequence of construction. They are also helpful for the erector to identify locations for study, or for detailers and engineers to visualize the locations of connections.

The shop drawings contain a list of steel called a "bill of material". This is a handy itemization in one, two, three fashion, of all the pieces that are on the drawings. It is a list, not a drawing.

Section 4 Subcontractors and suppliers

Mills

A steel mill manufactures structural steel and sells large orders to steel service centers. Mills and other suppliers of structural steel sell steel shapes from the steel manual and do not usually sell steel joists, steel decks, or miscellaneous steel. Contractors make smaller purchases to steel suppliers called warehouses or service centers. Mills and steel warehouses do not erect steel or provide shop drawings.

Steel contractors

The steel company that bids to the prime contractor must have the administrative and estimating ability to put together a package bid that includes purchasing and erecting the structural steel and often some nonstructural steel. They may employ subcontractors to complete portions of the work. The steel subcontractor may be a fabricator and subcontract the field erection of structural steel. Or, they may subcontract or exclude nonstructural portions of Division 5, such as the steel joists and steel deck.

A specialty contractor may be employed by either the steel subcontractor or GC/CM to install the steel stairs that climb up ten stories. Or, perhaps the steel contractor will build the stairs but subcontract, or exclude, the handrails, which falls under the category of miscellaneous steel. Handrails and other miscellaneous steel fabrications are subcontracted to specialty firms by either the steel subcontractor or the GC.

Sometimes all of these divisions of the steel trade are bid under the umbrella of one steel subcontractor who is employed by the prime contractor. However, there is no general rule about this, and the GC may contract with them separately. For example, one company may complete the structural steel and another the joists and steel deck; a third subcontractor may install steel stairs and landings, and a local shop may be used to fabricate some angles and bollards.

Fabricators, fabrication

Fabrication is commenced after shop drawing approval by the structural engineer. Structural steel is fabricated offsite before delivery. The first part of fabrication is cutting steel to size, and cutting a column or beam to length is the first order of business. Next, drilling or punching holes must be precisely accomplished. Automated equipment may also be used.

Shop welding and assembly is completed on as much steel as possible to save costs and so that field erection can proceed quickly. This includes prime painting at the end of the fabrication process. Before delivery, each piece is labeled with a "piece mark", the erector's method of identification, which varies among fabricators.

Sketch 1 below describes the shop fabrication of a steel HSS column.

See the online resources for diagram 514.1

A "fabricator" in the construction industry is one who takes delivery of manufactured products from a mill, not a manufacturer. The fabricator then creates an "assembly" by drilling holes, cutting and shaping, and often welding products together. "Assemblies" are made in a shop and then delivered to the site. The steel subcontractor is a fabricator during the time their shop is fabricating the column in Sketch 1.

By comparison, fabrication is also found in the hollow metal door trade. The manufacturer of doors and frames sends them to a door supplier, who cuts holes in the frames and doors for locks and other assembly, and then delivers them to the site for more installation.

Section 5 Structural steel products

Nomenclature

Every size of every product has a name. A product's shape is combined with its depth and weight per linear foot to provide a singular name. This is a combination of three factors, and these product descriptions are hereafter referred to in this text as a "designation". The first part of the product designation is the letter used to define its shape, for example W, M, S, or H. The remainder of the designation is two numbers, its nominal depth and its weight per foot. A W24X162 steel product is a W-shape with a nominal depth of 24″ and a weight of 162 pounds per linear foot.

Of these three factors, the depth of the product must be defined further. While the shape and weight factors are "as is" (what it says), liberty is taken with the depth. For example, there are 21 W-shape products in the manual with W24 as the beginning of its designation, *but none of them have a depth of exactly 24 inches.* They range from 23.6 inches to 28 inches. The depth of the product name is a nominal dimension, not an actual depth. The depth of a steel shape, shown as part of the name of a product, is rounded off!

Structural steel products

The following pages describe nineteen structural steel products from Part 1 of the manual. This is almost a complete list, as it only excludes crane rails, bolts, and some hardware such as turnbuckles. Of these nineteen, only eleven are stand-alone "original shapes". The ever-efficient steel industry uses some of these products back to back and cuts others in half to make other shapes. However, the original shapes, and their doubles and halves, are made in a multitude of sizes.

Most of the products are called "shapes" and are designated by a letter, such as the W-shapes. Unfortunately, the contour of the product does not usually resemble its letter name, for example the "W-shapes" looks like an "I" in profile. The first nine products listed in this chapter are called "shapes", but none of them resemble the letter names given to them, which is confusing.

To help the student better remember these products, fifteen of them that do resemble letters are grouped here by letter shape. Four of them resemble geometric shapes, and that's the category they are in here. These letter and geometric categories have no significance in the steel industry and are simply a means of putting them in a group of like-shaped products for study in this chapter.

Products 10 and 11 are angles, and they do in fact resemble their letter designation "L", as angles and double angles have designations beginning with "L" and "2L". The only other products that resemble their letter designation are C-shapes and 2C shapes, which are channels and double channels.

Each steel product is made in a variety of sizes, and products vary greatly in size as the following sketches show. For example, the W-shapes are made up to a depth of 44″, but the largest depth of an M-shape is only 12.5″. Concerning the quantity of products, there are several hundred sizes each made of the W-shapes, WT-shapes, and rectangular HSS products. However, there are about twenty sizes of the M-shapes (unfortunately shaped like a "T") and forty of the channels.

The typical lengths that structural steel products are manufactured in are 20′, 40′, and 60′.

Following are three methods used in this text to help describe structural steel products. They are a combination of sketches and tables provided to give some familiarity with the products:

1 An architectural "sketch" is shown of an average sized product at 1/16th scale. The sketches show what the products look like, and a comparison of size can be made between them. There is quite a range in both departments – their contour and size vary greatly. Each sketch is of an example product approximately midway in the manual between large and small (using size and weight). The designation of the product is given at the bottom of the sketch.
2 A structural "profile" is shown for some of the products with dimensions and properties from the steel manual concerning the exact contour of the product.
3 A table is provided listing the designations of three of the smallest and three of the largest products. Note that *all* the designations of every steel shape are shown in Part 1 of the manual, as well as their dimensions and properties, which makes quite a long list. The tables below, which also include the "average sized product" from the sketches, are a mini-list of the designations in the steel manual and show the range of products instead of listing them all.

"I"-shaped products – W, M, S, HP

The full names of the "I"-shaped products are:

1 W-shapes.
2 M-shapes.
3 S-shapes.
4 HP-shapes.

The architectural Sketch 1 below uses an "average size" of each shape from the AISC manual and compares it in scale to the other shapes. The name of this average sized product is given in the sketch. See profile of "I" shapes, diagram 515.2, for the steel manual properties.

See the online resources for diagrams 515.1 & 515.2

Where tw = web thickness.
tf = flange thickness.
d = depth.
w = width.

As shown in the structural profile, the W-, M-, and H-shapes conform to the profile on the left and the S-shapes to the flange profile on the right. Note that the term I-beam is *not* used to name any of the four structural steel shapes, but all four are shaped like an "I". If you haven't figured it out by now, there is no such thing as an "I" beam! The answer to many a crossword question is fiction!

The S-shapes, which have sloped (approximately 2 in 12) flanges, are known as American standard beams.

Table 1 shows the range in size that each shape is made, with the total number of product sizes at the bottom. Note how much larger the W-shapes are than all the other structural shapes. The designations used by the AISC do not leave a "space" between the letters and numbers as shown in the tables below. The designation of the largest W-shape is written W44X335. However, sometimes a space is used on building plans, and these tables use the spaces for clarity.

See the online resources for diagram 515.3

The W-shapes are the most-used products of all the shapes. They are doubly symmetric with parallel flange and web sides, and when used as beams are often called "wide flange" beams, their former nomenclature. The AISC has dropped the "flange" part of this description, but the term wide flange can still show up on some plans. W-shapes are also used as columns.

The HP-shapes have equal web and flange lengths. *This "equal leg" length is the best shape to use for vertical pilings in the ground underneath a building*, and this is what HP-shapes are used for.

"T"-shaped products – WT, MT, ST, plus W- and S-shapes with cap channels

The first three "T"-shaped products are:

5. WT-shapes.
6. MT-shapes.
7. ST-shapes.

These three shapes are called "Structural T's" and are made by cutting shapes 1, 2, and 3 in half. The WT-shape is made from a W-shape. The MT-shape is cut from an M-shape and the ST-shape is cut from an S-shape. *They are simply cut in half. The "I" shape cut in half makes a "T".* They are approximately half the size of the product cut from.

The architectural Sketch 2 below, diagram 515.4, uses an "average size" of each shape from the AISC manual and compares it in scale to the other shapes. The name of this average sized product is given in the sketch. See Table 2 for a range of sizes for each product.

Table 2 shows the size range for each Structural T shape, with the total number of product sizes at the bottom. Note how small the MT-shapes are. The smallest MT and ST sizes weigh only about 3 lbs per linear foot.

See the online resources for diagrams 515.4 & 515.6

There are two profile versions, with the WT- and MT-shapes shown in the left profile. The ST-shape is shown on the right and has a flange slope of 2 in 12.

See the online resources for diagram 515.6

The next two structural steel products shaped like a "T" are also combinations of shapes already made. They have a C-shape on top of either a W- or S-shape product. They are known as:

8 W-shapes with cap channels. Combine product 1 with 12.
9 S-shapes with cap channels. Combine product 3 with 12.

The architectural Sketch 4 below uses an "average size" of each shape from the AISC manual and compares it in scale to the other shapes. The name of this average sized product is given in the sketch. See Table 4 for a range of sizes for each product.

Table 3 shows the size range of cap channel shapes, with the total number of product sizes at the bottom. Note how much larger the W-shape cap channels are than the cap channels made with S-shapes.

See the online resources for diagram 515.7, 515.8

See the online resources for diagram 515.8

"C"-shaped steel products – C, 2C, MC, 2MC

The full name of these products are:

10 C-shapes. Known as channels.
11 2C-shapes. Known as double channels.
12 MC-shapes. Known as miscellaneous channels.
13 2MC-shapes. Known as double channels.

The architectural Sketch 3 below uses an "average size" of each shape from the AISC manual and compares it in scale to the other shapes. When two C-shapes are placed back to back they become 2C-shapes, and when two MC-shapes are placed back to back they become 2MC-shapes. See above architectural Sketch 3, diagram 515.9.

See the online resources for diagram 515.9, 515.11

The C-shapes have 2 in 12 sloped flanges similar to the S-shapes. The MC-shape is a similar version, the only difference being that the flange is *different* from 2 in 12. The two shapes, C and MC, are made singly or back to back as doubles. The back-to-back channels are known as 2C-shapes and 2MC-shapes, and are called double channels.

The C-shape channels are also known as American shape standard channels. The MC channels are known as miscellaneous channels.

Table 3 shows the size range for each C- and MC-shape, with the total number of product sizes at the bottom. These four products are made in similar sizes and quantities. There is very little difference between them, only the flange slope. These are relatively small products.

See the online resources for diagram 515.10

The group of C-shapes are the last products in this text that combine the three factors of shape, depth, and weight to define a designation.

"L"-shaped steel products – L, 2L

The full names of these products are:

14 L-shapes. Also known as angles.
15 2L-shapes. Also known as double angles.

The architectural Sketch 5 below uses an "average size" of each shape from the AISC manual and compares it in scale to the other shapes. The name of this average sized product is given in the sketch. See Table 5 for the size range of angles and double angles. The total number of product sizes are at the bottom.

See the online resources for diagram 515.12

See the online resources for diagram 515.13

There are 137 sizes of angles manufactured, the third most numerous of steel shapes. Some have equal leg lengths and are called "equal leg angles". The longest side of an angle is 12″. There are more unequal leg angles than equal. *The two sides of all angles are always the same thickness.*

The designation for angles is different from the prior steel shapes. The description of an angle begins with an "L" and is followed by three numbers – the length of both sides and the thickness.

The following acronyms are used to describe angles:

LLBB: Long legs back to back.
SLBB: Short legs back to back.
LLV: Long legs vertical.
SLV: Short legs vertical.

Geometric-shaped products – rectangle, square, and circle

The full names of these products are:

16 Rectangular HSS.
17 Square HSS.
18 Round HSS.

These products are hollow and their wall thickness is contained in its designation. These shapes are known as "hollow structural sections", or HSS. They have a uniform wall thickness.

Designation: The first number is a width.

The second number is a depth.
The third number is a thickness.

The architectural Sketch 6 below uses an "average size" of each shape from the AISC manual and compares it in scale to the other shapes. The name of this average sized product is given in the sketch. See Table 6 for a range of sizes for each product

See the online resources for diagram 515.14

Table 6 shows the size range for each HSS shape, with the total number of product sizes at the bottom. The rectangular HSS shape is made in as many sizes as the W-shapes!

See the online resources for diagram 515.15

Plate steel

Plate steel describes flat sheets of steel in various thicknesses and weights per square foot. Plate thickness can be specified in inches or by weight per square foot. It is manufactured smooth or in a raised pattern. Depending on thickness and width, it is known as bars, strips, or sheets. When a rectangular piece of steel is needed for a connection, it is cut from flat plate. If a 1/2″ thick piece of steel is needed, 6″ L × 9″ W, or 12″ square and 3/8″ thick, it is cut from flat plate. Patterned floor plates are different; they have a texture roughed into the surface to prevent it from being slippery. Patterned floor plate provides a safe walking surface at catwalks and mezzanines.

"Bent plate" is a product made from flat plate. If an engineer needs a shape that is 16″ long and turns up 4″ at the end (shaped like an angle but bigger than a standard angle size), call it bent plate and get the fabricator to make it in the shop.

Summary of products

This text has included nineteen total structural steel products from the steel manual. Not covered are crane rails, bolts and anchors, and pipe. Of the nineteen, several are made from others. The list in the left column is of the 12 stand-alone "original shapes" and the list in the middle is of doubled-up products, or products cut in half, from already manufactured products.

See the online resources for diagram 515.16

Section 6 Steel connections

Introduction

The connection, or attachment, of two steel members is an important consideration for all concerned –structural engineer, estimator, shop drawing detailer, shop fabricator, and field erector. *When a failure in structural steel occurs, it is usually at a connection.*

Whether the connection is bolted or welded is a prime consideration for the estimator, as is the repetition of connections, which reduces labor. Generally, welding is more expensive than bolting because it takes a higher degree of labor. Field welding is expensive and shop welding is preferred. Bolting requires precise drilling and punching in the shop, but field bolting is cheaper than field welding. Connection parts (e.g. an angle) are often shop-attached to one structural member by welding and later bolted to the second structural member in the field.

Review the following drawings of Sections 4, 5, and 6 for examples of steel connections.

Steel beam connection to steel column

Section 4 views two HSS columns supporting W18X30 beams. Both columns have 3/4″ steel cap plates bolted to the bottom of beam flanges. The column on the right also indicates a weld on the right end. The bottom flanges of the beams are fixed in place above the columns.

The webs of the two beams connect with a plate, or beam splice, that does not occur above a column. The splice joins the webs together but the beam can rotate here.

A steel "stiffener" made of rectangular flat plate is attached to the web and flanges of the beams above each column. The purpose of these stiffeners is to fix the beam in place and not allow the flanges to rotate. Web stiffeners are welded in place during fabrication.

See the online resources for diagram 516.1

Photo Of A Rusty Beam With A Web Stiffener

Steel beam connection to concrete

Section 5 shows a steel beam attached to an "embed plate". The 10″ high embed is made of flat steel plate with four "studs" attached during fabrication. The embed plate is sent "loose" to the field and turned over to the concrete subcontractor. It is fixed in place to a concrete form. After the concrete pour, the inside face of the concrete is flush with the embedded plate, which provides a steel surface to weld the beam and connection plate.

During fabrication, the 8″ high connection plate is loosely bolted to the W16X40 beam. In the field, the beam connector is welded to the embed.

See the online resources for diagram 516.3

Steel beam connection to steel girder

At the intersection of beam to girder in Section 6, their top flanges conflict. The top flange of the W14X22 beam is cut off during fabrication, called "coping". Holes are drilled into the beam and plate during fabrication. During field erection, the plate is welded to the W18X46 girder, which is called a girder because it is carrying the weight of (smaller) beams. In another situation, the W18X46 could be used as a beam instead of a girder.

See the online resources for diagram 516.4

Section 7 Construction techniques

Delivery

The site layout, delivery, and storage of structural steel are pre-planned by the GC and erector. The GC has other work and trades to consider, and sites are always busy. If there is not enough "lay-down" area, the steel may have to be lifted from the truck (and other truck deliveries delayed) by crane and erected immediately. If unloaded to the ground, wood blocking is placed under the steel.

Layout

The responsibility for the layout of columns is often split between the GC and concrete subcontractor (not the steel contractors) with surveying "offsets". The GC hires a surveyor who locates the corners of the property and from there places stakes near the corners of the building. These stakes are "offset" from the corners of the building exactly a certain distance, say 5′. The concrete sub then has a good target to work from, being able to measure from the offset stake, which is far enough away that batter boards and footing excavations will not disturb it. The key to layout accuracy is to have surveyors handle the long-range measurements.

Anchor bolts

Often the first onsite interaction between the prime contractor and the steel contractor is about the bolts that columns sit on. Early on, at the foundation stage, the steel contractor has provided the GC the locations (via shop drawings) of these bolts and sometimes even provides "templates" of wood or metal with punched out holes. The bolts, furnished by the steel subcontractor, are given to the GC. From there on, the GC is responsible for "laying them out" on footings and slabs. The GC takes the bolts and templates and, before concrete is poured, positions the bolts by hanging them in the air. *The bolts must be located with the accurateness of surveying.* It is not easy. They may be placed in footings way down in excavations where both lateral and vertical locations are difficult to determine and maintain. Or, bolt locations may be in slabs with multiple elevations that extend across a couple of acres. Regardless, it must be right, because bolts are hard to tear out of hard concrete and replace.

Dallas photo of large concrete column with base plate, Texas size! The weight of the column is fully supported by leveling nuts.

Vertical anchor bolts are several feet long and embedded into the concrete footing. The bottom row of "leveling" nuts can be adjusted up and down after the concrete pour to stand the column vertical.

The GC can expect the steel subcontractor to trot out to the site prior to the delivery of columns to check the layout of bolts that the GC has positioned in the concrete. With tape measure in hand and a grin on his/her face (anticipating a mistake), anchor rod locations are thoroughly checked. If any of them are misplaced, demand is made for them to be corrected before the steel sub (who is no doubt on the critical path) will come to the site, thank you. Since the anchor bolts are embedded in concrete, correction is no easy task. Epoxy is usually used to anchor a new bolt.

The columns and their base plates are positioned over anchor bolts and loosely attached. The beams and girders are erected next and temporarily braced; connections are loose. Beams and girders are then bolted or welded into place while columns are plumbed. Minor horizontal members such as bar joists, purlins (roof support), lintels (masonry support), and steel deck are attached. This sequence is repeated per floor. The steel business is repetition – order long lead materials; prepare shop drawings, erection plans, and diagrams; fabricate, deliver, and erect.

Bolts are not tightened all the way in some connections. Bolts may be loosely tightened so that the building frame can expand or contract ever so slightly. *This is called snug tightening.* There are other connections that are "slip critical" and must be securely tightened.

Erection

A crane can be provided by either the GC or the subcontractor. If one job crane is enough, sometimes the prime contractor provides it for the duration of the job, letting all subs use it. This may prove to be more economical than cranes coming and going for various subs, which is similar to the situation with scaffolding. Sometimes the steel subcontractor will include crane rental when bidding the job. If it is determined later that the prime contractor is going to provide a crane, they will issue a credit to the GC. Whatever the case, the important thing is for there to be a clear understanding.

Sections 1, 2, and 3 describe the phases of erecting a typical column. The interaction onsite between GC and steel subcontractor usually begins with columns and the placement of anchor bolts.

See the online resources for diagrams 517.2, 517.3, & 517.4

Section 1

Position a plywood "anchor bolt template" with four holes at the top of the concrete (see Part 3 Concrete) and drop the four anchor bolts through them. Brace the template beyond the excavation. Pour the concrete with the bolts (rods) in position.

Section 2

Stand the fabricated column (the base plate has already been welded to the column in the shop) and base plate over the bolts, resting the column on the bottom leveling nuts, which determines the elevation of the connecting beam(s) above. Top leveling nuts are snug, not tightened. Temporarily brace the column to a rough degree of vertical alignment.

Erect the connecting beam(s) at the top of the column. To make the top connections fit, sometimes it is necessary for the column to be moved a bit back and forth. The loose leveling nuts allow this slight movement. After the top connections are made, the leveling nuts at the base are tightened, the connections at the top of the column are tightened or welded, and the column is fixed in place.

Section 3

Place non-shrink grout under the base plate. The concrete subcontractor is usually tasked with this job of placing grout under base plates and around the anchor rods. The slab pour, which occurs after the grouting, would not accomplish this because of the large size of the aggregate in the concrete.

Section 8 Estimating

For the general contractor basing a budget price on tonnage, connections are often ignored because most of the weight is in the large members. A CM or GC can count the linear feet of beams and columns from the plans, and multiply their length by their weight per foot. A total tonnage for the job can be determined. Then, using an overall "price per ton", a budget can be determined, hopefully accurate to within ten or fifteen percent. This kind of budget estimating is similar to a concrete budget that is based not on specifics but just overall cubic yardage.

The bidder that takes on the risk for the job must be more accurate than the CM or GC. If a fifteen percent markup is being used by a bidder, the subcontractor can't afford to be wrong by five or ten percent.

Everything must be figured, including welding supplies and equipment. Bolts are counted and priced. Crane rentals are included. Connections are carefully taken off and accounted for. Columns and beams may be separated per floor or area, and the costs for them might be extended over several pages of work.

Section 9 A short history of American steel

Some of the early characters in the steel business are still heard about and left legacies that deserve mention.

Andrew Carnegie, of philanthropic fame, became the richest citizen of the United States when he sold his steel business to JP Morgan in about 1900. The Scottish American gave away his money in a professional and humanitarian way, studying the subject and even writing about philanthropy. He financed over 2,500 libraries built in communities across the country and some more abroad. Keeping with his philosophy of always requiring something of those receiving aid, part of his formula for giving a town the money for building a library was their commitment to buy the books and run the library. If he was going to build the building, they had to commit to paying for the expenses of keeping the library alive!

His contribution to the progress of the steel industry included plant efficiency using overhead cranes and hoists.

One of Carnegie's partners, Henry Frick, who had a near monopoly on turning coal into coke (used to heat the furnaces that melted the iron ore, the "smelting" process), became embroiled in one of the biggest labor union stories of all time. When workers went on strike, Frick sent 300 Pinkerton detectives to the Thomson Steel Works (still in operation today in Pittsburgh, Pennsylvania). Nine strikers were killed in the resulting confrontation. This union-busting attempt is an important event in the history of labor relations with big business in the U.S. Despite his hard stance against labor, Frick left us the Frick Collection, an art museum in New York City. And Carnegie's philanthropy is still with us in the arts.

JP Morgan, the financial giant who owned parts of many companies, was instrumental in the formation of General Electric, International Harvester, and AT+T (American Telephone and Telegraph). He helped bail the U.S. government out of financial trouble in 1893 and 1907. He merged several steel companies to form the U.S. Steel Company, which became the largest corporation in the world.

Retired Judge Elbert Gary was running Carnegie Steel when it was merged with other companies and became U.S. Steel. Gary, well respected for his integrity, was named chairman in 1901 and ran U.S. Steel until his death in 1927. The steel towns of Gary in both Indiana and West Virginia are named after him. In 1917, BC Forbes, founder of *Forbes Magazine*, said that the biggest job in America, next to the presidency of the United States, was the chairmanship of the U.S. Steel Corporation. He concluded that Elbert Gary headed an empire greater in income, resources, and area than the average European nation.

Charles Michael Schwab (not the stockbroker) began his career at Carnegie's Thomson Steel Works and in 1897 became president of the Carnegie Steel Company. At U.S. Steel, he clashed with Morgan and Gary and left to run the Bethlehem Shipbuilding and Steel Company. In 1908, the "H" beam was developed there, which helped usher in the age of the skyscraper.

2

STEEL JOISTS AND STEEL DECKS

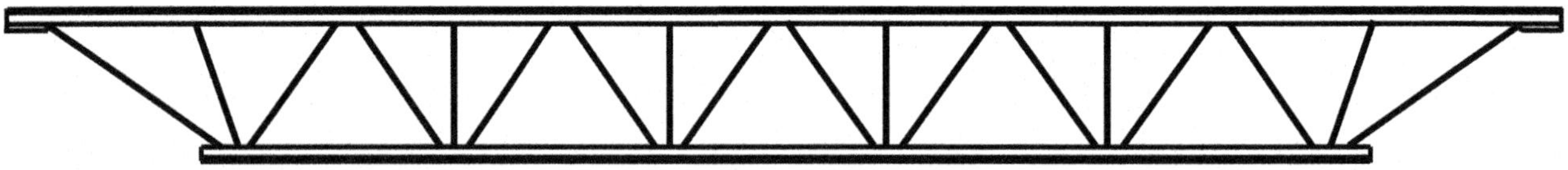

Section 7 Joist and deck plans

Joist and deck structural plans
Joist and deck shop drawings

Section 8 Steel joist bearing, bridging, and extensions

Joist bearing
Deep bearing seat
Bridging
Girder stabilizer
Joist extension

Section 9 Construction techniques

Steel joists
Steel decks

Section 10 Joist and deck fireproofing
Section 11 Estimating

Section 1 Introduction

Steel joists

Chapter 2 covers "joists and decks", two steel products used for many floors and roofs. Steel joists are spaced from 2′ to 12′ o.c. and are covered with steel decking perpendicular to the joists. Joists and decks are in Division 5 of CSI formatted specifications. This chapter covers products and profiles, nomenclature, subcontractors and suppliers, and some construction techniques.

Steel joists are designed to span from one bearing point to another with uniformly distributed loads. Steel joist girders are similarly shaped, but are designed to carry in-plane concentrated loads, such as a series of steel joists perpendicular to the steel girder.

Steel joists are made from structural steel products listed in the steel manual. They have a top chord and bottom chord, both commonly made of steel angles with a space (gap) between them to receive a web that is welded to the chords. The angles can be as small as 1-1/2″ × 1-1/2″. The webs, which run up and down in a triangular pattern, are typically made, for the smaller joists, of solid round steel bars. Single or double angles or other steel products might be used on the larger joists.

The open webs of a steel joist permit ready passage and concealment of pipes, ducts, and conduits within the space under a floor or roof, hence the name "open web" steel joists. The designer can use the same space for multiple purposes. Concrete, masonry, and steel beams are all solid materials that require mechanical and electrical work to travel under or over. Wood trusses of similar shape do not carry as much load and are limited in span. Only the triangular open web construction of a steel joist has the versatility of passing ductwork through it while still spanning a far distance.

Steel joists are lightweight, compared to the weight they can carry, but are flimsy to handle and erect until they are laterally braced perpendicular to their span. This bracing is called "bridging" and is a big part of steel joist design and construction. Strict safety rules (OSHA regulations) require a strict sequence of attachment. First, a worker must be supported, then a bit of construction material, then the full design load. If a steel joist is loaded prematurely, it will "buckle" or "roll over".

The Steel Joist Institute recognizes five kinds of joists, plus one girder, as industry standard shapes. Extensive testing has been done on these products and the SJI publishes load and span tables for them. These tables are used by engineers to select joists and joist girders for floor and roof design. In addition to these joists, some of the larger manufacturers also make custom shapes (for example, arches, bowstring, or scissor trusses) and provide their own testing and engineering.

The repetitive use of the words "joists and joist girders" is not always made here, and the use of the word "joists" in this text also includes the topic of "joist girders".

Steel decks

Steel decking (steel decking is the industry-preferred term, not metal decking) is made by cold forming (bending without the use of heat) sheet steel into a repeating pattern of parallel indentations called "ribs". The long panels can be made of different widths, such as 1′ or 2′, but full-width panels are 24″ or 36″ wide.

Steel decking is a versatile product used in building floors and roofs. It is placed on top of steel joists or on structural steel beams. Steel decking is used by structural engineers to create a diaphragm that resists horizontal movement, similar to plywood on a wood floor.

Section 2 Ruling bodies, SJI and SDI

Steel Joist Institute

The Steel Joist Institute (SJI) governs the manufacture and installation of steel joists in the United States by setting standards and developing regulations for the steel joist industry.

The following is a partial listing of publications by the Steel Joist Institute:

Standard Specification for K-Series, LH-Series, and DLH-Series Open Web Steel Joists and for Joist Girders.
Code of Standard Practice for Steel Joists and Joist Girders.
Standard Specification for CJ-Series Composite Steel Joists.
Code of Standard Practice for CJ-Series Composite Steel Joists.
Handling and Erection of Steel Joists and Joist Girders.

Steel joists shall be erected in accordance with the Occupational Safety and Health Administration (OSHA), U.S. Department of Labor, *Code of Federal Regulations 29 CFR Part 1926, Safety Standards for Steel Erection, Subpart R – Steel Erection.*

Steel Deck Institute

Since its founding in 1939, the Steel Deck Institute (SDI) has provided uniform industry standards for the engineering, design, manufacture, and field usage of steel floor and roof decks. Members of the SDI are manufacturers of steel roof and floor decks. The products that they manufacture are primarily used to support overlying roofing materials or to serve as a permanent form and/or positive reinforcement for concrete floor slabs.

Some of the publications by the Steel Deck Institute include:

SDI Code of Standard Practice.
SDI MOC, Manual of Construction with Steel Deck.
SDI Roof Deck Design Manual (RDDM).
SDI Floor Deck Design Manual (FDDM).
SDI Diaphragm Design Manual (DDM).
SDI Steal Deck on Cold Formed Steel Framing Design Manual (SDCFSFDM).

Section 3 Steel joist products and profiles

K-series steel joists

The K-series open web steel joist is a popular item. There are many building structures that can utilize joist spans of less than sixty feet (their maximum span). Their depths of 10″ to 30″ (made in 2″ increments) can provide enough open space to tuck away a lot of small sized ducts, plumbing pipes, and electrical conduit. The K-series joist is an economical and practical choice for the bays of many buildings.

K-series joists are "secondary" members, which means they span from end to end without receiving intersecting loads from other members. They are designed to carry uniformly distributed loads, not point loads.

KCS steel joists

The KCS joist is not a series but a type of K-series joist that can be designed to support concentrated loads as well as a uniform load. They can be selected from the same sizes as the K-series, with the same depths of 10″ to 30″, and a maximum span of sixty feet.

Longspan LH-series, and deep longspan DLH-series steel joists

The LH- and DLH-series are open web joists and *secondary* members – not designed to carry intersecting joists or beams. They are for carrying uniformly distributed loads, not concentrated loads.

The maximum span for an LH-series joist is 96 LF, and they are made with depths of 18 inches through 48 inches.

The maximum span for a DLH-series joist is 240 LF and they are made with depths of 52 inches through 120 inches.

Composite steel joists, CJ-series

Composite steel joists are open web, parallel chord, load-carrying steel members. They are a singular kind of *secondary* member, structurally purposed to act with a concrete slab to make a composite product. They are used to support floors or roof slabs composed of steel decks, plus concrete and "shear studs". Shear studs provide the connection between slab and joist. After concrete is poured, the slab and joists are joined structurally and make a "composite" slab. Steel joists designed for use with *concrete composite slabs* are called *composite steel joists*.

CJ-series joists are provided unpainted to facilitate installation of welded shear studs.

CJ-series joists must be cambered. A cambered joist, even after being loaded, can have a different elevation than an adjoining flat floor. For this reason, the placement of a joist directly against a flat area is often avoided by placing the steel joist at least one joist spacing away. Foretelling the eventual flatness of a cambered composite joist is not an exact science. It is important for the contractor to document that construction is proceeding "according to design", in case levelness does not match the designers' expectations.

The maximum span for a composite joist is 120 LF, and they are made with depths of 10 inches through 96 inches.

Joist girders

Joist girders are open web steel joists and are *primary* steel members carrying concentrated loads. They are designed to support intersecting joists bearing on the top chord of the joist girder.

The maximum span of a joist girder is 120 LF, and they are made with depths of 20 inches to 10 feet.

Profiles

The following are sketches of six SJI industry standard profiles for K-series, LH-series, and DLH-series joists and KCS and joist girders. Steel joists or joist girders shall have parallel chords or a top chord pitch of up to 1/2″ per foot.

A. Underslung parallel chords, top chord bearing

See the online resources for diagram 523.1

This profile is the most standard shape of a steel joist.

The following joists and joist girders are manufactured with this profile:

K-series KCS, LH-series, DLH-series, CJ-series, JG-series

B. Underslung, pitch one way, top chord bearing

See the online resources for diagram 523.2

The following joists and girders are manufactured with this profile:

LH-series, DLH-series, JG-series

C. Underslung, pitch two ways, top chord bearing

See the online resources for diagram 523.3

The following joists and joist girders are manufactured with this profile:

LH-series, DLH-series, JG-series

D. Square ends, parallel chords, bottom chord bearing

See the online resources for diagram 523.4

The following joists and girders are manufactured with this profile:

K-series KCS, LH-series, DLH-series, CJ-series, JG-series

E. Square ends, pitch one way, bottom chord bearing

See the online resources for diagram 523.5

The following joists and joist girders are manufactured with this profile:

LH-series, DLH-series, JG-series

F. Square ends, pitch two ways, bottom chord bearing

See the online resources for diagram 523.6

The following joists and girders are manufactured with this profile:

LH-series, DLH-series, JG-series

Section 4 Steel deck products and profiles

Typical steel decks

While there is no standard profile, there are several typical profiles. The one below is a type B, the most commonly manufactured steel roof deck profile.

See the online resources for diagram 524.1

The thickness of a steel deck is specified in inches or a gauge, given that the manufacturer defines how thick the gauge is in inches! There was confusion in the past regarding exactly how thick various gauges were, and that is the reason for the stipulation. The SDI has developed a consensus table for the industry to use that defines the thickness of various gauges. The most commonly used thicknesses for floor and roof decks are:

Gauge	*inches*
22	.0295
20	.0358
18	.0474
16	.0598

Composite form deck

The word "composite" in this description refers to a concrete slab acting integrally with a steel deck. A composite deck is one that contains shear studs, which is the steel reinforcement a concrete slab requires.

Non-composite form deck

This term describes any floor or roof deck used as a concrete form. Form deck is furnished galvanized, prime painted, or uncoated. Galvanized roof deck must be used if carrying a lightweight insulating concrete fill.

Roof deck

Roof decks are not designed to act compositely with other materials. Roof deck rib openings are usually narrower than floor decks. Roof deck rib openings are usually narrower than floor deck rib openings. Standard roof deck finishes are galvanized or prime painted (not intended for prolonged exposure to weather).

Cellular decks

These decks are a "two piece" system because they are composed of a flat sheet on the bottom and a ribbed panel on top. They can be used as floor deck topped with concrete or roof deck covered with roofing and insulation products. Cellular deck is always furnished galvanized or painted over galvanized.

Acoustical decks

Sound-absorbing material can be placed within the ribs of a deck. It can be installed in the field by the roofing contractor by placing the absorbent material into the open webs. This is sometimes called an open rib fluted acoustical deck. It can also be used as a two-layered "sandwich" product with voids that wiring is run through, or these voids can be used to place sound-deadening material in to create an acoustical product.

Or, the acoustical material can be factory installed into a cellular deck through holes placed in the bottom sheet. This is called a closed rib cellular type acoustical deck.

Section 5 Steel joist designations and characteristics

Joist designations

The word designation is the industry nomenclature for naming a specific joist. When the depth is used as part of the designation for a steel joist or girder, it is taken from the mid-span.

The first three joist types (K, LH, DLH) are described with three numbers or letters (symbols):

First, the depth of the joist in shown in inches, such as 10 or 20 or 30.
Second, letter(s) describe the joist series – K, LH, or DLH.
Third, a number from 01 to 25 is added, which relates to the chord size.

The designation for a KCS joist is as follows:

Depth of the joist in inches.
The letters, KCS.
A section number from 1 to 5, which has to do with the chord size.

An example of a CJ-series joist, 30CJ2188/1168/420, is where:

First symbol inches: depth at mid-span 30 inches.
Second symbol joist series letter: CJ indicates a composite joist.
Third symbol design load: 2,188 lbs per linear foot.
Fourth symbol live load: 1,168 lbs per linear foot.
Fifth symbol dead load: 420 lbs per linear foot.

The sixth joist type, joist girders, take five numbers or letters (symbols) for their description. An example of a joist girder designation is 72G10N7K, where:

First symbol inches: 72 beam depth (at mid-span).
Second symbol joist series letter: G for joist girder.
Third symbol quantity of joist spacings: 10 means there are ten joist spacings (which corresponds to nine joists bearing on the joist girder).
Fourth symbol N7: the load in kips at panel points.
Fifth symbol: K or F.

The designation symbols of the joists in Chart 1 have been separated by a space for clarity. The designations shown on each row of the chart are found midway in the tables of products of average size and weight.

See the online resources for diagram 525.1

Joist characteristics

Chart 2, describing joist characteristics, lists the first and last entries in the span and load tables for each product. This illustrates the range in size and weight of each product. Note that the longest and heaviest single joist product made is a DLH-series joist that can span 240′ and weighs over 36,000 pounds!

See the online resources for diagram 525.2

Section 6 Joist and deck suppliers and contractors

Joist and deck manufacturers

The Steel Joist Institute controls the specifications for steel joists. Companies that *manufacture* steel joists are required to submit their design data to the SJI (or an independent agency approved by the SJI) for verification of compliance with SJI specifications. Periodic inspections at the manufacturing plant are made by an independent agency approved by the SJI. The frequency, manner of inspection, and manner of reporting is determined by the SJI.

Joist and deck suppliers

There are many companies that specialize in selling joists and decking that are not manufacturers. They have a purchasing relationship with a manufacturer(s) and sell to contractors. Steel contractors send these joist and deck suppliers the structural plans of projects for pricing. Suppliers have knowledgeable sales representatives that can interpret the plans and interact with estimators, project managers, and manufacturers. And, very importantly, they employ drafters and provide shop drawings.

Joist and deck subcontractors

Joist and deck erection is a quick operation. Crews that do this routinely are knowledgeable about safety, installation, and proper techniques and are quick with their task. These are lightweight products and production can proceed rapidly, but their handling and erection can be dangerous.

Section 7 Joist and deck plans

Joist and deck structural plans

The structural plans, not architectural sheets, are where the estimator and supplier find steel joists and decks in the bidding documents. Each floor of a building is typically described with a framing plan. This plan locates the supporting structure beneath the joists, whether it is walls, beams, or columns, as well as any openings in the floor or roof, such as elevator shafts, roof hatches, skylights, mechanical chases, etc.

The location of loads, including walls, will be shown on the plans, as will joist designations. Joist sizing is repeated in entire "bays" where the span remains constant and bearing conditions on each end of the joists remain the same. The areas to receive steel deck are noted. Plan sections and details reveal bearing elevations, thicknesses of slabs, and the construction and thickness of the roof. The thickness of the steel deck will be either on the plans or in the specifications.

Openings in floor and roof slabs are a common occurrence. If they are for a dimension less than the joist spacing, as in Section 8, the cross framing is often done with angles. The exact placement of the steel angles, in this case depending on duct sizing, is determined by the A/C subcontractor verifying the exact duct size. The steel deck supplier will request, on the deck shop drawings, what the decking cuts are going to be. The GC would coordinate this between the trades, and all participants would account for this work in their estimates. Field cutting of floor and roof decks is made as late as possible in the construction sequence because of OSHA safety requirements.

See the note in Section 8, "Approx. 3′ square, see mechanical drawings". Locating this steel opening and determining its exact size is one purpose of the shop drawings. It requires coordination between the trades managed by the prime contractor. Section 8 is a typical section that might be found on a set of bidding plans where some dimensions are approximate. The mechanical plans would not have the exact dimensions for angle to angle either, although it says on Section 8 to go there for the answer. The mechanical plans may have more information, but the answer to the question lies with the subcontractor responsible for that work – the A/C subcontractor.

Joist and deck shop drawings

A joist and deck supplier provides shop drawings when a purchase order is received from a subcontractor. The drawing is made using *nominal* dimensions from the structural plans. A question about the duct size would be placed on the shop drawing as shown below and transmitted from sub to GC. (For this example, assume the joist and deck installer is also furnishing and installing the steel angles.) The drawing below is not large enough to show it, but *overall lengths and widths of the roof deck would also be questioned.*

See the online resources for diagrams 527.1 & 527.2

There is no requirement for joist and deck shop drawings to be signed and sealed by an engineer. The engineer does the design calculations, designates the sizes of all the joists, and provides bridging on the structural plans.

For erection, the joist and deck supplier provides installation drawings called "placement plans". Each joist is labeled with a "piece mark" at one end, which becomes known as the "tagged end". The placement plans for the decking clearly show the location of all deck sheets. These plans include erection and bridging. The design of the shear studs is by the structural engineer, but the joist manufacturer locates them on a drawing. The size, quantity, and layout of shear studs is drawn on the "placement plans". This field set of shop drawings are used exactly for what they are called – placement.

Section 8 Steel joist bearing, bridging, and extensions

Joist bearing

The bearing condition of a steel joist is very specific. An underslung top chord bearing joist may span for 100′ then rest on an 8″ masonry or concrete wall with a steel bearing plate only a few inches across buried in the top of the wall. This is why the verification of shop drawing dimensions is so important; there is no room for error. Steel and concrete do not stretch, everything must fit. If it doesn′t, the manufacturer must be called for help!

Section 1 shows how a steel joist bears on concrete, masonry, and steel walls. The joist cannot sit directly on the masonry or concrete, the *connection must be steel to steel*, and the usual practice is for steel flat plate to be embedded at the top of the masonry or concrete wall. There is only a small amount of contact between joist and the steel plate embedded in the concrete, and this surface area is clearly defined on the drawings depending of the type of joist. This plate is often called an "embed plate" by the contractor, and is *not* designed and furnished by the joist manufacturer or supplier. It is designed by the structural engineer and furnished by the structural steel or miscellaneous steel subcontractor.

There is no typical bearing seat at the bottom chord for a squared end joist. For bottom chord bearing squared end joists, see profiles D, E, and F.

See the online resources for diagram 528.1

Section 2 shows a steel joist bearing on a column, left side of the section, and a beam on the right. At the column, a connection piece is used called a "joist seat". This seat, only 8″ long, is welded to the side of the column flange, and the joist sits on top of the seat. Above the joist, there is a 5″ deep angle; angles are typically placed at the perimeter of a steel deck and act as a form for the concrete.

See the online resources for diagram 528.2

On the right side of Section 2, the steel joist is bearing on a W16X57 beam. The beam is connected to the column with bolted angles.

Deep bearing seat

Many joists have a steel plate attached for end bearing as indicated in Sections 1 and 2. However, long span joists and joist girders require a deeper end condition. The standard depth at the bearing ends for all joist girders is 7-1/2″. Composite steel joists vary in bearing depth from 2-1/2″ to 7-1/2″.

See Section 3 below for a deep bearing seat.

See the online resources for diagram 528.3

Bridging

Bracing of steel joists consists of bridging perpendicular to the trusses. This bracing prevents lateral movement, sometimes called "roll-over". Bridging can be horizontal, see Section 4, or diagonal × bracing, Section 5. It is important for the bridging to culminate at a side wall or beam (or other structural element) for rigidity. This anchor point for a line of bridging is called a "bridging terminus point" and *must be determined before joist bridging is installed.*

With bottom chord bearing joists, there is a tendency to "roll over" and collapse. For this reason, diagonal bridging is placed near the end, or at the end, of joist bearing.

Given a joist sixty feet long and its depth and loading, bridging requirements can be determined from the SJI tables. There are some joists that span thirty feet that require erection bridging before releasing the crane hoisting cable, whereas another joist spanning fifty feet will not need erection bridging.

Typical bridging is made of small angles from 1″ to 2″. It is welded to chords, not webs. When bridging installation is complete, this marks the point at which joists are aligned vertically and properly spaced.

See the online resources for diagram 528.4 & 528.5

Girder stabilizer

Bridging does not have to be used with joist girders. However, the bottom chord must be stabilized at its end to brace the girder from overturning. See the bottom chord stabilization in Section 6. If the supporting structure is steel, *the ends of the bottom chord of a joist girder are restrained from lateral movement with the use of a "stabilizer plate", see Section 6.* This braces the girder from overturning.

No loads shall be placed on a joist girder until its ends are secured *and* the steel joists bearing on the girder are in place and welded to the girder.

See the online resources for diagram 528.6

Joist extension

Often there is a need for the top or bottom of a joist to be extended. An extended top chord can be designed to carry loads, and when properly designed it can support some mansard framing, a canopy, or a marquee.

The plans will define any extensions to the "ends" of the joists, which are a special condition. At the bottom of a joist, these extensions are called either "ceiling extensions" or "extended bottom chords".

At the top of joists, these pieces are called "extended ends", "full depth cantilevers", or "top chord extensions" (TCX). See Section 7.

See the online resources for diagram 528.7

Section 9 Construction techniques

Steel joists

Joists and deck are lightweight products, and much about their erection concerns safety. Until joists are braced, they can be dangerous to handle, set in place, or add construction loads to. Joist installation is a step-by-step process. There are many stages in the stability of joists, and this text does not attempt to explain all of the sequencing and safety requirements.

Steel joists are tagged (a numbering system that matches installation drawings) and delivered upside down. Their top chords are longer than their bottom chords and are often built out of thicker and stronger members. They are transported on flatbed trailers in "bundles" of four to six each. They have one coat of shop paint on them, except for composite joists (because of the welding to shear studs that is going to occur), but this coat of paint is only expected to protect the steel for a short time. They are unloaded by crane or forklift and placed upside down on wood "dunnage" (dunnage is a term that describes material, often scrap, used in storage and handling). If 4 × 4 wood is used, for example, on the ground perpendicular to the joists, the 4 × 4 spacing is preplanned so that the joists are supported equidistantly and preferably at panel points. Joists should not be laid on their side.

Steel joists are moved through the air suspended from a crane and hoisting cable. With long joists, multiple cranes may be used. The joist is held in a vertical position by the crane(s) with the hoist always attached to the top chord, never the webs. Each stage of setting a joist provides a bit more stability, and safety is paramount.

The first load that will be placed on a joist is a worker, then small construction loads such as bridging and steel deck bundles, then larger construction loads, and finally the full design load. These incremental increases in loading all require a certain stage of attachment and completeness of installation.

If the joist is going to bear on a continuous wall or beam, the initial goal is to get "one end attached both sides." The attachment at one end is the point at which a worker is allowed on the joist. If the joist is column bearing, the work is more dangerous because a joist can more easily fall off a column than a wall. *A column-bearing joist must be connected at both ends before a worker is allowed on the joist.* Note that joist girders often bear on columns and joist girder installation is approached with more care than other steel joists.

Other considerations that factor into the incremental loading of joists includes erection bridging, completion of *all* bridging, stabilization of the bottom chord for joist girders, and installation of end bridging for bottom chord bearing joists.

"Erection bridging" is the minimum amount of bridging that must be installed before the hoisting cable can be released. The manufacturer denotes this on a placement plan. There may be additional bridging required beyond erection bridging, but its completion, and the release of the hoisting cable, represents an important stage in the stabilization of a joist. For bottom-bearing steel joists, the erection bridging includes bridging placed at the end spans of the joists at or near bearing.

Joists are usually lifted and set in place one at a time, but they can be grouped together in a process called "panelization". This method constructs an assembly of a few joists on the ground. Two dummy beams are set in place a few feet off the ground the same distance apart as the overhead beam span. The spacing of joists is marked on the dummy beams. A group of joists are then placed on the dummy beams and bridging *is permanently installed.* The group of joists are then lifted by crane and set in place on the beams above.

Some long joists are delivered in pieces. This requires them to be "spliced" onsite before erection. A suitable location on the jobsite must be found where the joist can be handled safely and the splice made. The splice must be defined by the manufacturer with instructions followed by the contractor. The joist supplier usually provides a bolted splice. The erector must then "match mates", which are marked such as 1A and 1B, and then bolt them together.

Joists are never modified, cut, or altered in the field to make them fit or attach. Any change to a joist must be investigated by the manufacturer, who will issue written instructions to the contractor.

Steel decks

Steel decking should be stored off the ground and sloped to provide drainage. It should be protected from weather with a waterproof covering and ventilated to avoid condensation. There are strict protocols for the loading of steel joists with bundles of decking, which are hoisted by crane to floors and roofs. A bundle of decking weighing less than 4,000 lbs, with this load spread over three joists, can be placed on joists if a minimum of one row of bridging is installed.

Decking is typically attached to the building frame with spot welds (puddle welds) or self-drilling screws, powder pins, or pneumatic pins. Sheet-to-sheet fastening is made with screws, by crimping, or by welds. Decking should not be installed with staggered "end laps". This does not increase the capacity of the deck, and a fall hazard is created when the leading edges of adjacent sheets are staggered and not in line.

Steel decks must be installed on lower floors prior to framing subsequent floors in accordance with OSHA Sub Part R, or alternate fall protection must be installed by the erector.

Decking is not a waterproof material. The SDI recommends that no sealant (or other material) be placed between the deck sheets at side or end laps.

Steel decks are designed to take on the loads of pouring fresh concrete for standard pours. However, the engineer and manufacturer do not control means and methods, and the contractor must utilize experience and judgment. Such things as "bulk dumping" of concrete, or the use of heavy power screeds and concrete buggies, need to be taken into consideration by the contractor in planning the pour.

"Shear studs" connect steel joists to a concrete slab. A shear connection is made between the slab and steel joist by embedding the studs in the slab. Shear studs are similar to short pieces of "rebar" a few inches long, have a diameter of between 3/8″ and 3/4″, and have a "head" on them like a bolt. Used with composite floor deck systems, they are placed directly above the top chords of the composite joists.

Shear studs are welded through the steel deck after the deck has been placed on top of the joists. They are embedded in the concrete when the slab is poured. After installation, *shear studs shall extend not less than 1-1/2″ above the top of the ribs of the steel deck. There must be at least 1/2″ of concrete cover above the top of the installed studs. When the top chords of joists are made of double angles, studs are placed on alternate angles.*

The welding of shear studs is carefully inspected as the work progresses by an inspector hired by the owner or engineer. One out of one hundred shear studs are chosen to be inspected by hitting with a hammer or bending with a hollow pipe lowered over the stud. Failure should not occur at the weld. If it does, there is a strict protocol on testing of additional studs.

Structural engineers are responsible for the design of shear studs (layout and sizing), not the manufacturer or supplier. Deck manufacturers do not typically make or furnish shear connectors.

Photo of Shear Studs And Metal Deck Before A Concrete Pour

Dashed lines indicate approximate location of steel beam top flanges, which the shear studs are welded to.

Shear studs welded to beams.

OSHA does not permit shear studs to be factory installed to steel beams except on steel bridges. This photo was taken on the Florida coast adjacent to new bridge construction.

Shear Studs And Steel Deck Isometric

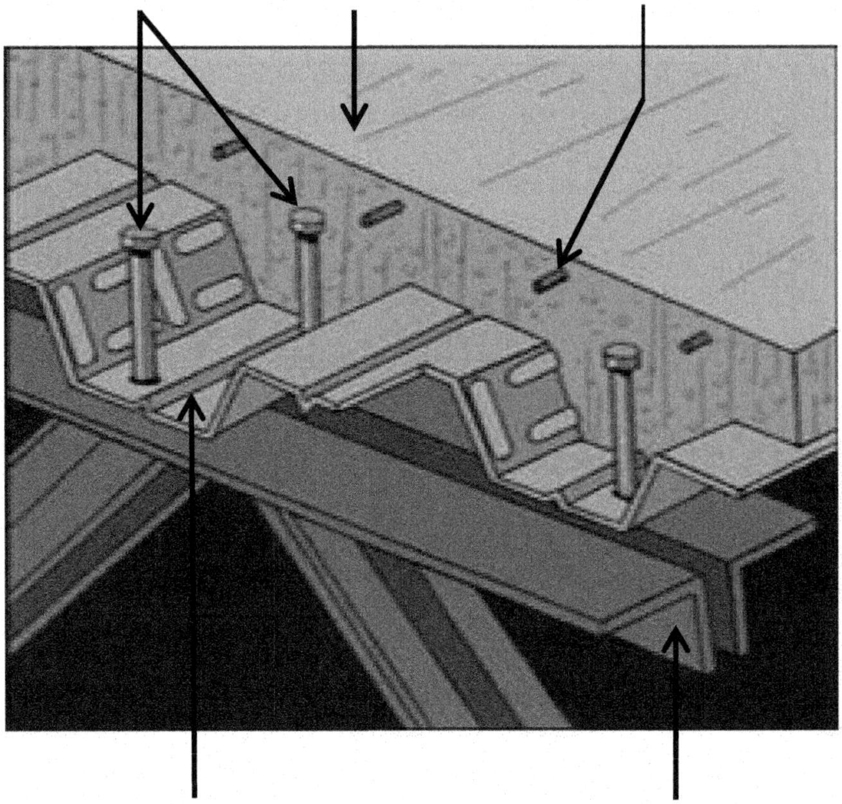

Section 10 Joist and deck fireproofing

The Underwriters Laboratories (UL) is the ruling body for fire-resistant designs. The UL has many approved details for fire rated "assemblies". Each assembly is given a number, and steel joists and decks are a part of many of them. If the plans reference a "UL assembly number 123", this instructs the contractor to follow the notes and instructions on the UL drawing (which may or may not be duplicated on the bid set of drawings), which is in the *UL Fire Resistance Directory*.

The UL assembly will specify what are acceptable deck finishes (for instance, galvanized, unpainted, or very specific paint products). This needs to be considered by the estimator.

The bottoms of steel decks are sometimes sprayed with fireproofing, and if that is the intent, the plans should clearly show it. The plans should clearly note if the bottom of the deck is to be left exposed. Sometimes the undersides of steel decks are used as finished ceilings, as in restaurants or big-box stores. Unless noted otherwise, the estimator assumes the deck is *not* exposed and aesthetics is not a consideration.

There are many fire rated assembles for floor decks, which are field applied and furnished by a separate trade than the joist and deck contractor. In combination with other materials, joists can be used in fire resistive assemblies, for both floors and roofs, for nearly any hourly rating.

The spraying of fireproofing can be expensive and affect the joist design. Fewer joists are sprayed if the joists are farther apart, and the selection of joist size and spacing might be increased in the design process to save money on the fireproofing.

Section 11 Estimating

The Steel Joist Institute recommends that bidders of steel joists include in their scope of work the following:

Joists.
Girders.

Joist substitutes.
Joist extended ends.
Ceiling extensions.
Extended bottom chords used as a strut.
Bridging and bridging anchors.
Joist girder bottom chord bracing.
Headers defined as members supported by and carrying open web steel joists, K-series.
If specified, one shop coat of paint.

3

MISCELLANEOUS STEEL

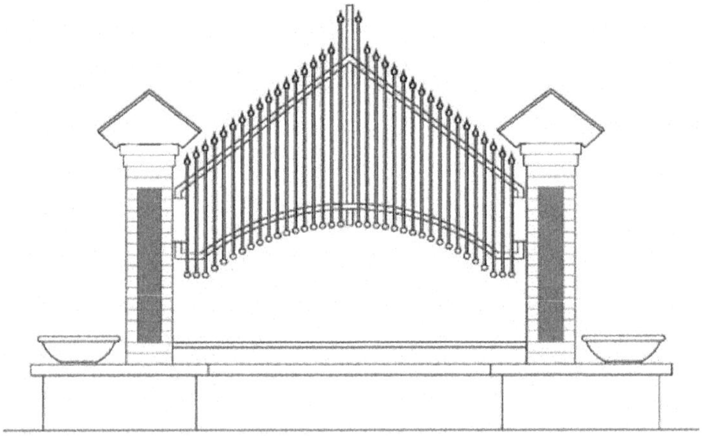

Section 1 Introduction

Chapter 3 of steel construction concerns miscellaneous steel, a collection of everything from stairwells to ladders, and has a specification division number beginning with the number 5. The word "miscellaneous" means an "assortment" of things, or "various" items. It is used in the steel trade to define a wide range of steel products. *The term "miscellaneous" is never used to name or describe structural steel or reinforcing steel (rebar).* The word "miscellaneous" is almost always seen abbreviated to "misc.". While there is only one category of structural steel, there are several types of misc. steel.

Steel stairs are *not* usually within Division 5 structural steel and are a separate division number between 5 and 6. Other types of misc. steel found within Division 5 include "ornamental metals" and "metal fabrications".

Some miscellaneous steel products are:

Angles at the perimeter of steel decks.
Bollards.
Gates.
Handrails and guardrails.
Lintels (brick shelf angles).
Metal pans.
Stairs.
Steel ladders.
Steel support on roofs for mechanical units, skylights, and roof hatches.

Miscellaneous steel can be found anywhere on the plans. Ten or more stories of steel stairs and landings may be shown on the architectural plans, not the structural drawings where the structural steel is found. The civil engineer, on the site drawings, may show a whole collection of steel bollards not shown anywhere else. The mechanical engineers may have misc. steel on their drawings, for example steel support for rooftop air conditioning equipment. Sometimes these drawing details overlap and conflicts occur. The estimator must review the entire set of plans, get everyone to fix their mistakes (it is the contractor's duty to point them out!), keep an eye on the clock (when is the deadline for bidding RFIs?), and "view the plans in their entirety".

This terminology about viewing plans means more than simply reading all the notes and looking at everything on the plans. The instruction to "view plans in their entirety" is especially pertinent concerning misc. steel because of the likelihood of it being anywhere on the plans. This contract language is from General Conditions such as AIA 201.

Even if a plumbing note is found on the electrical plans, the contractor is responsible for it. But, this is *not* what viewing plans in their entirety is about. The point here is that the plans and specifications may touch on some subjects many times, that a set of plans and the specifications contain hundreds of instructions, and since drawings show the work in various positions (plan view and sections), it will happen that, given all the commentary and drawings, that contradictory and confusing statements will be made. However, the contractor should look at all the notes, read the plans and specifications, and although there may be some confusion, when all the plans and specifications are "viewed in their entirety", the contractor should realize the designer's "intent".

Of course, this concept of interpretation leads to argument, with the designers seeking a wide interpretation and contractors sticking to a more literal one. The law on this subject generally is that large notes and drawings govern over small ones, and that specific language (whether on the plans or in the specs) always governs over general. The more detailed version of instructions governs over more vague directions. Intent gives way to specificity.

The steel subcontractor often provides misc. steel to other trades. These "furnished only" items might include connection plates given to the concrete subcontractor to be embedded in concrete for attachment to a steel beam. Or, a piece of misc. steel called a shelf angle might be furnished to the brick mason to support brick above a door or window opening. The connection plates should appear on structural plans; the shelf angle may only appear on architectural sheets.

The subcontractors and suppliers in the misc. steel trade are a diverse group of players, from the typical structural steel subcontractor to the local welding shop. Often it is a specialty contractor that does this work, not the "rough iron" guys, the structural steel folks.

There is no ruling body that governs miscellaneous steel.

Section 2 Shop drawings

The shop drawings are completed by the subcontractor that performs the work. It is typical for *all* steel in a building to be shown on shop drawings and be approved by the general contractor and architect/engineer. However, if the steel is being furnished to other trades, for example the above shelf angle, sometimes the shop drawing will be sent to the receiving trade for verification of length or other dimensions.

Some drawings for misc. steel can be quite involved because of the detail required when using small pieces of steel close together. Very exact dimensioning is required to describe a set of winding handrails proceeding up a stair tower.

Shop drawings are required for renovations as well as new buildings. A block and brick library that is being enlarged may need holes punched in the walls at dozens of locations for new doorways and reconfiguring of walls. The engineer may make use of two- or three-inch diameter vertical steel posts at the jambs of these openings, with horizontal angles and steel beams spanning between them (supporting masonry overhead), and a drawing with dimensions to be confirmed for every location!

Section 3 Wrought iron architecture and ornamental metals

Wrought iron has a low carbon content compared to modern steel and is more malleable. It is no longer produced on a commercial scale but was used for several hundred years. Its use reached its peak (in volume) in the 1860s for warships and rails for railroads. The stronger steel of the following decades, together with more efficient manufacturing, ushered in the use of steel in high-rise buildings.

Many of the fancy gates and fences that are now in historic sections of American cities were built of wrought iron in the decades around the turn of the century. Today we still build decorative architectural features out of steel products, but it is with different and stronger steel. Sometimes a modern-day construction of this former style is still referred to as "wrought iron" work, but of course it is not wrought iron steel. When you see the term "wrought iron" today, it is usually to describe an *architectural design style*, not a type of steel.

Another term that is used to describe decorative steel work is "ornamental iron". It, like wrought iron, is a term that more often refers to design than the composition of the steel material.

Section 4 Fabrication

Fabricating small pieces of steel in a shop is difficult, such as putting together the winding course of a handrail on incline. The work is for layout artists that have the tools and flat steel tables and computer programs that carve out the pieces. Less complicated are steel angles at the perimeters of floor decks, or angles above a door or window that support brick, which might only need to be cut to length.

Misc. steel products can be fabricated in a factory, mass-produced, and sold to contractors or the public. There are many metal fabrications, from "spear tips" (placed on top of fence railings) to spheres (round balls of various diameters), fancy handrails, and even lettering. These "architectural metals" can be seen in catalogs or online.

Section 5 Handrails and guardrails

There are many versions of handrails and guardrails. Some are:

- a Second-story balconies with 42″ high guardrails, a full set of vertical balusters, and no more than 2″ from the bottom rail to the balcony floor.
- b Handicap-compliant concrete sidewalks sloped less than one in twelve with a guardrail to 42″ high and a handrail stuck on the side at 36″ high. The requirement for balusters depends on how far the rail is above grade.
- c A city sidewalk with a foot drop on one side, and a simple two-line rail to keep bicycles from falling off the sidewalk.

A basic decision concerning the posts for handrails is whether they are going to be sunk into concrete or surface mounted with a metal plate and bolted down. Section E, which illustrates a collection of rails and balusters, shows the difference in construction between the two, with a plate bolted on top of the concrete shown on the left. To set the post on the right, the sidewalk has been "core bored", which means a round hole has been cut into it slightly larger than the post, drilled by a core bore machine operated by one person. After the post is positioned in the sidewalk opening, non-shrink grout is placed between the post and concrete.

If a base plate is used and the sidewalk is sloped or unlevel, the base plate must be built on an angle. Getting this precisely correct with uneven terrain often makes core boring a better option. The post grouted in concrete is a good choice for many locations because it more easily allows vertical alignment of the post.

A "two-line" or "three-line" handrail is often used in locations where there is no more than an 18″ drop and balusters are not required. These horizontal rails are shown on the left side of the section. On the right side an example of a handrail with balusters, or pickets, is shown. The note stating to use a spacing of 4″ o.c. guarantees passing the building code because of the thickness of the balusters.

See the online resources for diagram 535.1

Section 6 Bollards

Steel bollards are a typical misc. steel item. They are often "furnished only" by a steel shop fabricator (the only fabrication is cutting the pipe to length) to a concrete subcontractor for placement. It is best if bollards are positioned into place after excavation and temporarily braced with wood or metal before the pour (instead of being placed into wet concrete). After the pour, the bracing is removed and the bollard filled with grout and then painted. That's three trades – concrete, steel, and painting – that work on bollards. The estimator makes sure that all three have the right count and scope of the work.

Bollards can be found anywhere on the plans, and sometimes their quantity count or details will be in conflict because three or more designers place them on the documents. There may be confusion in the initial bid documents. The civil engineer may locate bollards on the site plans protecting the corners of dumpsters. The architect may show a few of them protecting the corners of a building, and the mechanical plans may use bollards to protect mechanical equipment. The estimator must search the entire set of plans looking for bollards, and be careful about which concrete footing details to follow because of conflicts when coordination has not been made between the designers.

See the online resources for diagram 536.1

Section 7 Stairwells and metal pans

Many buildings contain a fire rated stair enclosure, often of block construction a dozen feet wide and twenty or thirty feet long. These extend upward floor after floor, each one having a rated door that enters from the floor *to* the stairs (in the direction of exit).

The metal pans for treads are flat on bottom and typically turn up 2″ at the sides. They are filled with 2″ of concrete and are supported by steel stair stringers. The pans for stair landings are a bit deeper, 3″–4″, and often are supported by mid-sized MC channels underneath for support. Landing bottoms are often made out of regular steel decking with ribs (see Chapter 2 Steel Joists and Steel Decks).

Since stairwells are so common, a lot of manufacturers cater to this market. They may build some of the components (producing several versions of handrails, a mesh type, a thin baluster type, a thick baluster type, etc.) and buy the stringers and pans from others. Regardless of how they put their package together, they gather all the materials it takes to build a stairwell. It becomes a set of stairs by manufacturer ABC, model one, two, or three, that an architect or engineer can specify. When this happens, the bidding plans may not be very detailed.

Engineers do not consider stairs to be structural steel, and stair sections and details are often found on architectural plans instead of the structural plans. Recall that steel stairs are specifically excluded from a structural steel element within Part 16 (16.3–6) of the steel manual. Since this steel is not a part of the structural steel frame, the engineer may exclude its design, leaving it to the architect. For the architect and engineer, it is a way of dividing up the work. To the contractor, the concern is how to build the stairwell and who is going to do the work. The distinction and line drawn between what is, and what is not, structural steel is an academic issue when a contractor is responsible for the whole.

Section 8 Campus stair tower plans

Stair tower floor plan

The stair tower here is like many stairwells, enclosed inside four fire rated walls, with similar landings and flights of treads and risers formed with metal pans. The door at the bottom of the stairs of the first floor plan is the exit to the exterior of the building. Note that the third, fourth, and fifth floor plans are similar to the second floor plan, all having a rated door entering *into* the stairwell. There is a lone bollard guarding one corner of the building.

See the online resources for diagram 538.1

Section A campus stair tower

Section A notes that metal pans are used for 2″ deep treads and 3″ deep landings. Per the note at the top, there are four each vertical posts in each handrail flight. The size of two framing members is given in this section, the MC 6X7 (miscellaneous channel) at the landings, and the HSS 12 × 2 × 3/16 (hollow structural steel) stringers. Later, at the shop drawing stage, the elevations of the floor levels and landings will be verified onsite, as well as the length and width of the room. At the bottom of the stairs, the stringers rest on the floor, and Detail C is referenced.

See the online resources for diagram 538.2

Section A-1 campus stair tower

The stairwell has light steel framing to support the pans at the landings. Refer to the M6X7 framing shown in Sections A-1 and B. At the treads, the pans are self-supporting and can extend across from one stringer to another with simple 3/4″ angle framing attached to the stringers. The guardrails sit on top of the winding two center stringers that are spaced 4″ apart. The guardrail, with handrail attached to the side, follows the footprint of the stringers. Imagine how careful and exact the shop drawings must be prepared!

The thin vertical "balusters", or "pickets", are 1″ round. *The building code rule for the spacing of balusters is that a 4″ sphere cannot pass between them.* In other words, 4″ is the minimum spacing between two balusters.

Notice the 7″ and 11″ dimensions for the treads and risers in Section A-1, and the 7″ rise shown in Section B. This is an important ratio; *remember the numbers 7/11. The building code rule for the rise and run of stairs is that the sum of two risers plus one tread must be between 24 and 25.* The use of seven-inch risers and eleven-inch treads in this equation is 14 + 11 = 25. These dimensions are used at many stairways.

See the online resources for diagram 538.3

Sections B and C campus stair tower

The handrail is offset 1-1/2″ with brackets on each post.

There is an important length shown here, the 4″ in the middle of the top dimension line (with 4′-2″ on either side). This is the same 4″ found on the first and second floor plans, the distance that the stringers are separated. There is nothing within this 4″ space (vertically) all of the way up the stairwell.

The steel bracket clip angles shown on Detail C are attached to both stringers sitting on the first floor. Detail C is referenced on Section A. The clip angles are bolted to the floor and welded to the stringer.

See the online resources for diagrams 538.4 & 538.5

Section 9 Stair tower takeoff

No.	Stairtower Takeoff	Detail	Sides	Each	Landings	Flights	L	539.1 LF
1	Stringers HSS12X2X3/16			2		9	10	180
2	MC6X7 framing at landings:							
3	at 30" o.c. shortway pieces	Section A-1, B		5	9		6	270
4	longway pieces			2	9		10	180
5	3/4 x 3/4 x 1/8 pan support:							
6	beneath treads	Section A-1		8		9	1	72
7	beneath risers	Section A-1		9		9	1	81
8	4 x 4 x 1/4 bracket first floor	Section C		2				
9	1-1/2 guardrail top and bottom	Sect. A, A-1, B		2		9	10	180
10	1-1/2 vert posts @ thirdpoints	Sect. A, A-1		5		9	4	180
11	1-1/2 handrail on block	Plan, Sect B				9	11	99
12	1-1/2 handrail at center	Plan, Sect A, A-1, B				9	11	99
13	1" o.d. vert balusters	Section A-1		20		9	3	540
14	Handrail brackets on block	Section B		36				
15	Handrail brackets at alum posts	Section B		45				
16	Metal pans	Lump sum by manufacturer						

Takeoff Notes:

Row 1: There are two stringers and nine flights. Each stringer will use approximately 9 LF of steel, rounded up to 10 LF. Structural steel is manufactured in lengths of 20′ and 40′.

Row 3: Placing the framing at 30″ o.c., which is the plan note, would put five pieces (since the stairwell is 8′-8″ wide) at 0″, 30″, 60″, 90″, and 104″. This is an unequal spacing. It is not known from the plans if the underside of the landings are exposed, but these kind of notes are where the engineer is just reading from load tables and giving an instruction that is structural. When the detailer interprets this "on center" instruction, the five pieces of framing will be placed across the landing *equidistantly* on the shop drawings, instead of an installation with uneven spacing. The framing will be symmetrical viewed from underneath the landing and will provide the best support because the load will be distributed evenly.

Row 4: The two framing pieces would be approximately 8′-8″ long, but the length is rounded up to 10′ in the takeoff because they would be cut from a 20′ length of steel. The pieces left over would be waste and recycled.

Rows 6, 7: The length of the angles, see Section A-1, would be slightly longer than the treads and risers. For takeoff purposes, these angles are figured 12″ long.

Row 13: From Section A-1, six vertical balusters can be seen between two vertical posts. There are three sections of these balusters, see Section A, so that is 18 balusters per flight. The estimator adds two more for a total of 20. By adding two to what can be seen, this accounts for the ends of the flights (at the landings), which aren't shown.

Row 14: Nine flights times four brackets each, see Section B.

Row 15: Nine flights times five brackets each.

Section 10 Photo of three misc. steel projects

This photo contains three different miscellaneous steel projects. One is a subdivision sign designed and built by the author, consisting of 3/4″ diameter vertical bars with 2″ round spheres at the bottom and spear shapes at top. These bars are on both sides of HSS 6 × 2 steel framing. The bottom piece of this 6 × 2 was bent into an arch, outsourced to a steel shop that specialized in bending steel. With the curved piece in hand, the steel frame and bars were fabricated in a local shop, then taken to another one for powdercoating. The assembly was then delivered to the site and bolted to the brick columns.

Beneath the center of the arched sign and in the distance are steel handrails attached to the cast-in-place concrete stairs of the Livingston Square Condominiums pictured on the cover page of this book.

Behind the sign is a steel fence made from prefabricated parts from a manufacturer's catalog and assembled onsite.

Subdivision Sign

The painted wood sign was attached after the above photo was taken.

PART 6
Carpentry

1

PRODUCTS AND METRICS

Section 1 Introduction

Building floors, walls, and roofs with dimensioned lumber is called rough framing. Installing baseboard, paneling, and molding is termed finish carpentry. The talents for these two differ and sometimes carpenters (and carpentry crews) perform one, or the other, but not both. A subcontractor that completes carpentry work is often called a "framing contractor". Sometimes a general contractor does carpentry work in-house. It is these contractors who count and estimate the lumber in a project.

Many minor decisions are made by a carpentry crew in laying out and framing a wood building. Often it is the best (and fastest) crew that uses and wastes the minimum amount of lumber.

The 2 × 4 is a workhorse, used over and over for various purposes in a building. Hardly ever is a full length of lumber used (with some exceptions, for example precut studs). When a carpenter picks up a board, the lumber is cut to fit a purpose. That is what carpenters do, and before lumber has met nail the estimator has lost 10%–15%.

Two separate carpentry crews can use a different amount of lumber to build a wall (or a roof). The lengths of blocking and nailers can vary, as well as waste and miscuts. Vertical studding can vary somewhat, depending on which end the wall is laid out from, because the on-center spacing of studs is kept for the full length of the wall.

Although carpentry may seem to be a straightforward trade, counting the lumber from a set of building plans can be a risky business. Whether the plans are drawn by architects and engineers or house plans by a drafter, they will usually fall far short in defining the extent of lumber required for a project. This is not because the designers do a poor job.

There is a great deal of wood bracing and blocking that can be missing on the plans. That is why contractors have sad stories of that large wooden structure that went over budget. "Where could all that lumber have gone to? It wasn't stolen! We bought 20% more lumber than we could find on the plans!"

Lumber has many nonstructural uses, and a whole host of these go unreported on the plans. Common dimensioned lumber is a handy product. There are nailers, blocks, cats, deadwood, bracing, and more used throughout a building. The estimator has to know where these items might be located.

Section 2 Wood components not shown on plans

Gypsum board nailers

Sheet goods, often sized 4′ × 8′ (or 4′ × 10′, or 4′ × 12′, etc.) include gypsum board, wood paneling, and wood or cement board. Sheet goods require support (something to nail to) every 16″ or 24″ on center, and sometimes even more often (roof plywood often 6″ o.c. at edges, and 12″ in the field).

The need for wood nailers in ceilings depends on the direction of overhead framing. If the overhead roof framing (trusses or rafters) is perpendicular to a supporting wall, then the gypsum board applied to the ceiling can be nailed at every truss or rafter. If the direction of overhead framing is parallel to the supporting wall, and the truss or rafter is, say, a foot away from the wall, nailers are placed in ladder-like fashion across the top of the wall (see the sketch in the carpentry glossary).

See the online resources for diagram 612.1

Truss bracing

Shop drawings by truss manufacturers, signed and sealed by structural engineers, often instruct that bracing needs to be added onsite after the trusses are set in place. The quantity of bracing can be considerable, but even if there is a small amount, the building department is going to inspect the bracing because of the engineering requirement. Truss bracing is a big deal for the estimator, superintendent, and building department. The truss shop drawings are often a part of the building permit application.

Truss bracing often consists of horizontal runners perpendicular to the truss span. Another typical type is vertical X bracing at the last truss. The purpose of bracing is often just to hold the truss rigidly in place. The result is a truss(es) with fewer webs, built with smaller sized lumber, and less cost. It certainly seems to the estimator that truss engineers are good at trading truss components for field-applied lumber, because there can be a lot of it!

The quantity of braces is problematic for the estimator because at the time of the bid, the drawings by the A/E (architect/engineer) are available *but the shop drawings by the manufacturer (who is simply bidding the job to the prime bidders) are not.* The manufacturer is not going to go to the expense of creating shop drawings until a purchase order is received by the successful contractor. Estimators at the time of the bid have to "throw in" hundreds of feet of 2 × 4 bracing, guessing that perhaps three or four or maybe six braces may be used running the full length of the building, plus some vertical X bracing. At the bidding stage, no one knows.

See the online resources for diagram 612.2

Bridging and firestopping

These wood components are similar in that they are short pieces perpendicular to floor joists, wall studs, ceiling joists, and roof rafters. With a 24″ spacing of joists, studs, or rafters, "blocking" is 22-1/2″ long; with 16″ spacing the blocking is 14-1/2″ long.

The national fire code requires horizontal 2 × 4 firestopping in a wall at the ceiling level. Review the fireblocking sketch below. Such blocking, often not shown on the plans, occurs in walls often ten to twenty feet high.

Horizontal wall blocks are sometimes called cats. In addition to horizontal blocks at the ceiling level, with *tall walls* building codes and engineers require horizontal blocking at 4′–5′ o.c. vertically for bracing.

See the online resources for diagram 612.3

Division 10 blocking

A multi-story public building with large bathrooms on each floor may have dozens of toilet partitions, grab bars, waste receptacles, mirrors, etc. Each bathroom needs wood or other blocking to fasten these items to (behind the wallboard). Grab bars have to withstand a couple of hundred pounds of weight; installing wood blocking at exactly the right locations is an important task.

There may be a note on the plans to provide blocking for Division 10 items, but typically not much in the way of details. However, the pieces of lumber required can be considerable. The labor required to install this lumber can be far greater per piece than it is in other locations due to the small lengths and precise locations.

Section 3 Lumber types and metrics

Common lumber

Board sizes range from 2 × 4 up to 4 × 12. These sizes are typically sold in 8′ to 16′ lengths in 2′ increments. Longer lengths may be available, but a 22′ long 2 × 12 might be a special order (the lumber yard may not have it in stock).

Common lumber and timber can be purchased either rough sawn or dressed. Rough sawn lumber (called 2 × 4s, 2 × 6s, etc.) has been sawn to *nominal* sizes of 2″, 4″, 6″, 8″, 10″, and 12″ and retains the rough cut of a large saw blade. When dressed (still called 2 × 4s, 2 × 6s, etc.), it has been sawn *again* and planed down to smooth *actual* sizes of 1-1/2″, 3-1/2″, 5-1/2″, 7-1/4″, 9-1/4″ and 11-1/4″. Lumber that is dressed on all four sides is known as S4S (surfaced four sides); S2S is surfaced on two sides, and S1S is surfaced on one side.

Timber

Large sized lumber is 4 × 4 up to 12 × 12 (for a maximum of about 24″ square). The typical lumber yard will have on hand 4 × 4s and 4 × 6s, for use as posts, but may not have in stock 4 × 8 or 6 × 6 and larger sizes. Timber can be purchased from mills. Timber is more likely to be purchased rough sawn than common lumber.

Round poles

This is another category of lumber used as columns and pilings and is available in lengths up to 50′ or so (for larger diameter poles). Poles are trees with the bark cut off, so they are sometimes left tapered, and have a "butt" end that is larger around than the other end.

Precut studs

These studs are often used in residential construction. They are 2 × 4s or 2 × 6s, high enough in demand that mills produce them. The standard 8′ high wall is actually built an inch or so taller than 8′. Two horizontal 4′ × 8′ sheets of gypsum board or plywood are 8′ high exactly, and this leaves no tolerance for unlevel floors, nor room for the ceiling board. The wall needs to be taller than these two sheets.

Some precut studs are 7′-8-5/8″ tall (92-5/8″). When 4-1/2″ of base and top plates are added that creates a wall 8′-1-1/8″ tall. Refer to the carpentry glossary at the end of the book for a sketch of a wall built with precut studs.

A building wall height tolerance beyond 8′ is needed for several reasons. One is to allow for the slab being up and down a little (say 1/8″ to 1/4″), and another is the good practice of not placing sheet goods directly on a slab because of moisture and decay prevention. Another reason is that the ceiling boards are installed on top of the wallboard, and the wallboard already takes up a full 8′. When a 1/2″ ceiling board is slipped over 8′ of wallboard, there is 8′-1/2″ of gypsum board product against the wall.

See the online resources for diagram 613.1

Manufactured wood trusses

Wood trusses are structural members, and truss drawings must be sealed by an engineer. The truss shop drawings become an integral part of the project plans. This is the providence of truss manufacturers, who hire engineers and build trusses. The plans for a house or other wooden structure may not even have a "roof framing" plan, no matter whether a designer, architect, or engineer drew the plans. The structure may have some wood or steel beams designed by an architect or engineer, who seals this portion of the plans. However, the roof design is by the truss manufacturer, and the A/E is often silent in regards to it. This brings to light a problem for the poor estimator, charged with pricing the work.

At the time of bidding, the truss drawings do not exist. The estimator only has a quote for the trusses, which includes future shop drawings. The estimator, quote in hand, has to make an educated guess of the amount of wood that will be required for field-installed "truss blocking".

See the online resources for diagram 613.2

Roof trusses are built with top chords at various slopes. The estimator uses the geometric factors in the following "Table of roof areas", multiplied by horizontal measurement, to determine roof sheathing or the lengths of rafters. Horizontal measurements are converted into a sloped distance, or a flat area into a sloped area. See the carpentry glossary for more truss terminology.

Glue-laminated products

Glu-lams are made by gluing multiple pieces of common lumber together, often for architectural effect, but they are also quite strong and used structurally (to carry loads). Horizontal beams and "A"-shaped arches are two glu-lam products.

Section 4 Units of measure

Two units of measure

Once the lumber is identified, whether from the plans or inferred, the estimator chooses between two "units of measure". They are "each" and "linear feet". The decision on how to initially count lumber keeps the purpose at hand and the least number of steps are taken.

"Each" as a unit of measure

If the length is important (for example, rafters need to be purchased as 12s because they are shown to be 11′ or so long), then the initial *unit of measure for counting rafters is "each".* Note that if the rafters are 11′ long, then obviously 10′ pieces will not do and a 14′ or longer board would create a lot of waste. Consider the following items: girders, floor joists, wall studs, ceiling joists, and roof rafters. It is often readily apparent from the plans that these items have to be a specific length. If 13′ is needed in the field, the takeoff needs to state that 14′ lengths are required because that is the length that should be purchased. Any other length will not do.

If the length is important *then it becomes a part of the description on the takeoff.* The item is not 2 × 8 rafters; *it is 2 × 8 × 12 rafters.* Adding the length to the description on the takeoff is a big distinction. It narrows the scope and further defines the work, allowing the use of the specific price (on the estimate) for 12′ boards.

Total length as a unit of measure

If it is not important to keep track of the lengths of boards, the task is to count the total "linear feet" of the product. Consider the following items: base plates, top plates, fascia, and baseboard. The lengths of these pieces of lumber are typically unimportant to the estimator. It does not matter if the boards are 10′ long or 14′ long. Later, at the buyout stage, the project manager or superintendent can decide lengths. The estimator's task is to use *linear feet* as unit of measure and add a waste factor. This is lumber being counted by the running foot.

The description of the work item lists the size and not the length – *2 × 8 rafters.*

A consequence of counting lumber by the linear foot is that, on the estimate (where the pricing is done), an "average" price for several lengths of lumber will have to be used instead of the price for a certain length. This may be of some consequence because *lumber prices are not proportional* – a 10′ board may cost $1.00 per foot and a 16′ board may cost $1.10 per foot, a 10% difference. To further complicate things for the estimator, the lumber market is quite volatile and prices don't stay the same for long.

Section 5 Conversion factors

"Factors" are sometimes used to determine quantities of lumber that are different from waste factors. A conversion factor is sometimes used to multiply times a horizontal distance to arrive at the inclined distance. This is not a waste factor; it is a conversion factor. It does not include waste.

The quantity of wall studs is approximated based on the length of a wall type; to try and count wall studs individually would take too long and still be inaccurate (unless there are only a couple of walls). A "factor" is used, depending on how "busy" the wall is, to estimate that, say, 1.1 studs per linear foot will be needed. This factor converts linear feet of wall into the number of studs per foot. It is a conversion factor, not a waste factor. However, in this case it includes waste. This may, of course, be confusing for the beginner.

One last "factor" should be mentioned that sometimes occurs when working with lumber. Sometimes the plans are vague and have limited details. The estimator may suspect there is more lumber in the project than can be directly counted and inferred from the plans. If the plans are sketchy, an estimator might add five or ten percent to the entire carpentry estimate. This is a *subjective judgment factor*, not a waste factor.

Section 6 Measuring areas and lengths of inclined surfaces

This table is used to determine the *net* area of a roof. Given the roof slope, the *flat area* is multiplied by a *factor*, which results in the quantity of *net roof area*. The roof area is used to determine the quantity of roof sheathing (decking), vapor barrier (roof felt), shingles, etc.

See the online resources for diagrams 616.1 & 616.2

These factors can also be used to determine the length of a rafter, see Chart 1, based on the dimension of the horizontal distance. Horizontal distance times factor = board length.

Section 7 Carpentry takeoffs

The difficulty of counting lumber

Compared to the exactness of concrete and the modular nature of masonry, carpentry is the "wild west" of quantity surveying. While concrete and masonry are solid in nature and their extent defined on the plans, lumber is not static. With so much often left unsaid in the documents, with building and fire codes referenced but detail not shown, with other standards and best practices expected but not defined, and with lumber being so convenient to use as bracing and blocking (with none of this shown on the plans!), the estimator has to take special care to accurately determine lumber quantities for a project. Indeed, lumber quantities seem elusive, a moving target.

Boards are like toothpicks when compared to other building materials, and in counting their quantity the estimator should use a wide lens, not a microscope with engineering scrutiny. Exterior wall lengths are counted corner to corner without regard for counting the corner twice as in masonry. When the width of a wall is only 4 to 6 inches thick, and the length is only counted to the nearest foot, why bother? The precision in wall lengths is not as necessary in wood framing as compared to concrete and masonry.

The carpentry takeoff form

The following format works for much of Division 6. Work left to right. Note that both "Length" and "Linear Feet" are column headings. The length is for when a specific board length is selected; the linear feet is an extended total.

(Company Name)											617.1
CARPENTRY TAKEOFF											
No.	**Description**	**Det**	**Ea**	**Pcs**	**o.c.**	**L**	**W**	**Ht**	**%**	**LF**	**SF**
1	2 x 4 base plates										
2	2 x 4 top plates										
3	2 x 4 x 10' studs										

Floor framing, wall framing, and roof framing should all be keep separate.

Keep different areas separate within each type of framing (for example, floor framing separated by rooms). Keep the framing separate for various floors and separate buildings and wings. With walls, separate by height. If it is different, keep it separate! Remember, the takeoff is not cost-coded. It simply lists items sequentially 1, 2, and 3. When the estimator decides to take off the components of the exterior wall, base plates are #1, top plates are #2, studs are #3, and then it's time to go to the next item.

A takeoff is not simply a lumber list. Since a 2 × 12 header takes longer (per LF) to install than a 2 × 12 floor joist, they are listed separately. These two purposes for 2 × 12s are kept completely separate *because the estimate needs them that way.* Note that a "lumber list" sent to a supplier to get material prices would combine them. This is an important distinction for the estimator and a primary reason that even architects and engineers may not understand the purpose of a takeoff. This is why *takeoff descriptions are so important.*

Section 8 Waste factors

More confusion is in store when waste is figured differently for the two units of measure (LF and each).

By definition, counting by the running foot (linear feet) is an "exact" number, but this does not mean counting down to the inch. When using a scale and blueprints, round off to the next linear foot when turning a corner. If a wall appears to be 12′-6″ long, count it as 13′ and go to the next one. There are a hundred walls and not enough time to count them down to the inch and it does not matter anyway. Get the wall length, or fascia or baseboard length, *covered*, as they say in estimator's language.

For roof framing counted by the linear foot, add at least 10% for waste.

For "running" trim lumber such as baseboard, ceiling cove molding, and chair rails, count by the linear foot, and add 15%–20% for a waste factor.

With trim lumber such as window sills and door/window casing, it is more accurate to count by the piece (each) than by the running foot. This kind of trim is installed in one piece and requires a certain length of trim lumber to be purchased.

With 2 × 12 × 12 rafters and other pieces that are length specific, some waste occurs when a carpenter cuts them to length, often 10%–15%. This factor is already accounted for simply choosing 12′ lengths to begin with. The waste factor only has to account for miscuts and discards of crooked or damaged lumber, which should be in the range of 5%.

2

FLOOR FRAMING

Section 1 Photos and drawing(s)

Photo 1 2nd Floor Wood Trusses

Single 2 x Floor Joists supported by metal joist hangers at band joist

Photo 2 - 1st Floor Porch Framing

Temporary Brace

Double Band Joist

Glu-lam beam, or girder

Photo 3 - 2nd Floor Rafters And Glu-lam

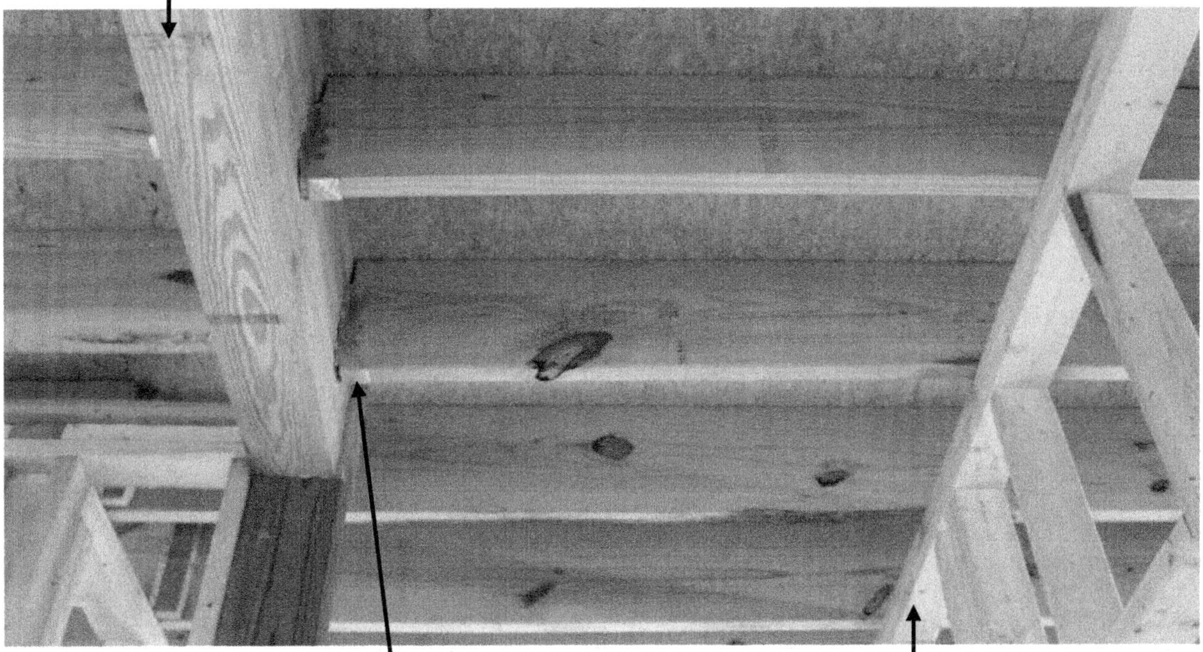

Joists supported by metal strap

Was this wall meant to be load bearing? Since the top plates have been built tight to the joists, it is a bearing wall, although the glu-lam is nearby.

Sometimes, carpenters inadvertently create bearing walls by building right upto rafters or trusses, when the design intent was non-load bearing (no footing underneath) with a single top plate.

Photo 4 - 2nd Floor Framing With Bridging

Section 2 Boardwalk

Mount Dora boardwalk plans

See the online resources for diagrams 622.1 & 622.2

Plan interpretation

The first thing to notice about this plan is the curious way the boardwalk is dimensioned. Perhaps the owner's instructions to the engineers were to give them a boardwalk 6′ across at the top, and they delivered with no other horizontal dimensions. To lay out the pilings requires starting at the wheel stop.

Round wood posts are used for the pilings. Each pair of pilings is held together with two each 2 × 10s. There are seven each 2 × 10 joists running continuously across the top of the beams. These joists continue for the length of the boardwalk. The two outer double joists are bolted to the pilings.

The decking consists of 2 × 6s with a space of 1/2″ between the boards. A double 2 × 4 wheel stop occurs on top of the decking.

The 220 linear foot length of the boardwalk is given in the note about pilings "10′ o.c. for 220lf". There are no turns in the boardwalk; it is a straight run.

Scope of the work

Include:

Furnish pilings, installation by others.
Provide labor and material for a complete job to install beams, joists, decking, and fasteners.

Exclude:

Concrete sidewalks.

Construction techniques

The pilings are purchased in this trade package, but a separate contractor will install them, using a "jetting operation", which is water under high pressure sprayed against the lake bottom to create a void for the piling to be dropped into.

The pilings will be left "high", by a few inches at the time of installation, with the top cut off later. The pilings will not be exactly plumb – the operation of installing pilings underwater is not an exact one. The boardwalk is shown to be straight above the water's edge. A site visit reveals the water to be about 1′ deep at the far edge of the boardwalk.

Bolting the transverse beams (perpendicular to the run of the boardwalk) in place will help to plumb the pilings. The joists can then be set. All of the work so far is done while workers are partially in water or on a raft or small boat. The decking can be placed from the top after the first few are installed. The wheel stops are the last boards to be installed.

Carpentry takeoff

When the length of a board is known, it becomes a part of the description. Note in this boardwalk example that determining a board's length is the first step in determining all of the quantities except for the wheel stop. A boardwalk does not have much lumber counted by the "running foot".

CARPENTRY TAKEOFF											622.3
job: Mt Dora Boardwalk			number						date		
No.	Description	Det.	Each	Pcs.	o.c.	L	W	Ht.	%	LF	SF
1	9" x 12' round wood pilings			46				12		**552**	
2	or		23	2				12		552	
3	Beams 2 x 10 x 14's		23	2		7				322	
4	Waste, one pair of beams			2		7				14	
5										**336**	
6	Joists 2 x 10 x 12		22	3		12				792	
7	Waste say two ea									24	
8										**816**	
9	Outer joists bolted		22	4		12				1056	
10	Waste say two ea									24	
11										**1080**	
12	2 x 6 x 14 decking			440		7			1.05	**3234**	
13	2 x 4 wheelstop			4		220			1.1	**968**	
14						Each	%	Each			
15	5/8" x 14" bolts		4	46		184	1.02	**188**			
16	2" x 5/8" washers					184	1.02	**188**			
17	5/8" nuts					184	1.02	**188**			

Row 1: The labor to jet and install pilings will be subcontracted. However, the material cost is in this bid so the pilings are counted.

When a configuration consists of a repeating pattern like a boardwalk, sometimes it helps to divide it into *sections* and *rows*. Here, there are 22 sections at 10′ long (see plan length of 220′), so there are 23 rows of pilings, or 46 total, measuring 9″ at the butt end. While doing the takeoff of individual items, there will be a certain number of them "per section" or "per row". Repeatedly using the sections and rows as a multiplier helps eliminate mistakes.

Assume no extra pilings will be purchased; there is no waste factor except in figuring the length (height), see below. The poles are expensive – it is not expected that one would be lost or cut up the wrong way, and a waste factor is not used in this example (a marine contractor may advise differently!). Forty-six pilings are estimated and forty-six will be used in this takeoff.

The drawings show a net piling height of 10′-6″ will be needed (30″ above the water, 8′ below). However, due to water fluctuations, the inexact nature of jetting the poles to any specific inch, and the poles being left a little "high", a foot or so should be added to the height of the piling for estimating purposes. Twelve-foot-high pilings are chosen for the takeoff.

Row 3: Some arithmetic and plan review must be made to determine the length of the beams because they are not dimensioned. At the edge of the crosswalk section the beam appears to be about 1-1/2″ beyond the wheel stop, making the length of the beams and decking to be 6′-3″. Therefore, the individual board lengths of each of these (beam and decking length) will need to be either 8′ or 14′. Select 14′ lengths, which will provide two pieces.

The quantity of beams is two times the number of rows (2 × 23), or 46 pieces total, the same count as the pilings. For the waste factor, use one pair of beams, a completely subjective factor.

Rows 6 and 9: To figure the length of the joists, the piling-to-piling spacing is 10′. Each joist will have to extend about another 8″ at each end in order to satisfy the requirement of each joist bearing on both beams. The minimum length of each joist is then 11′-4″. Twelve-foot lengths of joists are selected.

At this point, the estimator earns his/her pay and insists that the seven joists (see the plans and their interpretation – there are seven joists!) are in fact two different types. The three joists running down the middle of the boardwalk, says the estimator, are faster to install than the four (two on each outer side) that are bolted through the pilings. The estimator realizes that the bolted outer joists have a different labor factor than the middle joists. If they are different, keep them separate!

For the waste factor, use one pair of joists, a completely subjective factor.

Row 12: Since the decking has a 1/2″ space between the boards, one board and one joint is 6″ wide in the direction of the boardwalk length. There are two boards per foot. The number of boards is found by doubling the boardwalk length, which means there are 440 boards. It would be best to purchase them in 14′ lengths, which provide two planks per length.

Row 13: There are four wheel-stop pieces that run the length of the boardwalk, which is 220 LF. Add typical rough carpentry waste of 10%.

Row 15: There are four bolts per piling. Each one goes through a 9″ post and two boards (3″), this being 12″ exact. Adding for enough length to place the nut and using an even number, the estimator selects 14″ lengths for the bolts, which becomes a part of the description.

Section 3 First floor framing

Boy Scout lodge plans

See the online resources for diagrams 623.1, 623.2, 623.3, 623.4, 623.5, 623.6, & 623.7

Plan interpretation

The framing plan makes use of a grid laid out using the numbers 1–8, see these numbers in parentheses. This is a common drawing aid, as reference can be made to "row 5 column 2", which spots a specific location on the floor.

The floor structure rests on 6 × 12 girders. The girders are supported by 3′ square concrete pads as noted on the plans. There is an isolated row of girders at Section B on the east side – they are not connected to the other girders.

The wood-framed floor structure is approximately 30′ × 40′. The joists are 2 × 12s except for one row of 2 × 10 joists on the east side. At the perimeter the joists are doubled, see sections A, C, D, E, and F. There are three cantilevers (one for 2′ and two for 1′) where the floor joists extend past the girders.

The floor framing plan is *diagrammatic*. Only one line is drawn at the perimeter band joists, but two are in the sections. *Quantities of lumber cannot always be counted accurately in "plan view"*. As long as the drafter has indicated that a second band joist is required in the sections, the plans are sufficient; it is up to the estimator to capture all the conditions based on the plans in their entirety. In this case, the plan is busy enough without adding more lines, so the drafter chose to leave them off the plan view. The sections are more specific.

The 2 × 10s are supported on one side by metal joist hangers; see Section F. All of the joists are covered with 3/4″ plywood. There is a double joist at Section F, but only one is shown in plan.

The perimeter walls at Sections D and E are supported by perpendicular joists underneath, which provide support for the base plate of the wall. *Where the joists are parallel to walls*, see Sections A and C, transverse blocking will be required to support the base plate. The two band joists have a combined thickness of only 3″, which is less than the thickness of the wall, and the blocking is needed to keep the wall from rotating. These blocking pieces would be less than 2′ long and span from joist to joist in ladder fashion. They are not shown on the floor plan. *Section details sometimes reveal more pieces than what can be counted "in plan".*

Scope of the work

Include:

Furnish girder material only, labor by others.
Provide and install all joists and bands.
Provide and install plywood and blocking.

Exclude:

joist hangers, nails, and fasteners.

Construction techniques

The girder placement is by others. The band joists at the perimeter are laid out and installed next. Batter boards and string would typically be used for the layout. The hangers and joists would be installed next, then the plywood.

Carpentry takeoff

CARPENTRY TAKEOFF											623.8
job: Boy Scout Lodge			number						date		
No.	Description	Det.	Each	Pcs.	o.c.	L	W	Ht.	%	LF	SF
1	6 x 12 x 12 girders col's (1,3,4)	A,B, C	3	3		12				**108**	
2	6 x 12 x 14 E/W girders west side	E		4		14				**56**	
3	6 x 12 x 16 E/W girders east side			4		16				**64**	
				17	girders compare to plan						
	East side										
4	2 x 10 x 10 joists run E/W	B		18		10				180	
5	2 x 10 x 12 band joist N/S	B		3		12				36	
6	Adder for waste			1		10				10	
7	2 x 10 total									**226**	
	Middle area - there are two joists at Section C!										
8	2 x 12 x 12 joists	C		20		12				240	
9	2 x 12 x 14 joists			10		14				140	
10	2 x 12 x 16 band joists	D		4		16				64	
										\	
	West side										
11	2 x 12 x 12 joists			24		12				288	

	CARPENTRY TAKEOFF										623.8
job: Boy Scout Lodge			number						date		
No.	Description	Det.	Each	Pcs.	o.c.	L	W	Ht.	%	LF	SF
12	2 x 12 x 16 band joists			4		16				64	
13	2 x 12 total									796	
14	waste 5%								0.05	40	
15	Total lineal feet of 2 x 12's									**836**	
16	2 x 6 blocking	A		17		2				34	
17	2 x 6 blocking	C		19		2				38	
18	Total 2 x 6 blocking									**72**	
19	2 x 10 joist hanger	C, F		14							
20	4 x 10 joist hanger	F		2							
21	3/4" tongue and groove ply					14.0	31.5		1.1		485
22						14.5	34.5		1.1		550
23						9.67	30.5		1.1		324
24											**1360**

Row 1: The N/S girders at column lines (1), (3), and (4) are counted first. Since all of the lengths are over 10′ but less than 11′, 12′ lengths are selected. At column line (2), there are no N/S girders.

Row 2: The girders on the west half (rows 5, 6, 7, and 8 running E/W) of the lodge have a dimension of 14′ long, but the N/S girders on row 1 were just counted full length; these girders running E/W will be 6″ shorter than 14′; 14′ lengths for the girders between column lines (1) and (2) will work just fine.

Row 3: The girders on the east half (rows 5, 6, 7, and 8 running E/W) of the lodge have a dimension of 14′-6″, and because they butt into the N/S girders will be 6″ shorter or right at 14′. However, since these are big timbers, the lengths should not be figured down to the inch, and 16′ lengths are figured.

The girders are 6 × 12 timbers. They are large pieces of heavy lumber, contain a lot of board footage, and are expensive. It is unlikely for the lumber yard to deliver a crooked one, or for the carpenters to miscut one. There is waste involved in selecting them in 2′ increments, but extra pieces are not ordered.

Row 4: Count the quantity of 2 × 10 joists on the right side of the floor plan. Include the doubling of the joist at Section F, and the doubling of the band joist at the north end, which is inferred. From a plan dimension of 9′-8″, 10′ lengths are selected. The waste factor is one each.

Row 5: These are the joists at column line (4) perpendicular to other 2 × 10s. They are single band joists on the east side in 12′ lengths.

Rows 8, 9: The joists in the bottom two runs (see column lines 6 to 7 and 7 to 8) are 12′ lengths; at the north end the joist length is 14′.

Rows 10, 12: Sixteen-foot lengths of 2 × 12 band joists are needed at the north and south ends of the middle and west areas.

Rows 13, 14: Approximately 5% of waste would be added to the lengths of rows 8, 9, and 10 when these quantities are sent to the estimate. It is usually best, when adding extra boards, to get them in the longest length, which in this example is 16s.

Row 16: Use a length, from north to south, of 32′ and divide by two, which is 16. Add one to make it 17 each.

Section 4 Second floor framing

Livingston Square Condos plans

See the online resources for diagrams 624.1, 624.2, & 624.3

Plan interpretation

There are three bays of second floor framing. The wood floor structure consists of floor trusses 18″ high bearing on double ledgers that are fastened to the block wall; see Sections C and D. The ledgers do not occur on the north and south ends – see Section D which shows that the south end lacks a ledger; the north end is interpreted to be similar.

The porch is attached to the building by a steel angle shown on the floor plan and Section D. There is a double header at the outside of the porch with column support. The top of the floor joists are covered with 2 × 6 tongue and groove boards.

At the mid-span of the porch joists there is a row of 2 × 12 bridging, each piece being approximately 14-1/2″ long. See the Floor Plan for joist spacing of 16″ o.c. There is no bridging for the porch framing on the east/west sides.

Scope of the work

Include:

Furnish and install floor trusses, ledgers, and plywood.
Furnish and install porch framing including joists, blocking, and 2 × 6 t & g.

Exclude:

Division 5 steel angle and Division 6 columns.

Construction techniques

The 2 × 12 ledgers with drilled holes into the block would be installed first. They would be installed level and the holes embedded with epoxy. The 2 × 6 portion of the ledgers would then be nailed to the 2 × 12s.

There would be some material handling involved with the trusses, and perhaps a crane. Nailing them in place is easy; all the work is getting them into place.

The steel angle at the porch would be installed next. The porch columns would then be stood and braced plumb. The header beam and joists would be installed, then the porch decking.

Carpentry takeoff

CARPENTRY TAKEOFF											624.4
job: Condos			number						date		
No.	Description	Det.	Each	Pcs.	o.c.	L	W	Ht.	%	LF	SF
	2 x 12 ledger perimeter:										
	Method 1 visual										
1	2 x 12 x 10 ledger ext wall			4		60			1.1	264	
2	2 x 12 x 10 ledger int wall			4		63			1.1	277	
3	2 x 12 x 10 ledgers									**541**	
4	2 x 6 x 10 ledgers									**541**	
5	5/8" x 6 bolts at 2 x 12 ledger		**246**		2	492			1.05		
			(3)		(2)	(1)					
	Exception to entering data from left to right, start with the 492' of 2 x 12's, then divide by 2' oc to get EACH (the number of bolts).										
					(each)						
6	Floor trusses	ABCD	3	31	**93**						

CARPENTRY TAKEOFF											624.4
job: Condos			number						date		
No.	Description	Det.	Each	Pcs.	o.c.	L	W	Ht.	%	LF	SF
7	2nd floor plywood					60	60		1.06		**3816**
8	2 x 12 porch ledger on stl angle	D				90			1.1	**99**	
9	Dbl 2 x 12 band	D		2		138			1.1	**303.6**	
10	2 x 12 x 12 joists south end			45.1	1.3	60					
11	Add two for doubling			2							
12	2 x 12 x 12 joists south end			47.1		12			1.05	**565.35**	
13	2 x 12 x 12 joists E/W sides			13		12				156	
14	extra			3		12				36	
										192	
15	2 x 12 bridging					60			1	**60**	
16	2 x 6 t & g decking total area										1036
17	factor convert SF to LF (12 / 5.5)								2.18		2260
18	waste								1.1		**2486**

Rows 1–4: The perimeter walls are about 60 linear feet each. The interior walls are near the same but have a 3′ jog, call them 63 linear feet.

Row 5: The net linear feet of ledgers is 492 LF, which excludes the 10% shown on rows 1 and 2.

Row 6: The trusses run E/W for a distance of 60 linear feet, so there are 31 each per unit, maybe 32 depending on how the truss manufacturer would treat the jog in the walls. The difference does not matter; the estimate can use an approximate number for trusses because the total material cost is going to be captured in a lump sum quote.

Note that since the "each" column has already been used; another one is needed, so the format heading is slightly modified by placing (each) above row 6.

Row 7: The count of 60 linear feet square already has a little waste in it because of the footprint of the walls, so just 6% of waste is added to the total.

Row 8: This is the ledger against the exterior block wall. Count the length where it touches the wall.

Row 9: The perimeter at the exterior of the porch band is 138 LF.

Row 10: For the quantity, take 60 linear feet, divide by 16″, and add two for the doubling at the east and west ends. *Note in this row the arithmetic is worked from right to left!*

Row 13: It is safe to use 12 linear feet for the length of the 2 × 12′ joists, because the 12′ dimension at the front porch includes 3″ for the double band at Section D and 1-1/2″ for the ledger against the building.

Row 16: The three rectangles that comprise the deck are added to get 1,136 sf.

The decking is made of 2 × 6s. Each board is 5-1/2″ wide. The factor is 12″ divided by 5.5″, or 2.18, which gives a net square footage.

3

WALL FRAMING

Section 5 Two-story platform framing

First and second floor condo wall framing plans
Plan interpretation
Scope of the work
Construction techniques
Carpentry takeoff

Section 6 Sloping walls

Zig-Zag Evaluation and Treatment Center plans
Plan interpretation
Scope of the work
Construction techniques
Carpentry takeoff

Section 1 Photos and drawing(s)

Photo 1 A Wall Is Built On A Flat Surface

Photo 2 - 2nd Floor Walls With Temporary Bracing

Photo 3 Interior Wall With Blocking

Photo 4 Exterior Bearing Wall With Header And Jack Studs

Photo 5 Exterior Wall With Blocking For Siding

Exterior Sheathing will be covered with siding.

Why the floor sheathing here? It's only a spacer. Beyond this photo are 2nd floor bedrooms with floor sheathing. With the same wall components, equal wall heights are easier to maintain.

What are the headers for? There is no opening below…perhaps they are continuous around the entire 1st floor.

King stud with jack studs both sides. Note metal strap ties 1st floor studs to 2nd floor studs.

Blocking at exterior siding joint is for nails.

Photo 6 Exterior Wall With PT Base Plate And Anchor Bolts

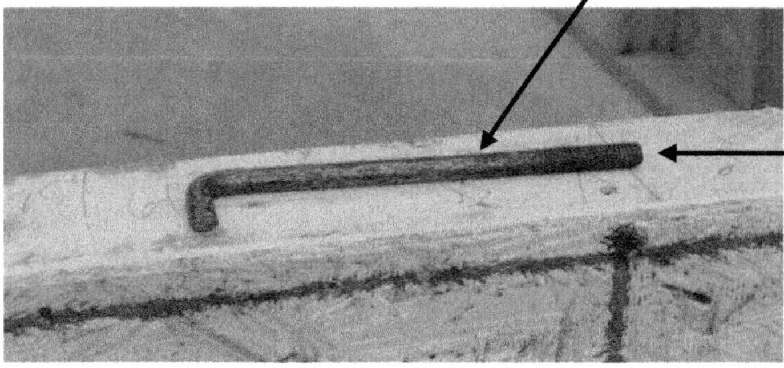

Photo 7 Busy Interior Wall In Texas With Stairs

This photo illustrates why it is impractical to "count studs" in a wall takeoff. It is best to use a factor applied to the wall length!

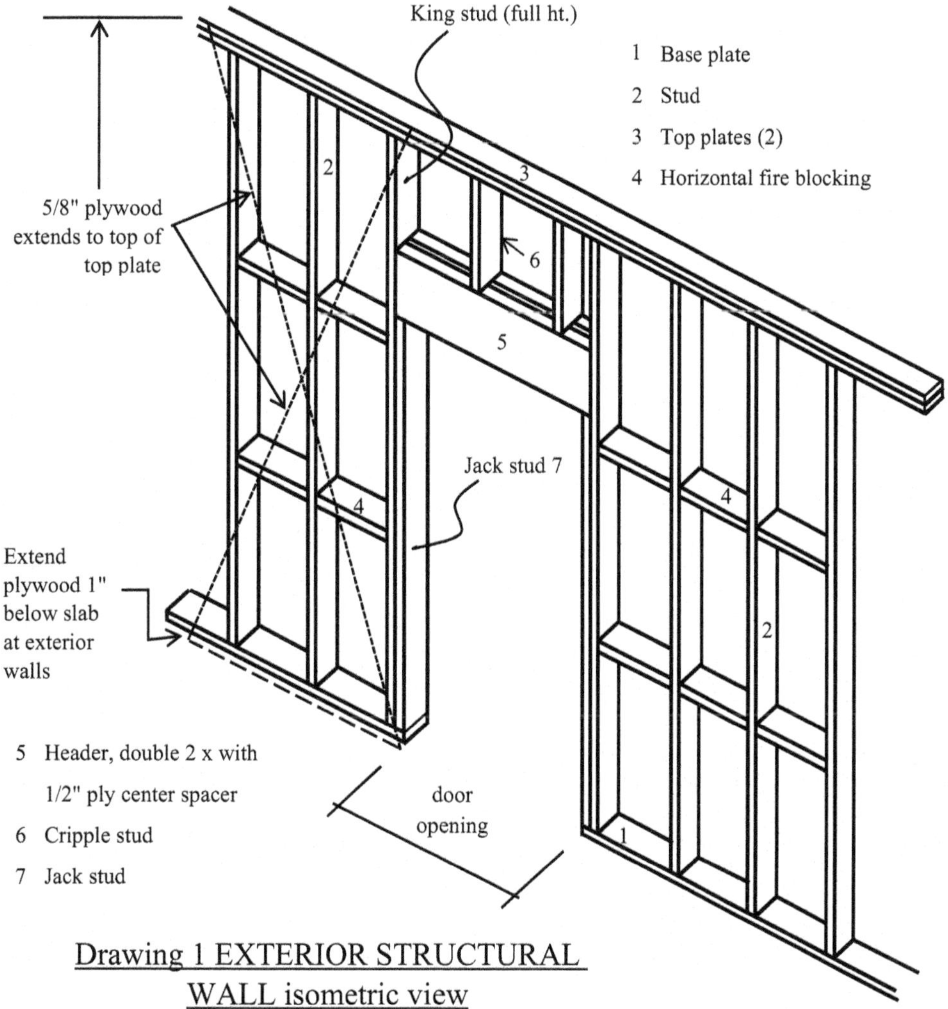

Drawing 1 EXTERIOR STRUCTURAL WALL isometric view

Section 2 Frame wall case study

Wall types

Bearing and nonbearing walls are usually 4″ or 6″ wide. The use of the terms "4 inch walls" or "6 inch walls" are generic and do not define the true width of a wall (a 2 × 4 is 3-1/2″ wide and a 2 × 6 is 5-1/2″ wide, and wallboard is usually at least 1/2″ thick on each side of the wall). What the terms mean is that nominal 2 × 4s and 2 × 6s are used to build the wall. However, architectural plans often dimension the width of walls as either 4″ wide or 6″ wide, which illustrates the point that most building plans have approximated dimensions.

Vertical wall studs are typically spaced 16″ or 24″ on center. They rest on a horizontal base plate of the same width. Base plates are taken off separately because they are "treated" with chemicals due to contact with concrete (which can cause decay). If the wall is bearing (structural), it will have two top plates. If it is nonbearing, it may only have one top plate.

All three wall components, base and top plates and studs, are quantified based on the length of the wall. First, walls are separated by type, such as:

Type 1: 4″ walls 8′ high
Type 2: 4″ walls 10′ high
Type 3: 4″ walls 12′ high
Type 4: 6″ walls 10′ high

Wall lengths

The lengths of all of these wall types are determined next, which is what is important for the takeoff. The length of each is used to determine the quantity of base and top plates and studs. The plates are easy; they can be counted with a *multiplier* (1.1 for base plates and 2.2 for top plates), and the studs are easy to count by using a *factor*, but the choice of the factor requires some thought, and is the purpose of this chapter.

How many studs per foot?

Wall studs are counted by assuming how many of them occur "per foot". A rule of thumb estimators often use is to count "one stud per foot" in a wall specified to have studs 16″ o.c. But how well does this "average" stand up under scrutiny? And what about the difference in walls – there are buildings with long straight runs of walls and then there are buildings with walls turning left and right every few feet (lots of corners) and with many windows and doors (lots of double and triple studded jambs).

It helps the estimator to study the range of studs required to build different kinds of walls. A minimum number of studs are needed to build long straight walls – place one every 16″ and add a couple for each end. A maximum number of studs are needed in busy walls with frequent turns and openings. This chapter reviews three of these walls and their "factors".

The following case study of a wall 100′ long is shown to provide three different factors for walls, all with studs at 16″ o.c. However, due to the number of wall intersections and doors and windows, all three require a different number of studs per linear feet of wall (quite a bit different, as will be shown).

For studs at 24″ o.c., see the 2 × 8 exterior walls of the Zig-Zag addition in Section 6 of this chapter.

Three cases of counting studs per foot

What determines the extra studs in a wall primarily consists of, first, the number of doors and windows in the wall, and second, the number of intersecting walls. The doors and windows have extra studding at their sides, and gypsum board nailers have to be added where there are intersecting walls.

Consider three different wall types, all of them in a straight line 100 LF long with studs spaced 16″ o.c.:

Wall type 1: No outs, no intersecting walls.
Wall type 2: Six doors or windows, three intersecting walls.
Wall type 3: Twelve doors or windows, six intersecting walls.
Review the elevation below of a wall 100 feet long.

Plans for a 100 LF wall

See the online resources for diagrams 632.1, 632.2, 632.3, 632.4, & 632.5

Plan interpretation

This is a single wall in a straight run. It is provided as an example to show the common components of a wall – studs, plates, and headers. It is also shown to find how many studs there are per linear foot in typical walls. Assume that the wall shown is a typical wall.

The wall has a single base plate and two top plates. All six openings have a double header; see Section A. All openings have a double stud at the sides; one of them, a jack stud, provides 1-1/2″ of bearing for the headers, and the other, a king stud, extends to the top plates.

The double top plate (one of them, the one on top) of the 100 LF wall is interrupted (not continuous) by the top plate of the intersecting walls, which laps into the 100 LF wall and ties the two walls together. The plate at the top of the 100 LF wall is in four pieces.

Scope of the work

Include:

In the takeoff, count the studs only.

Exclude:

Bottom and top plates, gypsum board, intersecting walls, headers, doors and windows, anchor bolts.

Construction techniques

Vertical frame walls are usually built horizontally on the slab they will stand on. The base plate and the lower top plate are first penciled on with stud locations at 16″ o.c. and window and door openings all laid out. *Stud locations every 16″ are maintained (even above and below openings) for the full length of the wall regardless of where openings occur.* The second top plate can be installed after the wall is stood.

Both sides of a header and a plywood spacer is assembled and installed as a unit.

The base plate runs continuous as the wall is built on the slab. It is drilled with holes that match the anchor bolts embedded in the slab. After the wall is stood, the plate is cut off at the door locations. It is easier to handle the wall and stand it up if the base plate is continuous.

Carpentry takeoff for three 100′ walls

Three stud counts are shown for wall types 1, 2, and 3. Headers and plates are not counted. Assume each door or window requires four vertical extra studs in the wall, two on each side. Assume each door or window requires 8′ (two pieces 4′ long) horizontal 2 × 4s to be placed above a door or below the window. That's five studs per window or door opening. Assume each intersecting wall requires two additional studs in the wall for gypsum board nailing.

Takeoff 1 is of a 100′ wall with the *fewest studs*, no door or window openings or intersecting walls.

Takeoff 2 is of the 100′ typical wall with an *average number of studs*, with six doors or windows and three intersecting walls.

Takeoff 3 is of a 100′ busy wall with the *most number of studs*, with twelve doors or windows and six intersecting walls.

For estimating purposes, if there are a lot of walls, it is not realistic to try and count all of the studs in a project, and a factor should be used to determine the number of wall studs. Using one stud per foot is a standard go-to formula for many estimators. These examples indicate that this factor is not high enough.

It is best to assume that when counting *studs beside openings* that none of them counted *at 16″ o.c.* will do double duty and also count towards the studs beside openings. This takeoff assumes each opening requires five additional studs, two each at each door or window jamb plus two pieces 4′ long either at the top of a door or the sill of a window.

It is best to assume that when counting added studs at the end of intersecting walls that none of the 16″ o.c. studs will be at the exact location that these nailers will be needed. This takeoff assumes two studs are needed at each intersecting wall.

It takes three studs plus some blocking to build a (d) *corner*. Use two per corner in a takeoff. When the intersecting wall is counted next, it will include the remainder.

632.6	**THREE DIFFERENT WALLS 100 lf LONG**	L	o.c.	Door or window openings	Wall intersect ions	Studs	Each
1	Wall type 1 NO DRS/WDWS						
2	Studs @ 16" o.c.	100	1.33				75
3	Add studs at doors/windows			0		0	0
4	Add studs at wall intersections			0		0	0
5	Add studs at corners						4
6							**79**
7							
8	Wall type 2 AVERAGE WALL						
9	Studs @ 16" o.c.	100	1.33				75
10	Add studs at doors/windows			6		5	30
11	Add studs at wall intersections				3	2	6
12	Add studs at corners						4
13							**115**
14							
15	Wall type 3 BUSY WALL						
16	Studs @ 16" o.c.	100	1.33				75
17	Add studs at doors/windows			12		5	60
18	Add studs at wall intersections				6	2	12
19	Add studs at corners						4
20							**151**

Wall type 1 takes 79 studs to build its 100′ length or 0.79 studs per linear foot.
Wall type 2 takes 115 studs to build its 100′ length or 1.15 studs per linear foot.
Wall type 3 takes 151 studs to build its 100′ length or 1.51 studs per linear foot.

The range shown in these examples is from 0.8 studs per foot to 1.5. The lower factor of 0.8 is unrealistic for takeoff purposes because this imaginary wall has no openings, no doors/windows, and no intersecting walls.

The author suggests a factor of 1.15 studs per foot, for estimating purposes, that will work for many walls (with spacing of 16″ o.c.). However, as seen in these examples, there is quite a range! The estimator should use judgment, review each wall type, and assign it a factor for use in determining studs.

The examples in this book use a factor of 1.15 studs per linear foot.

Section 3 4″, 6″, and 12″ stud walls

Plans for maintenance T building

See the online resources for diagram 633.1, 633.2, & 633.3

Plan interpretation

The floor plan has a Maintenance Room and Office with frame walls of three different widths, 4″, 6″, and 12″.

The 2 × 4 wall between the two rooms is shown on the floor plan and in Section 9. It is 8′ high and about 15′ long.

The 2 × 6 walls are shown on the floor plan at the office. These three walls are 9′ high per Section 9.

Another 2 × 6 wall is at the maintenance room, shown 12′ high on the floor plan. Section 9 also shows this wall "beyond" on the left side of the drawing.

The 12″ wide wall is at the garage door and its height of 12′ is given on the floor plan.

In Section 9, all openings are shown to be 7′ high, which means doors and windows. Note that door number 1 is only 6′-8″ high, which conflicts with this note. This kind of conflict is common, and the estimator can note this in an RFI if time permits, or ignore it since it is a minor consideration. As always, the estimator balances these options with the knowledge that contract language makes it the duty of the contractor to report conflicts in the documents.

There are three walk doors, two exterior and one interior. The header schedule reveals an instruction about the bearing width (the number of jack studs) beside the openings. Opening widths of 3′ must have a double 2 × 8 header with bearing of 1-1/2″ (one jack stud) at each end of the header. Two studs will be at each side of the opening (one jack stud and one king stud that extends to the top plate). At opening widths greater than 4′, the header must have bearing of 3″ (requiring two jack studs and one king stud at each end of the header).

The sheathing is "typical" on the exterior side of the wall in Section 9, which means it is on the exterior side of all of the exterior walls.

Scope of the work

Include:

Furnish and install all wall framing – studs, plates, and headers.
Furnish and install walk doors and windows.

Exclude:

Garage door.

Construction techniques

Except for the second top plate, all of the walls can be assembled on the flat slab.

When walls of the same height but different widths intersect (see the east 12″ garage wall and the intersecting 6″ walls), it does not matter which top plate extends past. The 12″ wall top plate can extend over the 6″ wall, or the top plate of the 6″ wall can extend across the 12″ wall.

Carpentry takeoff

Counting the length of a frame wall is different from counting the length of a block wall. There is no need to deduct for counting corners twice – with 4″ and 6″ walls it hardly matters. It can make a difference, however, if walls are 8″ or greater in width.

CARPENTRY TAKEOFF											633.4
job: T Building			number						date		
No.	Description	Det.	Each	Pcs.	o.c.	L	W	Ht.	%	LF	SF
	2 x 4 wall 8' high										
1	2 x 4 base plate (8's or 16)			1		16				**16**	
2	2 x 4 x 16 top plates			2		16				**32**	
3	2 x 8 x 8 header			1		8				**8**	
4	5/8" plywood spacer					4		1			**4**
5	2 x 4 x 8' studs			19				8		**152**	
	2 x 6 exterior wall 10' high										
6	2 x 6 base plate			1		54			1.1	**59**	
7	2 x 6 top plates			2		54			1.1	**119**	
8	2 x 8 x 8 header dr 2, wdw A			2	8					**16**	

CARPENTRY TAKEOFF											633.4
job: T Building			number						date		
No.	Description	Det.	Each	Pcs.	o.c.	L	W	Ht.	%	LF	SF
9	2 x 10 x 10 header wdw B			1		10				**10**	
10	2 x 12 x 14 wdw C			1		14				**14**	
11	1/2" plywood spacer					20		1			**20**
12	2 x 6 x 12 studs			62				10		**620**	
	2 x 6 exterior wall 12' high										
13	2 x 6 base plate					82			1.1	**90**	
14	2 x 6 top plates			2		82			1.1	**180**	
15	2 x 8 x 10 header dr 2			1		8				**8**	
16	2 x 12 x 14 header wdw D			2		14				**28**	
17	1/2" plywood spacer					17		1			**17**
18	2 x 6 x 12 studs			90				12		**1080**	
	2 x 12 wall 12' high										
19	2 x 12 base plate			2		8				**16**	
20	2 x 12 top plates *			2		24			1.1	**53**	
21	2 x12 x 14 top plates *			2		14				**28**	
22	2 x 12 x 12 top plates *			2		12				**24**	
23	2 x 12 x 12 header			2		12				**24**	
24	1/2" plywood spacer					11		1			**11**
25	2 x 12 x 12 studs			16				12		**192**	
*	Row 20 counts the top plates by a continous count and then adding 10% waste. Rows 21 and 22 counts the top plates by observing the specific lengths that will be needed, which are 14's and 12's.										

Row 1: The bottom and top plates of the interior 2 × 4 wall can be 8′ or 16′, since the length of the wall is 15′.

Rows 5, 12, 18: Add 15% to the length of the wall to get the number of studs.

Row 13: The base plate of the 2 × 12 wall would not be figured for the full length of the wall, as is usually the case with base plates. Since the garage door opening is 10′ and the walls on either side are only 7′, it doesn't make sense to buy the additional lumber just to assemble the wall, and have to remove it after the wall is stood. For the short wall in this example, that would be too much waste.

Row 18: The wall is not very busy so a waste factor of (only) 10% is selected for the number of studs (82 plus 8).

Section 4 Three-wall addition

Public library plans

See the online resources for diagrams 634.1, 634.2, 634.3, & 634.4

Plan interpretation

This is a three-wall addition to an existing library. Two openings are being created in the exterior wall for passage between existing and new construction. Several wall types are noted on the floor plan. The dashed line indicates the roof slope.

Section 1 is of the exterior 2 × 6 wall which is 9′-8″ high. The base plate is anchored to the slab with anchor bolts. Horizontal 2 × 6 blocking is at 4′ and 8′. A triple 2 × 8 header is at 8′-8″. The exterior side of the wall is covered with plywood sheathing and felt. The wall is finished with horizontal siding with a 5/4 band board at the top.

Section 3 occurs where the new walls intersect with the existing south wall. Since two studs need to be installed as blocking within the existing wall, the gypsum board inside the library has to be removed.

The "wall partition" schedule describes three interior wall types. Two of the walls are built up to the roof deck (E, G) and wall C is built 6″ above the ceiling. Wall C is the only wall with horizontal top plates, non-sloping. All three walls have horizontal fire blocking. In the wall partition type G note, reference is made to UL design U-305. It is common for plans to contain instructions referring to separate documents, which then become a part of the job plans and specifications.

Scope of the work

Include:

- Exterior and interior frame walls.
- Plates (including silicone), studs, and headers.
- Demolition of door openings, headers.
- Removal of gypsum wallboard at Section 3.

Exclude:

- Gypsum board, sound batts.
- Siding and trim.
- Anchor bolts.
- Doors and windows.

Construction techniques

The exterior walls would be built first. Since some of the interior walls are built to the underside of the roof deck (which needs to be in place), the trusses would be installed next and the roof dried in before the interior walls are constructed. If the demolition of the door openings between the two structures are created after the new addition is dried in, the openings won't have to be boarded up and sealed against rain.

About two linear feet of interior gypsum board would be removed from the inside face of the wall at Section 3. This is needed to place the double stud blocking for the new walls.

The east/west portions of walls E and G will slope up to the roof deck. The north/south portion of wall G will be horizontal.

Carpentry takeoff

Count the length of each wall; keep each wall separate.

CARPENTRY TAKEOFF											634.5
job: Library			number						date		
No.	Description	Det.	Each	Pcs.	o.c.	L	W	Ht.	%	LF	SF
	Exterior 2 x 6 wall 81' long										
1	2 x 6 base plate	Sect 1				81			1.1	**89**	
2	2 x 6 top plate			2		81			1.1	**178**	
3	2 x 6 x 10 studs			89				10		**890**	
4	2 x 6 horizontal blocking			2		65			1.0	**130**	
5	2 x 12 x 10 headers wdws B			3		10				**30**	
6	2 x 12 x 12 header wdw A			3		10				**30**	
7	1/2" plywood spacer			2		20		1			**40**
8	5/8" cdx plywood sheathing					81		10	1.1		891
9	less outs B			-2		4		5			-40
10	less outs A			-1		8		5			-40
11	s/t cdx plywood										**811**
12	15 lb felt same as plywood										**811**
	Int. 2 x 4 wall C 16' long	Partition types									
13	2 x 4 base plate					16			1.1	**18**	
14	2 x 4 top plates			2		16			1.1	**35**	

CARPENTRY TAKEOFF											634.5
job: Library			number					date			
No.	Description	Det.	Each	Pcs.	o.c.	L	W	Ht.	%	LF	SF
15	2 x 4 x 12 studs			18				10		**180**	
16	2 x 4 horiz fire blking @ clg.									**16**	
	Int. 2 x 6 wall E 18' long										
17	2 x 6 base plate					18			1.1	**20**	
18	2 x 6 top plate			2		18			1.1	**40**	
19	2 x 6 x 12 studs			6				12		**72**	
20	2 x 6 x 14 studs			6				14		**84**	
21	2 x 6 x 16 studs			9				16		**144**	
22	2 x 6 header			1		8				**8**	
23	2 x 6 horiz fire blking @ clg.									**22**	
	Int. 2 x 4 wall G 20 lf										
24	2 x 4 base plate			1		16			1.1	**18**	
25	2 x 4 top plates			2		16			1.1	**35**	
26	2 x 4 stud 12' ht			7				12		**84**	
27	2 x 4 stud 14' ht			17				14		**238**	
28	2 x 6 header			1		8				**8**	
29	2 x 6 horiz fire blking @ clg.									**16**	
30	Demo door openings say 4 x 8		2			4		8			**64**
31	New headers 2 x 12			3		8				**24**	
32	New 2 x 6 x 8 door jambs		4	2				8		**64**	

Row 1: The three walls combined are 81 linear feet, the board lengths do not matter, and 10% is added for a waste factor.

Row 2: There are two top plates and the length of the wall is from row 1, 81 LF.

Row 3: Since this wall has no doors and is not very busy, a factor of 1.1 is multiplied times the wall length of 81 to get 89 studs. The arithmetic for this does not occur on the takeoff, just the number 89.

Row 4: The 2 × 6 horizontal blocking shown in Section 1, at a height of 4′, will be interrupted at all three openings because the sill height of these openings is 2′-8″. Fire blocking from stud to stud (16″) is figured by the linear feet of wall *but without a waste factor* because the thickness of the studs already accounts for a deduction in length of 1-1/2″ every 16″, approximately a 10% waste factor. The length of wall blocking is 81 − 4 − 4 − 8 or 65′.

Row 5: Window "B" openings scale 4′ long. Adding a few inches on each side for the header to bear on jack studs, use 5′ for the takeoff length of the one piece of the header. Since there are two "B" openings and three header pieces needed for each one per the exterior wall section, a total of six headers 5′ long will be needed. Two of these pieces can be cut from a ten-foot length. Ten-foot lengths or sixteen-foot lengths would work.

Row 6: Window A scales 8′ long so 10′ pieces are selected as a length.

Row 7: For plywood spacers, count the takeoff lengths (5′+5′+10′) of the headers times two.

Row 8: The wall is 9′-8″ high but 10′ is used for the height of the plywood because in plywood dimensioning 4″ can't be used for anything. When figuring the area of "plywood, use "plywood dimensions" by rounding up to the nearest foot, simply because the product is a 4′x 8′ size.

Rows 9, 10: Sometimes estimators do not deduct for "outs" when counting exterior sheathing. They trade the outs for a waste factor. The problem with this is, while it makes the takeoff easier, the waste factor can be a wide range. This is not recommended unless the outs are a small size, less than a door opening. The quantity and size of the outs should be figured, and after they are deducted from the total square footage of the wall, an appropriate waste factor can be added. The window outs on the south wall are shown on the following takeoff.

Row 12: Repeat row 8.

Row 13: Base plates do not need to be a specific length. Add 10% to the length of the wall.

Row 14: Top plates do not need to be a specific length.

Row 15: Add 10% to the wall length of 16′ to get the number 18, which becomes the quantity of studs. The height of wall C is 9′-6″ (required to be 6″ above the ceiling; see partition types and Section 1), so 10′ studs are used.

Rows 17: The length of the wall is 18′.

Rows 19, 20, 21: See wall heights in the sketch below.

See the online resources for diagram 634.6

There are two ways to count the studs:

Method 1: Average height method.

Use 14′ as the average height of the studs, and use 18 × 1.15 (factor) = 21 each total studs. So, 14′ × 21 = 294, or *about 300 LF* for a total length of wall studs.

Method 2: Draw a sketch and figure out the stud heights!

Use a combination of 12s, 14s, and 16s, and choose 7 each of the 12′ and 14′ lengths and 9 each of the 16′s. There are a *total of 326 LF* counting this way.

Row 24: The length of the wall is 20′ and is composed of three different wall heights.

Row 26: See Sketch 1.

Row 30: The scope of the work in this package includes the demolition of the door openings, which is to be done after the dry-in of the new structure is completed. Note that the unit of measure for this demolition is square feet, a composite area made up of studding, drywall, and siding. That is how it will be treated on the estimate. The scope does not include patching drywall, but it does include the rough framing of new headers and jambs. No door size is given, nor is the size of the header. Since the openings scale 3′ wide, a header length of 4′ can be assumed and 2 × 12s can be assumed for the header because they are shown in Section 1. The estimator further assumes a 2 × 6 wall instead of a 2 × 4 wall, and three header pieces are estimated for each door opening. Two studs are estimated for each side of the door jambs.

Section 5 Two-story platform framing

First and second floor condo wall framing plans

See the online resources for diagrams 635.1, 635.2, 635.3, 635.4, 635.5, 635.6, 635.7, 635.8, 635.9, & 635.10

Plan interpretation

The plans describe a two-story building with 2 × 6 exterior and interior walls 9′ high. Three sides of the building have the same elevation. There are six units – three upstairs and three downstairs.

Two window openings plus a center jamb are 6′-10″ wide and 5′ tall per the floor plan and per Section A and F. This window pattern is repeated around the building. All of the doors are 3070.

There are smooth cement panels (cut from 4′ × 8′ sheets) at the center jambs of the windows and at the single location where two doors are adjacent (see Section F and the front elevation). There are various trim boards at the doors, windows, and corners (see Sections C through F). The horizontal siding runs into (not behind) the thicker trim boards. The trim and siding pieces are installed over the vapor barrier and plywood sheathing.

Section B indicates the party walls (walls between tenants) have 2 × 6 horizontal blocking. Sections C and D show 1 × 8 horizontal cement board siding with 6″ exposure over felt and plywood. Section E shows triple 2 × 12 headers. Section D shows the vertical 5/4″ trim at the corners.

The second floor flat trusses are 18″ high. The gypsum floor poured on the plywood is poured against the walls, not underneath.

Scope of the work

Include:

Furnish and install exterior first and second floor wall framing.
Include plates, studs, and headers.
Include plywood sheathing, siding, cement board, and vapor barrier.
Include trim lumber at doors and windows, 5/4″ boards, and 2 × 10 sill.
Include horizontal banding and galvanized flashing.

Exclude:

Gypsum board.
Trusses.
Windows and exterior doors.
Interior window sills.
Interior walls.

Construction techniques

All of the first floor walls are to be built concurrently since they are all load bearing. Before the second floor trusses are set, the walls would be braced plumb (2 × 4 temporary diagonal bracing). Installing the exterior sheathing to the first story would accomplish the same thing, but the crew may want to install all of the sheathing at the same time. Regardless of the sequence here, the walls should be plumb before the weight of the trusses is placed on the walls, which will hinder wall alignment.

The second floor trusses and 3/4″ plywood subfloor are installed next.

The second floor walls would be installed and plumbed with temporary bracing.

The roof trusses, fascia, and sheathing would be installed, and once the roof sheathing is covered with a vapor barrier (typically 15 or 30 lb felt), it would be "dried in".

The exterior sheathing and vapor barrier are installed, and then windows and the exterior door frames. This would be followed with installation of the siding and trim.

Carpentry takeoff

The following observations can be made about the quantities:

1 The first and second floor exterior walls are identical in their count of plates, studs and headers, and window and door trim.
2 It is faster to build headers above the windows that extend across the pairs of windows than building separate headers above each individual opening.
3 When counting the plywood sheathing (or any sheet product sized 4′ × 8′), round off to sheet dimensions. With a wall height of 19′-6″, use 20′ for the height of the sheathing on the takeoff. The door and window openings are treated as outs.

CARPENTRY TAKEOFF											635.11
job: Condos			number						date		
No.	Description	Det.	Each	Pcs.	o.c.	L	W	Ht.	%	LF	SF
	Exterior walls 226'										
1	2 x 6 **pt** base plate 1st floor					226			1.1	**248.6**	
2	2 x 6 base plate 2nd floor					226			1.1	**248.6**	
3	2 x 6 top plates		2	2		226			1.1	**994.4**	
4	2 x 6 x 10 studs		2	260				10		**5198**	
5	2 x 12 x 8 **wdw** headers		30	3		8				**720**	
6	2 x 12 x 8 **door** headers		6	3		4				**72**	
7	1/2" plywood spacer		2	15		16		1			480
8	1/2" plywood spacer		2	6		4		1.00			48
	s/t ply spacer										**528**
9	5/8" ply sheathing 2 stories					226		20			4520
10	less wdw outs			60		3		5.00			-900
11	less dr outs			6		3		7.00			-126
12	net area										3494
13	add waste								0.1		349.4
14	total ply sheathing										**3843.4**
15	1 x 8 siding…net area										3494
16	times 2 plus 10%								2.2		**7687**
17	30 lb felt (400sf rolls so round up)										**4000**
18	2 x 12 ledger at porch	C				57			1.1	**63**	
19	Galv flashing porch	C				57			1.05	**60**	
20	5/4 x 6 vertical :	D									
21	Ext 4 corners	D	4	2				19.5	1.1	172	
22	Front @ porch	D	4	2				18.5	1.1	163	
23	5/4 x 4 wdw head	E		30		10				300	
24	5/4 x 8 wdw head	E		30		10				300	
	s/t 5/4 x 8's									**934**	
25	Galv Z flashing	E		30		10				**300**	
26	5/4 x 5 1/2 x 14	F	2	30				7		**420**	
27	5/4 x 2	F	60	3				6		**1080**	
28	Smooth cement panel 4 x 8	F		30			0.5	6	1		**108**
29	2 x 10 wdw sill	G		30		8				**240**	
30	5/4 x 4 door trim			6		20				**120**	

Row 1: The length of the exterior frame wall, including the doorways, is 226 LF.
The arithmetic of adding all the walls together to arrive at 226 LF is *not* shown on the takeoff.

Row 3: The 2 in the "Each" column is the number of floors. There are 2 top plates, 2 pcs.

Row 4: The quantity of 260 studs is arrived at by multiplying the length of the wall times 1.15 (see factors in the 100′ wall case study) or 226 × 1.15.

Row 5: There are 3 each headers per opening 8′ long. Counting both floors, there are 30 openings.

Row 6: Count each header as 4′ long (above a 3′ door plus a 2 × 4 on each side of the door). The best lengths to use are either 8′s or 12′s.

Row 7: Each spacer is 8′ long, each opening requiring 16 LF of plywood figured at 1′ high. There are 15 openings per floor, 2 floors.

Row 9: The plywood is counted at 20′ high instead of 19′-6″, using sheet dimensions. The windows and doors are counted as outs, and then 10% is added to the net area. Note that the ending quantity of 3,711 sf is far less than if plywood had been figured straight through the openings (not deducted).

Row 12: See total amount of sheathing. Felt is sold in rolls of 200 sf each.

Rows 15–16: Section C states that the siding has a "6" exposure, or coverage on the wall. It doesn't matter what the actual size of the siding board is, it could be 8″ high or 12″ high; it is only exposed for six inches in height for every one linear foot. It takes 2 LF of siding to install 1 SF of siding wall surface.
Double the SF and add a waste factor (row 16).

Row 19: A low waste factor of 5% can be used for several reasons. It is a straight run across the front of the building (no bends), flashing lap only needs to be 4″–6″, and the material is purchased in 10′ pieces.

Row 25: This product is made in 10′ pieces, and is placed at the top of each pair of windows, including the jambs widths. Each pair of windows will use one piece of Z flashing, and there are 30 window openings.

Row 26: There are two pieces per opening and 30 total openings of two windows each. Each piece is figured as 7′, since they scale to be about 6′ from the elevations. The most economical length to use is 14′, and this length becomes part of the description.

Row 30: See front elevation note, "5/4 × 4 head and jamb trim". That's three sides of each door, which will take two 8′ and one 4′ pieces of trim to accomplish, or 20 LF per door (one face). There are six doors.

Section 6 Sloping walls

Zig-Zag Evaluation and Treatment Center plans

This project has walls that are sloped; the wall ends have different heights. With sloped walls, estimators often have to rely on a sketch to determine various heights. This can often be done by hand or computer utilizing an existing section(s). These will result in a helpful aid such as Sketch 1 shown below, called the "Wall Height Study".

Another aid for the estimator is a sketch that numbers the walls to ensure that they are all accounted for in the takeoff. See Sketch 2, provided for the purpose of giving each wall a sequential number and grouping them into like kinds.

Three "factors" are introduced in this chapter. They are determining "studs per foot" for walls *with studding 24″ o.c.*

Refer to Chapter 3 Wall Framing, Section 2, for the study of determining studs per foot in a wall 100′ long, *with a spacing of 16″ o.c.* Three factors were determined for walls with this spacing:

Wall type 1 with no doors or windows and no intersecting walls requires 0.79 studs per linear foot.

Wall type 2 has a typical number of doors, windows, and intersecting walls, and requires 1.15 studs per linear foot.

Wall type 3 is a busy wall with many doors, windows and wall intersections, and requires 1.5 studs per linear foot.

Using these same wall types, to count studs in walls *with a spacing of 24″ o.c.*, use these factors:

Wall type 1 with no doors or windows and no intersecting walls requires *0.56 studs per linear foot.*

Wall type 2 has a typical number of doors, windows, and intersecting walls, and requires *0.75 studs per linear foot.*

Wall type 3 is a busy wall with many doors, windows and wall intersections, and requires *1.0 stud per linear foot.*

Use a factor of 0.75 studs per linear feet in the Zig-Zag takeoff.

Another consideration with sloping walls is that the top plate length is not the same as the horizontal length. If the slope is less than 3 in 12, the estimator may wish to do so; however, with slopes greater than this the inclined length should be determined. Either the slope length can be scaled by the use of drawings like Sketch 1, or a table can be used such as that shown in Chapter 1, Section 6.

In the following takeoff of the Zig-Zag Treatment Center, the roof slope is 10′ over a distance of 40′, which is a 3 in 12 slope. The factor from the roof table in Chapter 1, Section 6 is 1.0308. See how this factor is used in the following takeoff.

See the online resources for diagrams 636.1, 636.2, 636.3, 636.4, 636.5, & 636.6

Plan interpretation

This is a three-wall addition to an existing building. The roof bearing is 20′ above finished floor on the west side and 10′ A.F.F. on the far east side. All other wall heights are in between; see Section B.

From the floor plan, the exterior walls are built with 2 × 8 studs spaced 24″ o.c., the doors are numbered with triangles, the ceiling heights are shown in each room, and the interior walls are type R or S, both with studs spaced 16″ o.c.

Section A shows the outline of the plywood (see X) on the shear walls labeled R. Two different ceiling heights are shown. The S walls are shown to be built 12″ above the ceiling.

In Section B, wall R is shown built up to the roof deck. Two adjacent ceiling heights are shown. This section is also cut through the front door and the 2 × 8 wall is shown with 5/8″ plywood clad on the exterior. The location of B1 is also shown in Section B.

Section B1 depicts the sloping top plates positioned in place with the use of solid shims. The four each 2 × 12 headers plus the plywood shown equals the width of a 2 × 8 (7-1/4″).

The door schedule shows that the headers for all of the interior doors are double 2 × 6.

A detailed description of walls R and S are given in the wall schedule. The R walls are built to the deck; the S walls are built 1′ above the ceiling. Note the use of the abbreviation U.O.N., "unless otherwise noted".

Scope of the work

Include:

All wall framing interior and exterior.
Base plates, top plates, studs, and headers.

Exclude:

Trusses and roof framing.
Ceilings, concrete, structural steel, gypsum board.
Doors and windows.

Construction techniques

One way to sequence the work is to get the steel beam (by others) installed against the building first to give an exact target for the top of construction. When the 10′ high exterior wall (NW wall) is then built and string is pulled from it to the steel beam, the remaining wall inclines are determined.

Only the nominal sizes of the windows and doors are shown on the plans. Their rough openings would be obtained from the "product data sheets" from the manufacturer. These are the same sheets turned in as "submittals" to the architect.

Although the plans state that type R walls "extend to the roof deck", a construction allowance is made because the roof will actually have some "give" to it; a small amount of deflection and movement may occur. It would be best to leave the walls, say, 1/2″ short of the roof decking. This space can be stuffed with fire rated insulation or foam.

Carpentry takeoff

The following takeoff observations can be made:

1. The exterior walls are built to the *bottom* of the trusses. The studs are 2′ o.c. – use 0.85 studs per linear foot of wall.
2. The R walls are built to the deck, the *top* of the trusses; *since they all run N/S they will be horizontal walls.*
3. Since the S walls are built 1′ above the ceiling; *they will all be horizontal.*
   The base plates can be counted as they have been in the other chapters; there is no reason to separate them by wall as with the studs.
4. The top plates do not need to be separated by wall, but they have two conditions – horizontal and inclined.

Sketch 1 is used to determine wall heights. The top line, shown dashed, is the roof deck.

See the online resources for diagram 636.7

Sketch 2 goes beyond wall numbering and has both the length of each wall and the number of studs in it, using a factor of 0.85 studs per linear foot. Grouping this information in one place allows the takeoff to be completed much faster.

See the online resources for diagram 636.8

Sketch 2 simply labels the walls with a sequential number. Like kind walls can be grouped together for counting.
The following wall lengths can be combined:

Walls 11 and 17, for a combined length of 19′. Type R walls to deck.
Walls 10 and 19, for a combined length of 22′. Type R walls to deck.
Walls 13 and 16, for a combined length of 20′. Type S walls 1′ above ceiling.

CARPENTRY TAKEOFF Part A										636.9
job: Zig-Zag Treatment Center			number					date		
No.	Description	Det.	Factor	o.c.	L	W	Ht.	%	LF	SF
	EXTERIOR WALLS									
1	All 2 x 8 base plate				168			1.1	185	
2	2 x 8 top plate:									
3	Horizontal walls 2,4,6,8				54			2.2	119	
4	Inclined walls 1,3,5,7,9		1.0308		114			2.2	259	
5	Total 2 x 8 top plate								377	
6										
7	Wall 1 studs 34 ea :				Each					
8	2 X 8 Studs 20' high	1/5th of 34 total studs			7		20		140	
9	Studs 18' high	1/5th of 34 total studs			7		18		126	
10	Studs 16' high	1/5th of 34 total studs			7		16		112	
11	Studs 14' high	1/5th of 34 total studs			7		14		98	
12	Studs 12' high	1/5th of 34 total studs			6		12		72	
13	Wall 2 - 2 X 8 studs 9 ea				9		10		90	
14	Wall 3 studs 8 ea				8		12		96	
15	Wall 4 studs 16 ea				16		12		192	
16	Wall 5 studs 10 ea				10		14		140	
17	Wall 6 studs 9 ea				9		16		144	
18	Wall 7 studs 14 ea				9		14		126	
19					5		12		60	
20	Wall 8 studs 13 ea				13		12		156	
21	Wall 9 - 2 X 8 studs 32 ea				4		12		48	
22					7		14		98	
23					7		16		112	
24					7		18		126	
25					7		20		140	
26					145				2076	

	CARPENTRY TAKEOFF Zig-Zag Treatment Center Part B										636.10
No.	Description	Det		Ea		L	W	Ht	%	LF	SF
1	EXTERIOR WALL HEADERS										
2	2 x 12 x 10 hdr wdw a			2		10				20	
3	2 x 12 x 10 hdr wdw b			4		10				40	
4	2 x 12 x 8 hdr wdw c			4		8				32	
5	2 x 12 x 14 hdr wdw d			4		14				56	
6	2 x 12 x 14 hdr door 4			2		14				28	
7	2 x 12 x 8 hdr wdw e			4		8				32	
8	2 x 12 x 10 hdr wdw f			2		10				20	
9	2 x 12 x 10 hdr wdw g			2		10				20	
10										**248**	
11	3/8" plywood spacer			2		54					**108**
12	1/2" plywood spacer			1		54					**54**
13											
14	INTERIOR WALLS										
15	All base plate					145			1.1	**160**	
16	All top plate					145			2.2	**319**	
17	STUDS										
18	Wall 10,19 R wall to deck			19				16		304	
19	Walls 11,17 R wall to deck			17				20		340	
20	Wall 12, S wall to 11'			19				12		228	
21	Wall 13,16 S wall to 11'			19				12		228	
22	Wall 14, R wall to deck			17				20		340	
23	Wall 15, S wall to 10'			9				10		90	
24	Wall 18, S wall to 11'			27				12		324	
25				127						**1854**	
26											
28	R WALL PLYWOOD			Sides							
29	1/2" plywood Walls 10, 19			2		22		16			704
30	1/2" plywood Walls 11, 17			2		17		16			544
31											**1248**
32											
33	2 X 6 DOOR HEADERS			EA	Pcs	L					
34	Doors 1, 2, 3, 5, 6, 7, 8, 9			8	2	4				**64**	
35											

4

CEILING AND ROOF FRAMING

Section 1 Photos and drawing(s)

Photo 1 Truncated Trusses

Photo 2 Gable End Truss

This "drop truss" is 3-1/2" lower than the other trusses not yet installed. Outlooks will be installed on top of and perpendicular to the drop truss, creating an overhang. Since drop trusses are built on top of walls, their chords can be vertical, carrying loads straight down.

Photo 3 Outlooks On Top Of Drop Gable Truss

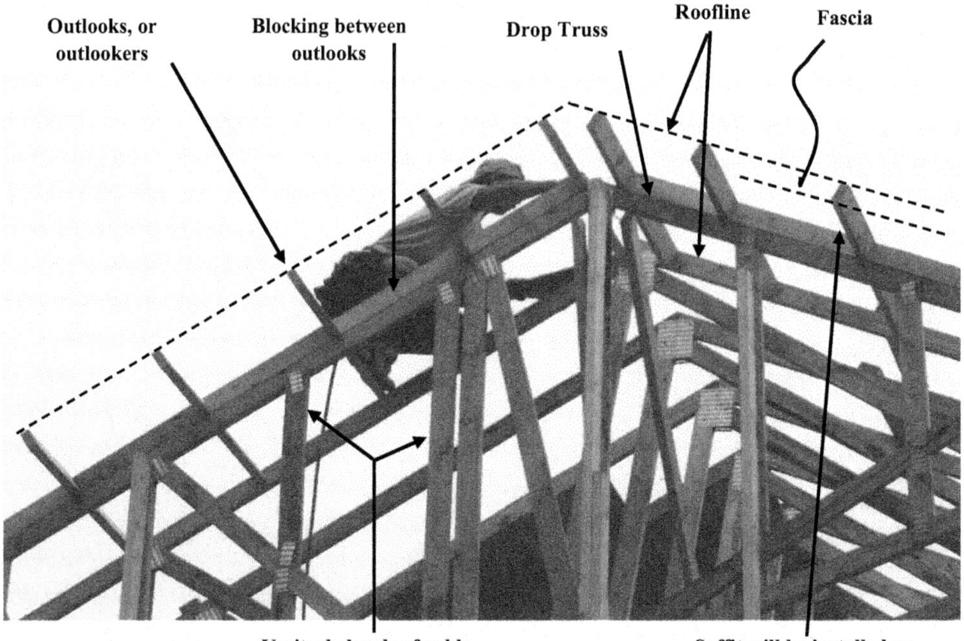

Photos 4 and 5 Hip End Trusses

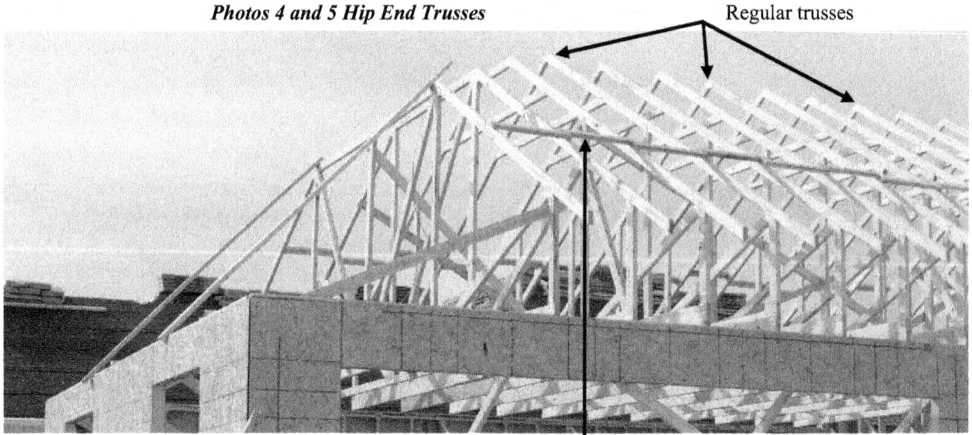

SIDE VIEW

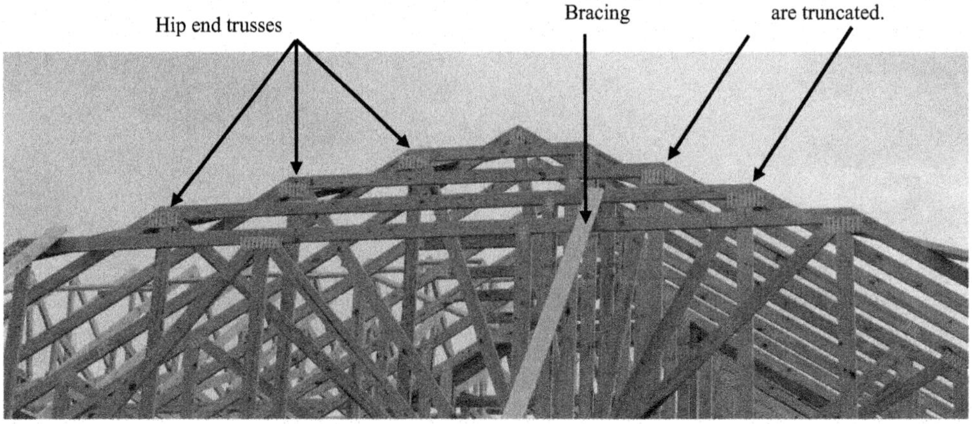

FRONT VIEW

Photo 6 Ceiling Framing At Two Heights

Photos 7 And 8 Porch Ceiling And Roof Framing

Section 2 Roof framing case study

Framing cube plans

See the online resources for diagrams 642.1, 642.2, 642.3, & 642.4

Plan interpretation

The roof framing plan is for a small roof structure 12′ square. Girder 1 is 12′ long (see floor plan) and horizontal at its bottom (Section A), and sits on walls that are 8′ high (Sections A and B). The girder rises to a peak at 14′ (front elevation and Section B), runs E/W, and splits the roof into north and south. Girder 2 is a rectangle (Section B), runs N/W, and splits the south half into quarters. It is 5′-10.5″ long and the top and bottom is horizontal.

All four frame walls are 8′ high (A, B).

The north framing members are trusses collectively known as a "hip set". The roof overhang is created by the extension of their top chords.

The roof of the southeast quadrant is made with rafters (a ceiling is not shown). The top end of the rafter sits on a ledger (Section A) attached to the side of girder truss 2. The roof overhang on the south side (of the SE quadrant) is created by the use of outlooks that extend past the exterior wall (roof plan). On the east side, the 2 × 6 rafters have a "birds-mouth" (Section A) cut in them at the bearing wall. The fascia here is a 2 × 8.

The roof members in the southwest corner are mono trusses (also called half trusses). They overhang the west exterior wall with a flat (horizontal) projection (A). On the south side (of the SW quadrant), 2 × 4 outlooks are used to create the overhang (B).

In Section B, the south exterior wall makes use of a "drop truss" to provide bearing for the outlooks. The drop truss is lower than the other trusses by the 2 × 4 width (see floor plan) of the outlook, and is flush with the face of the exterior wall. The front elevation has the shape of the 8′ wall plus the drop truss above it.

Scope of the work

Include:

Trusses, girders, and metal hangers.
Ledgers and outlooks.
Rafters and fascia.

Exclude:

Wall framing.

Construction techniques

Girder 1 would be installed first, then girder 2, which would stabilize and plumb girder 1. Any one of the other three quadrants could be built next.

The metal truss seats would be attached to girder 2 before the mono trusses are installed.

The 2 × 6 ledger would be attached to girder 1 before the rafters are installed.

The drop truss on the south wall would be installed before the outlooks can be installed. The fascia would be the last item put into place.

Carpentry takeoff

Rows 1–4:

Some estimators would not itemize trusses by the piece on the takeoff. On the estimate, their entire *material* price will be entered lump sum. *Labor*, too, can be entered in its entirety, such as 80 hours to set all of the trusses. This makes a further itemization of the trusses unnecessary. Items such as crane time and temporary and permanent bracing are figured separately, but sometimes the trusses themselves are simply listed on one row with a quantity of 1 each or LS (lump sum).

CARPENTRY TAKEOFF											642.5
job: The Framing Cube			number						date		
No.	Description	Det.	Each	Pcs.	o.c.	L	W	Ht.	%	LF	SF
1	Girder truss 1, 12' span		**1**								
2	Girder truss 2, 6' span		**1**								
3	Hip truss set		**1**								
4	Mono trusses, drop truss		**3**								
5	2 x 6 ledger at rafters	A		1		6					
6	2 x 6 rafters	A		4		8				**32**	
7	2 x 4 x 10 outlooks	B		4		5				**20**	
8	2 x 6 x 12 outlooks	A		6		4				**24**	
9	2 x 6 x 8 clg joists	B		4		4				**16**	
10	2 x 6 x 8 fascia			2		8				**16**	
11	2 x 8 x 8 fascia			2		8				**16**	
12	2 x 8 x 16 fascia			2		16				**32**	
13	Metal truss seat	A		**3**							

Section 3 Trusses, sheathing, and soffit

Storage building plans

See the online resources for diagrams 643.1, 643.2, & 643.3

Plan interpretation

The roof is fully "trussed" (entire roof covered with trusses) at a slope of 3-1/2 in 12; there are no rafters.

Girder trusses (see double truss on roof framing plan and Section A) that carry loads from jack trusses are often two trusses resembling the other full-length trusses. All of the trusses are secured to the exterior wall with truss straps embedded in a tie beam. The jack trusses sit in a "truss seat" (designed by the truss manufacturer) against the girder trusses.

From a center ridge at the roof peak, there are four ridges that slope down to the corners of the building. In plan, these ridges are at 45 degrees to the exterior walls (which occurs when the slope is the same on the sides and ends of the roof). These four ridges also have 2 × 4 blocking, which provides nailing for the plywood sheathing edges on top of the trusses.

The soffit and fascia framing is with common lumber. An 18″ overhang is created with a 2 × 12 perimeter fascia perpendicular to the trusses. It is braced against the building with 2 × 4s and the soffit is covered with plywood.

Scope of the work

Include:

- Trusses and sheathing.
- Soffit and fascia including A/C plywood and 1 × 4 trim.
- Blocks and nailers.
- 15 lb felt vapor barrier.

Exclude:

- Block walls.
- Shingles.
- Concrete tie beam.

Construction techniques

The full-sized trusses would be set first, including the girder trusses. The metal hangers (sometimes called truss seats) are then fastened to the side of the girder trusses. The jack trusses could then be installed, which would fix and align the girder trusses in place.

The 22-1/2″ roof blocks along the ridge (fitting between trusses) would be installed, which would align the trusses vertically. The fascia would then be installed. Since the fascia is a 2 × 12 and the truss overhang is only a 2 × 4, the soffit nailer and ledger would be installed to brace the fascia vertically before the roof sheathing is installed. The vapor barrier on top of the sheathing would be installed before the soffit plywood to keep the soffit from being damaged.

Carpentry takeoff

To determine the quantity of roof sheathing, determine the flat area (the footprint dimensions) and then use a slope factor to convert the flat area to a net sloped area. A waste factor is then used. Waste factors for simple roofed areas are in the 6%–8% range. This method (using flat areas times a factor) will work well even when there are multiple sloped areas.

For busy roofs, calculating the sheathing areas may not be this simple. With multiple roof planes and perhaps various slopes, the elevation sheets of the drawings, and sometimes sectional views, are used to measure lengths and widths of roof areas instead of by flat area.

Row 2: Determine the flat area by length and width, then multiply times the factor for this slope. See the roof slope table within Chapter 1, Section 3 Lumber types and metrics. Using the slope factor results in a quantity of net square feet, then a waste factor is added.

Row 3: Use 112 LF for the length of the ledger, which is the perimeter of the block wall.

Row 4: Nailers are 2′ o.c. all the way around the soffit, which has an inside length and an outside length. Divide the length by two to get the number of nailers, but which length is used? The inside perimeter is the block wall (112 LF) and the outside perimeter is the fascia (124 LF). When placing nailers or blocking that have an "inside" and "outside" perimeter, use the *longer* one when counting pieces on center (this is for modest overhangs of 3′ to 4′).

Row 6: A length of 118′ is used, halfway between the inside and outside perimeter, i.e. the length for the plywood is counted by its centerline. For the width, the exact dimension is 18″ but this is not a "plywood dimension". Use 2′, which prevents having to use a large waste factor and more accurately represents carpentry cuts.

CARPENTRY TAKEOFF											643.4
job: Storage Building			number						date		
No.	Description	Det.	Each	Pcs.	o.c.	L	W	Ht.	%	LF	SF
1	All trusses		1								
2	Roof sheathing net area	**A**				35	27		1.042		984
3	add waste factor								1.075		**1058**
4	2 x 4 ledger to wall			1		112			1.1	**123**	
5	2 x 4 soffit nailer 2' oc			62		1.5			1.2	**112**	
6	2 x 4 ridge blocking					40			1.1	44	
7	2 x 4 blks at attic access					4			1.1	4	
8	S/T 2 x 4 blking top chord trusses									**48**	
9	2 x 12 fascia			1		124			1.1	**136**	
10	3/8" A/C plywood					118	2		1.05		**248**
11	1 x 4 trim			1		112			1.1	**123**	

Section 4 Ceiling joists and chases

Public library addition plans

See the online resources for diagrams 644.1, 644.2, 644.3, 644.4, & 644.5

Plan interpretation

This roof structure is planned to "run into" and match the existing library roof. Per the note on the floor plan and Section 3, the existing overhang will be removed, and dimensions taken of the existing profile (rise and run) of the roof trusses. These dimensions would be shown on shop drawings to define the new trusses (the slope shown on the original building plans would not be used, nor would the slope shown on these plans).

Note that 4′ of existing sheathing is to be removed north of the south exterior wall (roof plan and Section 3). The new sheathing, spanning across both new and existing trusses, will help to tie the two roofs together.

The sections indicate the roof bearing is 9′. The ceiling is lowered to 8′-1″ in the Tech Services room with 2 × 8 ceiling joists (Sections 2 and 3). The joists are supported by joist hangers on ledgers. Note how the "area" of floor joists is shown in plan by the use of diagonal dashed lines, which indicates the Tech Services room gets a lowered ceiling. The joists are not shown in plan, but appear in Sections 2 and 3.

The trusses do not have an overhang – their span is from exterior wall to exterior wall. Rafters are used to create the overhang in order to match the existing library, which has a tongue and groove "v joint" soffit; see Section 1. This soffit is exposed, can be viewed from below, and is on all three sides of the structure. There is also some tongue and groove soffit shown in Section 4.

There is a duct chase shown in the Director and Reference rooms. Section 2 locates it at 7′ high and built with 2 × 4s. This much detail is not usually shown on the plans. Often an arrow pointing to a dotted line is all that is given!

Scope of the work

Include:

Trusses and decking.
Rafters and fascia.
Ceiling joists and ledgers.
Duct chase.
Joist hangers.

Exclude:

Wall framing.
Metal roofing, peel and stick, and eavesdrip.
Demolition.
Truss hangers.

Construction techniques

The demolition of the roof overhang would be one of the first activities, and would be carefully dried in to keep rain out of the existing building.

The main trusses will be set with the blocking at the ridge helping to stabilize and plumb the trusses. The metal truss hangers are to be fastened to the girder truss, and then the hip end trusses installed.

The rafters and blocking (top of the exterior wall) are installed, as are the outlooks in Section 4. The fascia would be next, both up top with 2 × 6′s as shown in Section 4, and down below with 2 × 8′s as shown in Section 1.

The roof decking would begin with the tongue and groove followed by the plywood sheathing.

Typically, the dry-in would occur before the drop ceiling and duct chase are built, because the dry-in is an important milestone to get to as soon as possible to keep the new construction dry.

Carpentry takeoff

CARPENTRY TAKEOFF											644.6
job: Public Library			number						date		
No.	Description	Det.	Each	Pcs.	o.c.	L	W	Ht.	%	LF	SF
1	Set trusses		LS								
2	5/8" 4 ply cdx sheathing :										
3	Flat area new roof footprint					35.5	29				1030
4	Add 4' north overlap					35.5	4		1.042		148
5	Add 12" overhang	4				21	1		1.042		22
6	total all flat area										1199
7	multiply times slope factor								1.042		**1249**
8	1 x 6 t and groove soffit E/W	1		2		29.0	3				174
9	1 x 6 t and groove soffit south	1				29.5	3		1.042		92
10	1 x 6 t and groove soffit abv	4				21.0	1		1.042		22
11	total all t and g										288
12	factor convert sf to lf								2.182		629
13	factor add waste								1.1		**691**
14	2 x 8 x 8 rafter	1		41		8				**328**	
15	extra rafters			2		8				**16**	
16	2 x 12 fascia	1				94			1.1	**103**	
17	2 x 6 fascia	4				21			1.1	**23**	
18	2 x 8 ledger	2,3		2		20			1.1	**44**	
19	2 x 8 x 10 ceiling joists	2,3		15		10				**150**	
20	2 x 8 rip blking @ perim.	1				88			1.1	**97**	
21	2 x 4 duct chase 40lf	2		7		40				**280**	
22	2 x 8 mtl joist hanger	3	32								

Row 4: See roof plan note, "Remove and replace 4′ of sheathing". The flat area is multiplied by the roof slope factor.

Row 5: Section 4 has a second fascia and soffit on top of the roof. The roof sheathing, viewed "in plan", exists on top and below (in two planes). The horizontal distance is multiplied from hip girder to hip girder and multiplied by the roof slope factor.

Row 8: Two sides of the soffit are flat (east and west) and its area is 2 × 29.

Row 9: The south soffit is on slope so is counted here instead of row 8.

Row 10: Add the 1 × 6 on top of the roof. The horizontal distance, from hip girder to hip girder, is multiplied by the roof slope factor.

Row 11: For the factor, divide 12″ by 5.5″.

Row 12: The 1 × 6 will be purchased by the linear foot, not sf, so the conversion is made here so that LF can be sent to the estimate. The factor of 2.182 is found by dividing 12inches by 5.5 inches (5.5″ is the width of a 1 × 6).

Row 14: The rafters "sistered" to the trusses will be 8′ long.

Row 17: Same location as rows 5 and 10.

Row 18: There are ledgers on both sides of the room against the side walls. The room is 20′ long.

Row 20: This is the blocking above the exterior wall shown in Section 1. Pieces will be less than 2′ L and can be cut from any length of lumber.

Row 21: Even with a good detail such as Section 2, there is often more lumber needed to accomplish the work than is shown. Items 3 and 7 below are 2 × 4s not shown on the section but might have to be added to properly build the chase. Chases have to be sturdy; after finish materials are installed, and perhaps carrying some ductwork load, there can be a lot of weight.

Using the shop drawing below, say that the seven pieces shown are used to build the chase. Pieces 1, 2, 5, and 7, are approx. two feet long and occur 2′ o.c. These four pieces account for four feet per linear foot. Pieces 3, 4, and 6 are all counted by the running foot also, and these seven pieces account for seven feet every linear foot.

See the online resources for diagram 644.7

It is often the case that unimportant details are half drawn but, with study, take a lot of lumber. When the details are sketchy, load up on the quantity of lumber. Where there is confusion there is money. Do not let it be yours!

Section 5 Roof framing

Church plans

See the online resources for diagrams 645.1, 645.2, 645.3, 645.4, 645.5

Plan interpretation

Four roofs surround a church "dome" that rises up higher than the roofs (see Section C), which butt into the dome.

The exterior walls are shown with a dotted line on the roof plan. The walls in Section A are four each N/S walls at 16′ high. There are four each E/W walls in Section C that are 15′ high.

Three of the end walls (E, W, and N) are gable walls that slope up to a mid-peak (Section B). Since the length of the north wall is longer than the two side walls, the north wall would be taller. All of them rise at a slope of 45 degrees.

The south end wall is under the roof as shown in Section D, unlike the other three end walls that rise higher than the framed roofs. The south wall does slope up to a mid-peak like the others, but outlooks bear on the wall. These outlooks extend past the wall to create an overhang.

All of the walls have a 2 × 12 cap on top of them. At Sections A, C, and D, rafters bear on top of these "sill plates". See Section B where the 2 × 12 is called a parapet cap.

The walls at Section A, north and south, are 16′ high and include a 2 × 12 sill plate for 2 × 10 rafters to bear on. The rafters slope at 12/12 and rise to a height of 25′-7″ where they meet a 2 × 12 ridge beam. The soffit is at 13′-7″ and the fascia is 3′ horizontally from the exterior wall. There is blocking between the rafters at the exterior wall line.

The four end walls slope at 45 degrees matching the roof rake (see Sections B and D). The rest of the walls are horizontal. The four north/south walls are 16′ high (see Section A), and the four E/W walls are 15′ high (see Section C).

The rafters shown in Section C rise to a height of 21′-7″ where they meet a 2 × 12 ridge beam. Section C reveals that the fascia and soffit alignment is maintained by having a 2′ overhang where the wall is only 15′ high. Without a change in the overhang distance, the fascia would not align where the roof overhang occurs at a 16′ wall.

Scope of the work

Include:

Rafters and ridge beams.
Sill plates, nailers, and blocking.
Fascia, drip, and soffit.
Roof sheathing and felt.

Exclude:

Roofing and flashing.
Dome and girder trusses.

Construction techniques

The construction of the church carpentry work has a straightforward sequence. Sill plates would be installed first. The ridge beams would be placed into position, supported with temporary posts and bracing. Rafters would then be installed, leaning against the ridge beams. The fascia and soffit would be installed, then the roof decking.

Carpentry takeoff

CARPENTRY TAKEOFF											645.6
job: Church			number						date		
No.	Description	Det.	Each	Pcs.	o.c.	L	W	Ht.	%	LF	SF
1	2 x 12 horiz. sill plate top of wall	A, C				128			1.1	**141**	
2	2 x 12 sloping sill top of wall	B, D				74			1.1	81	
3	times slope factor								1.414	**115**	
4	2 x 10 x 18 rafters on slope	A		36		18				**648**	
5	2 x 10 outlooks horizontal	D		11		10				**110**	
6	2 x 8 rafters on slope	C		28		12				**336**	
7	2 x 12 ridge beam	A, C				68			1.1	**75**	
8	2 x 8 horiz.blk at perimeter	A, C				128			no	**128**	
9	2 x 10 blks on slope	D				18			1.414	**25**	
10	2 x 12 horiz fascia	A, C				114			1.1	125.4	
11	2 x 12 fascia on slope	D				24			1.414	34	
12	Total fascia									**159**	
13	1 x 4 pt drip	A,C, D								**159**	
14	2 x 4 2' oc horiz soffit blk	A		44		3				**132**	
15	2 x 4 2' oc horiz soffit blk	C		28		2				**56**	
16	2 x 4 continuous ledger	A, C				128			1.1	**141**	
17	2 x 6 nailer against block	B				56			1.1	62	
18	Add slope factor								1.414	**87**	
19	2 x 6 nailer at dome	plan				60			1.1	66	
20	Add slope factor								1.414	**93**	

Row 1: The sill plate on top of the walls is horizontal at Sections A and C. The horizontal measurements add to 128 LF.

Row 2: The sill plate sits on sloping walls at Sections B and D. The horizontal measurement is 74′. It has to be multiplied by a waste factor and a slope factor.

Row 4: From the roof plan or Section A, the horizontal distance of the rafter is 12′; when multiplied by the slope factor for a 12/12 roof (1.414), this is 17′. The rafter length is then 18′. Count the quantity from the plan. A rafter is not needed against the dome; a 2 × 6 nailer is noted to be placed there.

Row 5: There are eight each 2 × 10 outlooks on the south end of the building, which excludes the one line in the middle, the ridge. (Note that the ridgeline is made of 2 × 12s, and it extends to the fascia; see row 12) The outlooks scale

8′ long, so figure 10′ lengths. Running the other direction, there are six each pieces about 4′–5′ long E/W, figure three more 10′ lengths for a total of 11.

Row 6: Figure the length of the 2 × 8 rafter first. The horizontal distance is 8′, and when multiplied by the slope factor of 1.414 it becomes 11.3, so select 12′ for the length. Count the quantity from the plan. A rafter is not needed against the dome; a 2 × 6 nailer is placed there.

Row 7: A 2 × 12 is used for the ridge beam for all of the joists (2 × 8 and 2 × 10). See Sections A and C. They are counted by the linear foot for a total distance of 68 LF.

Row 8: These 2 × 8 blocks occur at the horizontal walls for a length of 128 feet. No waste factor is needed because every 2′ the depth of a rafter is accounted for but is not needed.

Row 9: These 2 × 10 blocks are on a slope and the horizontal distance must be multiplied by the slope factor; no waste factor is needed because blocking lengths are diminished by the width of the outlooks.

Row 10: There is no fascia at the B walls.

Row 11: This is only for the fascia figured on the slope at south end.

Row 13: Same as row 12.

Row 14: The blocks at Sections A and C are different in length so are shown separately. For Section A, figure each piece 3′ long including waste. The total run along the horizontal soffit is 80 LF. Spaced 2′ o.c., this would be 40 pieces, but add 4 for each run of soffit for a total of 44 each.

Row 15: Figure each block takes 2′of lumber for a length. The total run along the horizontal soffit is 48 LF. Spaced 2′ o.c. this is 24 pieces, but add 4 for each run of soffit for a total of 28 each.

Row 16: The 2 × 4 ledger is against the block wall as shown in Sections A and C.

Row 17: This nailer goes up and down the rake at Section B so that plywood can be nailed to it. It is found on the E/W and N ends.

Row 19: This nailer is at the perimeter of the dome so that plywood can be nailed to it.

PART 7

Thermal and moisture protection

1

ASPHALT SHINGLES

Section 1 Introduction

Ruling body, the NRCA
Allowable slope
Wind resistance

Section 2 The shingle product

Shapes and styles
Composition of asphalt shingles
Self-sealing strips and roofing cement

Section 3 Construction techniques

Underlayment
Nailing
Starter course
Six-inch offset pattern
Hip and ridge cap shingles
Venting the roof

Section 4 Estimating

Restaurant roof plan
Plan interpretation
Scope of the work
Construction techniques
Shingle takeoff
Roofing products furnished by one trade, installed by others

Section 1 Introduction

Ruling body, the NRCA

The National Roofing Contractors Association is the largest trade group that represents the roofing industry. Their technical publication *Asphalt Shingle Roof Systems, 2017 edition*, was used for much of the information in Part 7.

Allowable slope

From a given point on a horizontal plane, a rise of 12″ and a run of 12″ ends at a point forty-five degrees from the starting point. In roofing terms, this describes a slope of 12 in 12. Shingles should not be used for slopes lower than 3 in 12. Shingles and other roofing products made for roof slopes exceeding 3 in 12 are "water-shedding" roof systems. Water does not drain fast and sure enough over shingle products, with slopes of less than 3 in 12, to prevent leakage in a consistent manner. Water-shedding roof systems are composed of many pieces and work with gravity to shed water, and include:

1. Asphalt shingle, clay, and concrete tile.
2. Clay and concrete tile.
3. Slate.
4. Metal shingles.
5. Some of the architectural metal panel types.
6. Wood shakes.

Wind resistance

Like doors and windows, roofing products must pass wind tests that meet the wind speed assigned to their locality by the building code. The wind speeds in building codes are different all across the country, with the coastal regions having to meet stricter standards. Shingles are tested according to UL and ASTM protocols, with fans used to create wind speed.

Manufacturers make shingles to meet five "classes" of wind speeds. They are:

1. Class A, 60 mph.
2. Class D, 90 mph.
3. Class F, 110 mph.
4. Class G, 120 mph.
5. Class H, 150 mph.

These classifications are applicable for buildings up to 60 feet tall. For taller buildings, and for some high occupancy buildings, NRCA suggests that designers seek more specific guidance than these classifications.

Section 2 The shingle product

Shapes and styles

The traditional shingle is made of asphalt and fiberglass and is three feet long by one foot high. Two "slits" are cut into each shingle creating three parts, or tabs, and "three-tab" shingles are called strip shingles. These tabs become the exposed portion of each shingle. Approximately half of each shingle, the top half above the tabs, is covered up by the next row of shingles, and each shingle has adhesive sealing strips that bond to the shingle above. Shingles are held in place with four "roofing" nails, six in high wind regions. These nails are not long (1″ or 1-1/4″ or so) and have a large round head. See Sketch 1.

"Laminated" shingles, also called dimensional or "architectural" shingles, are also a strip shingle. They are heavier (thicker) than traditional shingles, and instead of three similar tabs have irregular shapes. Laminated shingles do not have cutouts. The various types and designs of architectural shingles are not covered in this textbook.

Composition of asphalt shingles

Asphalt shingles are manufactured from four components: asphalt, fillers, reinforcing mat, and surfacing.

Asphalt is a petroleum product, and fillers are primarily minerals that are added to the asphalt. These mineral additives help bind the asphalt and add to its fire resistance. Manufacturers use various minerals for this purpose.

Reinforcing mats are a layer that supports the asphalt and are composed of thin glass or polyester fibers bound together with plastic binders or resin. The added thickness of laminated shingles comes from a thicker reinforcing mat and thicker asphalt.

On the top of the shingle, the most common surfacing material is ceramic-coated mineral granules. These have the purpose of:

1 Protecting the asphalt from sunlight and weathering.
2 Increasing fire resistance.
3 Expanding design flexibility because of their range of color and texture.
4 Adding weight, which increases wind resistance.

Self-sealing strips and roofing cement

Adhesives are placed in strips on both three-tab and architectural shingles. They bind the shingles together after placement on a roof and help keep the shingles in place during high winds. After installation, the adhesive is activated by heat (the sun will do it) and forms a bond with adjacent shingles. See Sketches 1 and 3.

Roofing cement, purchased in one-gallon cans and five-gallon pails, is manually applied in strategic locations for the following purposes:

1 To seal shingles to starter strips.
2 To seal tabs of shingles in high slopes or some cold-weather applications.
3 To use as a bedding system for underneath flashings and bases of accessories.

There are two common types of roofing cement – flashing cement and lap cement. Flashing cement is put on vertical surfaces with a trowel. Lap cements are for bonding asphaltic materials together and are placed with trowels or brushes.

Section 3 Construction techniques

Underlayment

Asphalt felt is a flexible sheet manufactured principally from wood pulp and vegetable fibers (organic felts), asbestos fibers, fiberglass fibers, or polyester fibers. They are saturated in asphalt.

A typical first layer of roofing is 15 or 30 pound felt, with the 15 lb weight made in 3′ widths and a roll long enough for a roof coverage of 400 sf (after lap), and the 30 lb weight made in 3′ widths and a roll long enough to cover 200 sf of the roof. If the surface of the roof is wood, it should be covered with a layer of felt as soon as possible so that rain does not damage the wood. This operation is called the "dry-in", because it prevents rain from leaking into the structure. At least that is the goal.

Felt is an initial layer of roofing called underlayment that is placed underneath all of the shingle and metal flashing components described below.

Underlayment is placed as shown in the following sketch before other valley construction. Underlayment from the two sides laps onto the valley underlayment by 6″ minimum.

See the online resources for diagram 713.1

Nailing

Four nails per shingle is typical for three-tab and laminated shingles. For most three-tab shingles, their location is about 5/8″ above the cutouts, and on the ends about 1″ from the shingle edge. For laminate shingles, they should be nailed about 5-1/2″ above the shingle edge and 1″ from the edge. Nails should penetrate the "head lap" of the shingle underneath.

Nails should penetrate at least 3/4″ into the wood. If the wood is less than 3/4″ thick, the nails should extend 1/8″ past the wood.

In high-wind regions, six nails may be required per code, *with an extra nail added at the center two cutouts.*

The NRCA does not recommend the use of staples to fasten shingles.

Starter course

Two layers of shingles are placed on the bottommost row, and the initial layer is called the starter course. It is installed over the underlayment and the shingle edge is laid to the end of T-Type drip edges (see drip edge configurations in Chapter 2 Metal Flashing), and from 1/4″–3/4″ past L-Type drip edges. This important first layer of shingles sheds water that may migrate through the joints and cutouts of the overlying first course above it and holds the leading edge of the first course with sealing strips.

Starter courses can be cut from regular shingles, or manufactured starter courses can be purchased. If made in the field, the leading edge of the shingle, the half that has the cutouts and three tabs, is cut off and discarded. The top half of the shingle, measuring approximately 6″ × 36″, is used for the starter course and laid on top of eavesdrip.

Another method is to use a 9″ or wider strip of asphalt roll roofing as a starter course. If it does not have adhesive strips, roofing cement can be "hand-tabbed" to the underlayment before the roll roofing is placed.

Sketch 2 is of both the starter course and the first and second rows of shingles on top of it.

See the online resources for diagram 713.2

Six-inch offset pattern

There are several offset or side-lap patterns that can be used, but the most common is the six-inch method. It establishes a pattern created at the bottom and continues up the rake. It applies only to three-tab strip shingles. Laminated shingles vary in appearance, and manufacturer guidelines vary.

The first courses are often laid with the following pattern, which provides the same appearance to left and right uphill rakes. This method provides for either a half or a full shingle tab to be shown at the ends of all the rows that traverse up a straight line. Any odd cuts of a shingle will occur in the middle of the roof where it is less likely to be seen.

See the online resources for diagram 713.3

Hip and ridge cap shingles

With laminate shingle roofs, hip shingles must be ordered from the manufacturer. They are sold by the bundle and usually cover about 35 LF of a hip.

Manufacturers of three-tab shingles do not make hip and ridge shingles. They are cut in the field from regular shingles; see Sketch 4.

Nails may need to be longer for hip and ridge caps. The same nailing rule applies – penetrate at least 3/4″ into solid wood or 1/8″ beyond the underside of decks less than 3/4″ thick.

See the online resources for diagram 713.4

At the ridge and hips of a roof, shingles are first placed all the way to the ridge from both sides. Shingles are then cut into thirds and placed across the ridge, lapping down on both sides. At hips, the individual tabs start at the low end then other tabs are placed "ladderstep" up the slope.

Venting the roof

An attic, or plenum, beneath a roof can become very hot. The roof should be interrupted with openings to allow air penetration and circulation. Two of these methods for shingle roofs are ridge vents and off-ridge vents. Off-ridge vents are located downhill from the ridge.

Venting can be continuous, such as a ridge vent that sits atop the roof extending for the length of the attic area. This is accomplished by installing vents, which are often 4′ or 8′ long, end to end. Or, they can be installed individually. They are often made of galvanized steel.

Section 4 Estimating

Restaurant roof plan

A bundle of shingles will cover 33 square feet, and three bundles make a "square", or 100 square feet. A square is the unit of measure for shingles and all roofing. If a roof measures 10,000 sf, then it will take 100 squares of roofing to cover it.

After counting the flat area of the roof, additional areas must be added to account for the starter course, ridge, and hips.

See the online resources for diagram 714.1

Plan interpretation

The roof areas, and quantity of shingles, can be calculated from this roof plan. No elevations are needed because all of the roof slopes are given. Note the explanation of the edge of roof area A given various slopes. The dotted lines represent where the valley would be located at various roof A slopes. When both roof plans have the same slope of 5 in 12, the resulting angle is 45 degrees as viewed in plan. The slope of the valley itself is somewhat less than 5 in 12.

The restaurant has two hip ends and two gable ends.

Scope of the work

Install 30 lb asphalt felt and architectural shingles to cover the entire roof over existing 5/8″ exterior grade CDX plywood. Include taking delivery and unloading of material. Install valleys, eavesdrip, starter course, and ridge shingles for a complete roof.

Construction techniques

Delivery will be made by a boom truck that will unload to the roof. Install four nails per shingle; extend nails 1/8″ beyond the thickness of the plywood.

Shingle takeoff

Review the lengths of ridges, valleys, and roof edges that an estimator might write on a roof plan. These numbers are used on the takeoff.

See the online resources for diagram 714.2

Figure areas of the same slope together, starting with horizontal dimensions and arriving at flat areas and then converting to sloped areas using a factor. Starting with roof A, see the rectangular area on row 1, and then the triangles that extend over roof C on row 2. The conversion factor for a 3/12 slope is used on row 4 to arrive at the sloped square footage; see row 5.

The unit of measure for shingles is the "square", or 100 sf. Three bundles of typical three-tab shingles have a coverage of 100 sf. However, the exposure of the shingle can vary by region, and the coverage of architectural shingles can vary. Regardless, the first task of the estimator is to properly figure the square footage.

Roofing products furnished by one trade, installed by others

Many trades of work can be involved in roof construction, and "who does what" can be confusing. One trade might furnish a product (buy it), and another trade might be responsible for the labor to install it. How it is handled follows generally accepted practices for the industries affected, subject to local preferences. This kind of co-mingling of products probably occurs more in roofing than in other divisions of work.

RESTAURANT SHINGLE TAKEOFF								714.3
No.	**Description**	**Det**	**Ea**	**Factor**	**L**	**W**	**SF**	**Squares**
	Roof area A @ 3 in 12 slope							
1	Rectangle				20	16	320	
2	Triangles above roof C		2	0.5	5	8	40	
3							360	
4	Multiply times 3/12 slope factor			0.0308			11	
5	Total area roof A						371	3.71
6								
7	**Roof area B @ 4 in 12 slope**							
8	Rectangle				14	20	280	
9	Triangles above roof C		2	0.5	8	10	80	
10							360	
11	Multiply times 4/12 slope factor			0.0541			19	
12	Total area roof B						379	3.79
13								
14	**Roof area C @ 5 in 12 slope**							
15	Rectangle				58	24	1392	
16	Less row 2						-40	
17	Less row 6						-80	
18							1272	
19	Multiply times 5/12 slope factor			0.0833			106	
20	Total area roof C						1378	13.78
21								**21.28**
	Net area without waste. Excludes the starter and ridge shingles below.							
22	**Starter shingles:**							
23	Roof A				20			
24	Roof B				28			
25	Roof C				196			
26					**244**	**LF**		
27								
28	**Ridge shingles:**	**LF**						
29	Roof A ridge is horizontal				25			
30	Roof B ridge is horizontal				22			
31	Roof C ridge is horizontal				36			
32	Roof C ridges on 5/12 slope	64		1.0833	69			
33					**152**	**LF**		
34								
35	**Valley metal:**							
36	Roof A				18			
37	Roof B				26			
38					**44**	**LF**		

A simple example of this is a plumbing vent stack, which is the roof component of an air vent through the roof and extending about a foot above it. Plumbers are responsible for the vertical piping extending up a building and reaching the level of the roof. At the roof penetration, a vent stack, sometimes called a "soil stack", finishes the job.

The plumbing trade counts the piping and vent stacks at the estimating stage, using the plumbing plans and floor plans. To estimate their cost, the number of vent stacks must be determined, but since they may not be drawn on the roofing plan, the roofing trade may not know how many there are.

Piping and vent stacks are defined in the plumbing specifications and typically supplied and paid for by the plumber. However, while the plumber will install piping up to the roof surface, the vent stack is often handed over to the roofing contractor. It is the roofing trade that sits the stack in place and places roofing material up to and around the vent stack. Vent stacks are not difficult to install but need to be handled at specific times during the course of roofing. So it is expedient for the roofing contractor to handle the labor, and it is important to have them when needed.

One trade buys the vent product, another installs it. This instruction is not part of the plans and specifications. It is how the contractors have decided to divide the work up. Another example would be skylights, which are a Division 8 item, contained in the same division of specifications as doors, windows, glass, and storefront. A steel contractor might, using dimensions from the skylight shop drawings, provide steel angles perpendicular to bar joists to surround a roof opening. The roofing contractor might build up a curb out of lumber around this opening six inches or a foot high above the roof and install base flashing around it. Then the skylight is installed on top of the curb by the Division 8 contractor, who has supplied shop drawings to the parties. Skylights usually have a lip around them that extends down like a drip edge, and this component acts as counterflashing.

In this example, the roofing company might have excluded carpentry work, which means the skylight curb would be framed by the carpentry subcontractor. So there's four trades affected – Division 5 steel, Division 6 carpentry, Division 7 roofing, and the Division 8 subcontractor who is installing the skylight.

Through-wall flashing is typically installed by masons; the material may or may not be furnished by the mason; it might be furnished to them by either the HVAC or roofing trade, who both can be in the sheet metal business.

Air conditioning roof top units (RTUs) are set up on curbs like skylights are. Carpenters may or may not build the framework for the curb, but the roofing contractor does the flashing and installs the roof membrane up the side of the curb. There is a small product that the A/C subcontractor might supply to be placed against the curb before the flashing and membrane is installed. This product is called a "cant strip" and is triangular in shape. The cant strip is given by the A/C sub to the roofing contractor – another example of one contractor buying a product and another being responsible for its installation.

2

METAL FLASHING

Section 1 Introduction

The NRCA, ruling body

The NRCA Roofing Manual: Architectural Metal Flashing and Condensation and Air Leakage Control, 2018 edition, was used as source material for this section.

Flashing and counterflashing defined

Flashing types can change in shape somewhat but still have the same design purpose. For example, base flashing, the first piece that goes on bottom, can be continuous for many feet along a horizontal surface such as the low side of a chimney or other wall. Where the wall turns and the roof is upslope, flashing of the same shape is used, but it is in short lengths, called step flashing, and is placed underneath each individual shingle as it traverses uphill.

Counterflashing, the piece directly above base flashing, has many possible standard shapes, but all are still called counterflashing, or sometimes a reglet. When used above membrane base flashing, it is better if counterflashing consists of two pieces, reglet and counterflashing. Where the base flashing is connected to the roof, counterflashing typically is fastened to the substrate above. See Sketch 1.

Counterflashing used on walls can be one or two pieces, but a two-piece design is better. If a single piece is used, it is called a receiver or reglet, and in two-piece designs the top piece is the receiver (same as reglet) and both pieces are termed counterflashing.

The NRCA's definition for receiver and reglet are the same: The top piece of two component counterflashing that can be surface mounted, inset into a raggle (see Sketch 1), or embedded behind cladding (siding or other veneer).

Counterflashings should extend over the top of the base flashings approximately 4″ and have a 1/2″ minimum drip edge at the bottom. End joints can be lapped, sealed, fastened, welded, or soldered.

Yes, counterflashing can consist of a piece of equipment or a skylight! Skylights and equipment often have edge flashing that drops down over curbs and flashing below. In this case, the base flashing is "counterflashed" by a product.

See the online resources for diagram 721.1

Metals used for flashing

The kinds of metals used for flashing material are shown in Chart 1.

See the online resources for diagram 721.2

Section 2 Common flashings

Drip edges

Drip edges are a type of flashing. The NRCA recommends the use of drip edge metal for asphalt shingle roof systems at horizontal eaves and uphill rakes. Drip edge metal should be installed directly above the deck at eaves, with underlayment on top. Drip edges should be installed on top of the underlayment at rakes. When perimeter edge metal is attached to wood blocking, the face of the metal should extend 1″ below the bottom of the blocking.

See Sketch 2.

See the online resources for diagram 722.1

Valleys

A valley is created at the intersection of two roof planes. If the two roofs are at the same slope, the resulting slope down the valley will be *less* than that of the intersecting roofs. For this reason, and because of the volume of water collected in them as water travels downward, they are more susceptible to leaking than the open "field" part of the roof.

Valley metal for use with shingles should be a minimum of 24″ wide. Obstructions, such as pipe penetrations and expansion joints, should not occur within a valley.

There are four basic types of valleys used with asphalt shingles:

1 Open valleys.
2 Closed-cut valleys.
3 No-cut valleys.
4 Woven valleys (not recommended by the NRCA).

These four types of valleys are constructed only after the specified layers of underlayment are installed.

Open valleys are metal flashings, usually made in 8′ and 10′ lengths, laid in the direction of the valley, with the first piece laid at the bottom low end and subsequent pieces lapped on top. They should have a width of at least 24″. Asphalt shingles are then lapped over the sides of the metal with their edges cut at an angle parallel with the valley flashing.

A valley's width, or the distance between shingles, should increase going down the valley, as there will be more water volume further downhill. In most climates, the width should increase by about 1/8″ per foot. For a valley 16′ long, this is 4″.

See the online resources for diagram 722.2

A good practice with open valley metal is to have an "inverted V" formed in the middle of the flashing. This acts as a diverter and water dam, turning water downward. The splash diverter should be at least 1″ high.

The shingles at *closed-cut* valleys are installed fully across the valley on one side, and the shingles on the other side are installed about 2″ short of the centerline of the valley

No-cut valleys are sometimes referred to as California valleys. After metal valley flashing is in place, two rows of shingles are placed in the direction of the valley, lapped over the edge of the flashing and leaving the middle portion of the flashing exposed. The exposure side of the shingles faces the center of the valley.

Penetration flashing

There are many small vertical penetrations that need to be flashed into asphalt shingle roof systems, such as vent pipes, exhaust vents, exhaust fans, furnace or water heater flue pipes, electrical piping, etc. This kind of flashing is often accomplished with a flat flange either square or rectangular. Shingles lap over the flange uphill of the vent, and the flange laps over shingles below the vent.

See the online resources for diagram 723.3

Section 3 Wall flashing

Through-wall flashing

It is difficult to flash a masonry wall, which is porous, without horizontally placing the flashing all the way through the wall. This is typically accomplished at mortar joints and is a common type of flashing. See Sketch 1 where the flashing extends under the brick.

Vertical wall flashing

Three types of vertical wall flashing are shown in Sketch 5 – apron flashing, counterflashing, and step flashing. On the high side of a vertical roof penetration, cricket flashing is used on walls *longer than 2′*. See Sketch 7. The term cricket is used to define both the "hump" in the roof and the associated flashing.

Apron, or headwall flashing, is located at the low sides of chimneys, dormers, curbs, and walls.

Short pieces of flashing that resemble the shape of apron flashing are used on the sides of walls such as chimneys. As the roof slopes up, individual pieces of flashing are laid at the end of each row of shingles. This is called step flashing.

See Sketch 5 for the term "let into". The top edge of this flashing has a 90-degree bend so that it can extend (or be "let") into a mortar joint. This flashing can be installed when the masonry is laid, but due to the precision-like nature of its placement, it is often installed later. The mortar joint can be sawn later and the flashing inserted and caulked.

See the online resources for diagram 723.1

On the high side of a vertical roof penetration, backer flashing is used on walls that are *less than 2′ long*. See Sketch 6.

See the online resources for diagram 723.2

On the high side of a vertical roof penetration, cricket flashing is used on walls *longer than 2′*. See Sketch 7.

See the online resources for diagram 723.3

Coping

Parapet walls form the perimeter of many low-slope roofs. The tops of these must be protected from moisture intrusion to prevent damage to the wall, other building components, and the interior of the building. The flashing that covers the tops of these walls is shaped like an inverted "U" and is called coping. See Sketch 8.

A cleat is a continuous metal strip, or angled piece, used to secure metal components.

Parapet walls can also be built on an incline, such as a gable end wall extending higher than the roof plane. The metal on top of these walls is also called coping.

See the online resources for diagram 723.4

Section 4 Curbs and gutters

Curb flashing

At roof locations such as air conditioning units and skylights, short "curbs" are built to raise the unit six inches or a foot above the roof surface. A roof membrane or base flashing surrounds the curb, and when the unit is set it usually acts as its own counterflashing, with perimeter metal at the base of the A/C unit or skylight that covers the base flashing. If the roof is sloped, the unit is flashed similar to chimney flashing consisting of apron, step, and backer flashing.

Gutters and downspouts

The design of gutters – their size, gauge, and fastening – including the quantity and placement of downspouts, is extremely important. Adequate drainage and strength is hard to constantly maintain without observation and maintenance. Otherwise, leaking gutters, downspouts, and joints can cause damage to a building's interior and exterior.

Gutters can be externally attached or built into the structure of the roof, called "built-in" gutters. Externally attached gutters are more common. Built-in gutters, are, of course, problematic, because they tempt Mother Nature. All it takes is a clogged drain or two and a roof can be flooded (and possibly overload the structure).

Externally attached gutters have two common types of support – *straps* and *support brackets*. Straps, or spacers, are used to keep the front rim of a gutter a set distance from the back of the gutter. They are spaced about 30″ o.c. and extend across the top of a gutter. Brackets, which support the weight of the gutters, should be installed between 12″ and 30″ o.c. Brackets are "U" shaped and fit around the three sides of a gutter. The side of the bracket to the rear of the gutter might be fastened to the building. The rear of the gutter may also be attached to the building, or the rear side of a gutter can extend up and over the edge of the roof for gutter support.

If both straps and brackets are used, they should be staggered. The front face of external gutters are usually lower than the back to allow for overflow.

Built-in gutters have the same "U" shape as external gutters but are located within the confines of the roof area. There are four types or conditions of built-in gutters located in the field of the roof, not at the edge.

1 Eaves.
2 Near eaves.
3 Valley.
4 Parapet wall.

These gutters recess below the surface of the adjoining roof to collect water, and the bottoms of built-in gutters slope towards a drain. If the gutter is located near the edge of the roof (see 1 and 2 above), it may discharge into external downspouts. If the built-in gutter is at the interior of a roof (see 3 and 4 above), drains are placed at locations where piping extends down into the building.

If water is not allowed to fall off the edge of a roof, or is only partially allowed to, and relies on roof drains dotted across the roof, it is said to be "internally drained".

Built-in gutters that are at or near an eave are inherently better than "valley" gutters or gutters against a parapet wall. If eave drains stop up and the gutters overflow, the water simply flows over the edge of the roof nearby. Gutter types 3 and 4 above should be avoided if at all possible. When they are used, secondary drainage, such as scuppers, becomes critical.

3

METAL ROOFING

Section 1 Introduction

Metal roofing can be purchased in several thickness, or gauges, be factory painted, and can last a half century. A residential thickness is 29 gauge, which is the thinnest metal roofing. This thickness has a tendency to show "oil canning", which is a term used to describe small wrinkles and bends in the metal.

Commercial gauges are often 24 and 20 gauge, thick enough to lie flat and not telegraph slight unevenness in the underlayment. Flat metal roofing is interrupted by ridges, called "ribs", that are at least 7/8″ high, every 16″ or 24″ o.c. Each manufacturer has their own shape of these ribs.

In former times, some metal roofing was not installed over solid sheathing, as it is now. Warehouses and farm buildings with fairly thin metal would span between wood supports, which could be rafters, joists, or horizontal ledgers on top of wood trusses. This support would typically be 16″ or 24″ o.c.

There are two categories of metal roof panels, "architectural" and "structural".

Section 2 Substrates for metal roofs

Architectural metal roof substrates

In new construction projects, the roofing contractor is responsible for accepting the surface of the structural roof substrate as suitable and compatible for the installation of the specified roof system.

Architectural metal roof panels require a continuous or near continuous surface, called the roofing substrate, underneath them. They are water shedding; the seams are not watertight, but they perform well on slopes greater than 3 in 12.

Steel decking is sometimes used as the substrate for architectural metal panels. Steel decks are usually installed over bar joists spaced 4′ o.c. Steel decks have many different rib configurations; see Part 5.

Plywood or oriented strand board (OSB) are two wood sheathings that provide a continuous substrate for architectural metal panels. They are typically installed over wood or metal trusses, but can also be the top layer in a roof composed of other products such as rigid insulation, which might be on top of a steel deck.

Other wood products used with architectural metal panels are wood planks and wood boards. Wood boards are common lumber pieces less than 2″ thick and up to 12″ wide. Planks describe larger pieces that a lumber yard might refer to as timber. The size of this lumber used for roof decking ranges from 2″ to 5″ thick and might have a tongue and groove edge, perhaps even a double tongue and groove (on each side of the board).

Structural metal roof substrates

Structural metal roof panels are waterproof and can span framing members that are perpendicular to the panels and spaced some distance apart. A minimum slope of 1/2″ per foot is recommended, although some manufacturers allow as little slope as 1/4″ per foot. Structural panels are mostly manufactured from steel, Galvalume, galvanized or stainless steel, or aluminum and are thicker than architectural panels.

Substrates for structural metal panels can be composed of various kinds of framing usually spaced a few feet apart, such as steel joists which are commonly spaced 4′ o.c. Wood or metal trusses can also be used, which are typically spaced 2′ apart. Structural metal panels span supports below running perpendicular to the panels. Sometimes the structural members are called purlins, which is a roofing and metal building term that in carpentry language might be called a roof rafter.

Section 3 Architectural metal panels

Configuration of architectural metal panels

The length of manufactured panels is generally limited by shipping constraints. For shipping lengths of approximately 30′ and less, roofing contractors can carefully measure a roof from ridge to eave and purchase panels that exactly fit a given roof. This eliminates field cutting of panels. *The center panel is placed first, with the centerline of the roof running either through the middle of the first panel or at one edge of a panel.* This is similar to many other layouts of rectangular building surfaces, whether an acoustical ceiling or a wall of paneling. Layout is from the middle, or center, with an eye towards having the two end pieces being the same size and appearance.

Standing seams have raised vertical legs. There are different designs, but the term "standing" means that the seam, which is the weak point concerning leakage, is raised above the plane of water drainage.

Profiles of architectural metal panels include the categories of single-lock, double-lock, snap-on cap, battens, T, snap-lock seam, and integral seam. Two of these are shown below.

See the online resources for diagram 733.1

Construction techniques of architectural metal panels

Underlayment should be installed on top of the substrate. One example of a substrate is plywood sheathing.

Architectural metal panels are attached to roof substrates with a series of individual clips. Once a section of the roof panels is secured with clips (see Sketch 2), the seaming process can begin. Seam types and seaming techniques vary, but they all are designed to interlock panels to form a continuous, water-shedding, and weatherproof system.

The legs of seams can be mechanically folded in the field. Other systems utilize a snap-on clip.

Using internal drains is not recommended with metal roofs. Exterior guttering should be used. When there are vertical penetrations in the roof, crickets should be used when the penetration is more than half the width of the roof panel.

Skylights installed on metal roofs should be placed on raised curbs so they are out of the plane of drainage.

Some typical features of eave and rake conditions for architectural metal roofing are shown in Sketch 2.

See the online resources for diagram 733.2

A typical ridge condition for architectural metal panels is shown in Sketch 3. Note how the metal panels are turned up at the top to prevent water from splashing or being wind driven beyond the end of the panel. This is termed a "secondary" means of stopping water passage, because the primary means is the Z flashing. The Z flashing is also how the ridge cap is held in place.

See the online resources for diagram 733.3

Section 4 Structural metal panels

Configuration of structural metal panels

Striations or longitudinal stiffening ribs are often used to minimize the effects of oil canning. Panel widths are wider than the one- and two-foot pieces of architectural panels. There are three common profiles of structural metal panels.

1 Trapezoidal. The rib has a trapezoidal or triangular shape, providing height and strength.
2 Intermediate rib. This is the use of a center rib, with a shape of the same profile of the end ribs. This profile allows the use of wider panels because of the increased strength provided by the center rib.
3 Vertical leg. This profile is a standing seam design similar to architectural panels.

See the online resources for diagrams 734.1, 734.2, & 734.3

Construction techniques of structural metal panels

Structural panels are attached to intermediate supports with individual "clips". Underlayments are not used because the substrate is not continuous; note the open area in Sketch 7. See how the metal panel extends past the drip edge at the eave. This projection is usually about 1″ and allows water to drip straight down from the edge of the panel. Note the use of cleats, a common flashing item.

See the online resources for diagram 734.4

Since there is usually no continuous substrate under structural metal panels, support is needed for the ridge cap and Z flashing. In Sketch 8, the heavy gauge ridge support runs longitudinally along the ridge to do this.

See the online resources for diagram 734.5

Section 5 Metals and metal problems

Some of the metals used for metal panels are:

1. *Aluminum.* Lightweight, and will expand and contract more than other roofing metals except zinc.
2. Aluminized steel.
3. *Copper.* Over time, changes color, referred to as a "patina", and can become brown or blue-green depending on geography and atmospheric conditions.
4. Copper-coated stainless steel (Copper Plus). Type 430 stainless steel has a small layer of copper on each side but is approximately 80% stainless steel. Provides a roofing or flashing product with the appearance of copper but the strength of stainless steel.
5. *Galvalume.* A proprietary aluminum-zinc-coated steel approximately 55% aluminum and 45% zinc by weight. Galvalume is used for many metal building roofs.
6. Galvanized steel. Thin zinc coating on steel.
7. *Lead.* Soft and workable, useful for flashing. Designated by weight, not gauge.
8. Lead-coated copper. Less cost and weight than lead alone and greater tear resistance.
9. Prefinished aluminum. Factory-applied protective coating such as paint.
10. Prefinished Galvalume. Factory-applied protective coating such as paint.
11. Prefinished galvanized steel. Factory-applied protective coating such as paint.
12. *Stainless steel.* 38 standard stainless-steel alloys! Types 302 and 304 flashing most common.
13. *Zinc.* Turns to blue-gray patina.
14. Zinc-tin-coated copper. Proprietary, referred to as FreedomGray.
15. Zinc-tin-coated stainless steel. Proprietary, referred to as TCS II/TCS II Satin/IndependenceGray. Type 304 stainless steel coated with zinc and tin.

Oil canning

Oil canning is an unavoidable phenomenon occurring in sheet metal products. It refers to distortions in the flatness of metal. It is an aesthetic problem but does not have any adverse effect on the capability of the metal. These "creases" in the metal can become apparent when, for instance, thin metal remains flat for too great a distance without bending into another plane. Temperature and an uneven substrate can also cause flat metal to deviate from a straight flat surface.

Oil canning can occur during the roll forming process, meaning that if that portion of the coil is used, the crease will be seen at installation.

Oil canning can often be prevented by using a heavier gauge of metal to begin with.

Dissimilar metals

When two dissimilar metals come into contact and moisture is present, one of the metals can show increased corrosion and the other decreased corrosion.

4

MOISTURE PROTECTION AND WATERPROOFING

Section 1 Introduction

Waterproofing and dampproofing are complicated subjects with specialists using different techniques around the country depending on subsurface conditions, the amount of rainfall, and other factors having to do with the building design. This textbook's approach to these subjects is that since moisture protection is a part of all buildings, the products and construction techniques for it are covered in more detail than the various waterproofing and dampproofing techniques, which are many. After the topics of basic moisture protection for walls and slabs are presented, the other two subjects are more generally covered, with information about a few products and some definitions.

Moisture protection is the first line of defense, dampproofing is the second, and waterproofing is the full defense against water and water pressure.

Section 2 Vapor barriers (retarders)

Introduction

Where to place "vapor barriers", using the terminology often seen on plans, is a design consideration for architects. The word barrier is not a good choice, but it has been used a long time. Vapor barriers retard the flow of moisture. Our concern here are a few building products that are much used in the industry that prevent rainwater from getting in while allowing excess interior moisture to get out.

The contractor shouldn't be deciding where the barrier is placed, but should understand its importance. Incorrect placement of vapor barriers (for example, the wrong side of the wall) can result in interior spaces that won't dry out properly or become subject to mold. Improper moisture control can cause deterioration of structural (often not visible) and finish components.

Polyethylene

Vapor barriers used in wall, floor, and roof construction can consist of a simple layer or layers of polyethylene, often known as "Visqueen", which is a brand made by British Polythene Industries Limited. It is also known as simply "poly". Polyethylene is *the most common plastic*, and it is also used to make plastic bags.

Polyethylene is a thermoplastic, a polymer that becomes pliable or moldable above a specific temperature and solidifies upon cooling. Polyethylene "film" is a much used product in building construction, being placed horizontally under concrete slabs and vertically against frame and block walls. It is also used for temporary purposes such as covering other materials for rain and dust protection. It is sometimes used temporarily on top of a concrete slab as it sets to minimize moisture release of the slab and to block wind, rain, and sun.

The product is a thin material, relatively inexpensive, and sold in rolls of 1,000 and 2,000 sf at a time. The most common thicknesses are 4 mils, 6 mils, 8 mils, and 10 mils (0.004 to 0.01 in.). It is available in clear, opaque, blue, and black.

Polyethylene is placed on top of the foundation before slabs on grade are poured. It is placed after fine grading is completed and the ground has been treated (sprayed) for termite protection. Wire mesh is placed over the top of the polyethylene "vapor barrier", and then the slab is poured. The purpose for using this product under slabs is to prevent water from escaping the concrete into the ground below. This greatly facilitates the proper curing of concrete. Common duct tape is often used at the joints where laps occur.

Asphalt felt paper

Another product that is commonly used as a vapor barrier is asphalt felt paper. It can be used under slabs and on walls and is a typical underlayment for shingles. It is often used in wall construction behind siding or other cladding. It comes in 3′ wide rolls, typically covering 200 sf and 400 sf. The product is usually applied horizontally.

Felt paper is black in color (it's made of asphalt) and sold by weight, not thickness, although it is a thin product like polyethylene. The two common weights used to be 15 lb and 30 lb, meaning that a square, or 100 sf of asphalt felt, weighs either 15 pounds or 30 pounds. The 15 lb felt rolls contain 400 sf, and the 30 lb rolls (twice as thick) contain 200 sf. They no longer weigh this much, but "15 lb felt" and "30 lb felt" are still terms used to describe them.

Building felt, called a vapor barrier just like polyethylene is, can be used vertically against frame or masonry walls. It can be used on roofs, sometimes in layers called "plies". Hot tar is often used at the joints and then another layer applied. This product can also be used under slabs, as a good protection against moisture rising due to capillary action.

Felt paper is usually fastened to plywood with "felt tabs", which are short nails with a large square head.

Housewrap

Housewraps are made from a synthetic material and have largely replaced asphalt felt as a vapor barrier in frame wall construction. The material is lighter in weight, comes in long rolls, and can be installed faster than asphalt felt can. One popular tradename for it is *Tyvek*, made by DuPont. It is difficult to tear but can easily be cut with scissors or a knife. Small amounts of water vapor can pass through Tyvek, but liquid water cannot.

Peel and stick underlayment

This product resembles several layers of building felt. It is a good choice for a moisture barrier underneath metal roofing, where a secondary protection against leaks is important. Metal roofs can sometimes have minor openings such as a rusty screw, and blowing rain can find a way to penetrate the metal laps or ridge.

Section 3 Waterproofing and dampproofing

Dampproofing and waterproofing defined

Much of the research for this section was enabled by *The NRCA Waterproofing Manual, 2011 edition.*

The NRCA defines *damp-proofing* as the treatment of a substrate to "resist the passage of moisture". *Waterproofing* is defined as the treatment of a substrate or structure to "prevent the passage of water under hydrostatic conditions" (which means the presence of water and water pressure).

Since most waterproofing products are covered up, there is often only one chance to install it correctly. This makes the surface inspection of prior work, and the implicit acceptance of it, if waterproofing proceeds without objection, of paramount importance. Of course, this is the case in all construction, and waterproofing is subjected to the same pressure to proceed that is inherent in a fast-paced schedule that all trades experience. However, since waterproofing materials often are applied directly to the substrate, any defects in the substrate can directly affect the performance of the waterproofing. Having a clean, dry, and smooth substrate, free of voids, can be hard to insist on during the rainy season and when the prime contractor is falling behind.

A waterproofing "assembly" consists of a substrate and a membrane and may incorporate a protection/drainage/insulation layer. Waterproofing systems are on either the *positive* or *negative* side of a wall. This is an important distinction. See Sketch 1.

Positive waterproofing systems are those placed on the exterior side facing the source or supply of water. Negative systems are those where the substrate is between the source of water and waterproofing, which is inherently not as preventative.

It is recommended that waterproofing be installed on the positive (exterior) side if at all possible. Exterior systems are best suited to resist moisture intrusion. Waterproofing is typically adhered or bonded directly to the substrate. Adhered materials are more desirable than those that are loose-laid (layer of water absorbing material) against the substrate, because adherence reduces the possibility of lateral water migration between the waterproofing material and substrate.

See the online resources for diagram 743.1

Substrates

Most waterproofing materials are applied to surfaces built by other trades. The substrate by others should be free from excessive cracks, holes, and projections and relatively smooth without sharp edges.

Masonry substrates include brick and block walls. Holes, joints, and voids in masonry substrates should be pointed (a masonry term meaning grouted) flush with the surface. Waterproofing over concrete substrates include cast-in-place concrete, precast concrete, and shotcrete. Lightweight insulating concrete is *not* an acceptable substrate for a waterproofing system.

Substrates composed of wood include marine-grade plywood and pressure-treated wood. OSB is not a good substrate for waterproofing. The masonry surface should be smooth and free from projections. If the masonry surface is irregular, it should be covered with 1/2″ thick parging, leaving a smooth steel trowel surface. Split faced block is very difficult to waterproof.

Concrete substrates should be properly cured and the surface dry before installing waterproofing materials. Form release agents and concrete curing compounds must be compatible with the waterproofing materials.

Shotcrete is mortar or concrete pneumatically projected at high velocity onto a surface. The NRCA recommends designers consult manufacturer requirements of the waterproofing products for instructions about shotcrete.

Marine-grade plywood is made entirely of Douglas fir or western larch. The grade of all plies is B and better, which allows knots but no knotholes. The surface must be smooth, and panel edges should bear on joists or blocking to reduce deflection.

At the exterior of a building, compacted earth and/or gravel can be used as a substrate for waterproofing. It should be free of sharp projections, voids, contours, and ridges. Eighty-five to ninety percent compaction should be obtained. Where the surface of the substrate is not appropriate for direct installation (e.g. drainage rock), installation of a protection board is recommended.

Walls below grade

On precast concrete and masonry below grade, waterproofing should only be applied to the positive exterior side.

Waterproofing on cast-in-place concrete below grade can be applied to either the positive or negative side. Sometimes the exterior face of cast-in-place concrete may not be accessible after a pour because of concrete placement against sheeting or shoring. In these cases, waterproofing is best applied against the sheeting or shoring before the placement of concrete.

Hydrostatic conditions include water pressure. In parts of the country, a big waterproofing consideration for basements in buildings built below grade is the amount of pressure exerted against the foundation wall by an adjacent column of water. Refer to Sketch 1. Water weighs 62.4 lbs per cubic foot, so at a depth of one foot, 62.4 lbs of pressure is against a wall, and double that for two feet of depth, and so on.

Section 4 Estimating

Gymnasium plans

Polyethylene is an inexpensive way to combat moisture. The following is an example of the method that a school board used to prevent moisture from rising up through a foundation crawl space and reaching a dry wood basketball floor, which was being replaced. The new wood flooring had a very low moisture content, 10% or so (ordinary moisture content of wood is about 19%), and the ground was about 3′ below the gym floor. As it is in Florida, the water level can be high, so moisture in the air must be prevented from reaching the wood.

See the online resources for diagrams 744.1, 744.2, & 744.3

Plan interpretation

Both sections indicate a vapor barrier in the crawl space on top of the ground. Section A/A-3 states that the vapor barrier is two layers of 15 mil polyethylene. The polyethylene extends up the sides of the foundation block on four sides of the gym, per the floor plan notes, as well as against the foundation block piers.

The existing gym flooring is being removed and new flooring is being installed, the purpose of the project. The wood joists are also being replaced, so once the flooring and joists are removed, the floor will be open and workers can walk around the crawl space.

Scope of the work

Include:

Furnish and install polyethylene.
Overlap edges 6″ and tape the joints.

Exclude:

Soil poison.
Hand grading.

Construction techniques

A large portion or all of the existing flooring and joists will have been removed before the barrier will be installed. The job will be sequenced so that much of the barrier will be installed over the same few days. Installing a small section at a time, say 20 LF × 40 LF, would not be productive.

There are three areas being covered with vapor barrier – the horizontal crawl space, four vertical sides of each block pier, and the four foundation walls. Of these surfaces, wrapping the block piers is the most tedious. Perhaps the best way to approach the job is to wrap the blocks first, tape them in place, and then install the polyethylene on the flat crawl space second.

The superintendent would consider the size of the rolls that the product comes in. His/her company has purchased the polyethylene in rolls 20′ wide, 100′ long, two thousand square feet in each roll. Since the spacing of the block piers is 10′ apart in both directions, he or she reasons that it won't matter in which direction the barrier is placed.

For the second layer of polyethylene, the superintendent has the barrier installed perpendicular to the first. This prevents the tape and laps from overlapping and makes for a better installation because the seams are in a crisscross pattern.

The takeoff

Polyethylene TAKEOFF								744.4
No.	**Description**	**Det**	**EA**	**L**	**W**	**Ht**	%	**SF**
1	Crawl space flat area	Plan		100	90			9000
2	Add factor for 6" lap						0.025	225
3	Wrap each block pier	A/A-3	88	5		2		880
4	Foundation walls	1/A-4	2	90		2		360
5			2	100		2		400
6								10865
7	Layers		2					**21730**

Row 2: Allowance has to be made for a 6″ overlap at the edges. To determine this factor, consider a 20′ wide roll laid flat 100 LF long. For every one foot length of it, there is one-half foot (six inches × one foot) of lap for every section 1′ × 20′. So 0.5/20 = 0.025, or 2.5%. Another way to figure the factor is to consider the entire roll with a six-inch lap at the edge. The lapped portion is 0.5′ × 100 LF, or 50 square feet, and 50/2000 = 0.025, or 2.5%.

Row 3: The size of the block piers is given on the floor plan at 8″ × 16″. The perimeter of them is 4′, and another 1′ is added for easy wrapping, so the perimeter used here is 5′. From Section A/A-3, the pier will be covered vertically with one foot of product; adding another foot for the barrier to spread on the ground horizontally, a height of 2′ is used. Therefore, the pieces that wrap the block piers are 5′ × 2′.

Row 3: The product covers 1′ vertically of the foundation wall per Section 1/A-4. One more foot is added is added for takeoff purposes to lap horizontally flat on the ground, so each piece is 2′. It wraps around the entire perimeter of the gym.

The final quantity on the takeoff is very close to an even eleven rolls of polyethylene, or 22,000 sf. There is not much to spare here, so the estimator could add another roll on the estimate and consider how the takeoff has been figured. Both rows 2 and 3 have a 1′ lap, and if this is conveyed to the superintendent so that these laps are perhaps trimmed down some, the 22,000 sf will work.

PART 8

Door and window openings

1

DOORS, FRAMES, AND HARDWARE

Section 1 Introduction

In commercial, office, and government buildings, the doors and frames are usually made of steel. The industry standard hollow metal door frame is made of 16 gauge steel, and the doors, called slabs, are 18 gauge. Doors can be hollow metal or wood and are almost always 1-3/4″ thick. The most common sized door is 3′ wide and 7′ tall, a 3070 door. The largest standard door that many manufacturers make is 4′ wide and 10′ tall; doors beyond this size are a special order and price. Hardware pieces can be numerous, with weird-sounding names like T astragals, electric strikes, and hospital stops. All of the parts of a door are called an "assembly" – the frame, door slab, and pieces of hardware.

The codes and product requirements found in Division 8 are very complicated. Doors have many life, safety, and handicap considerations, such as having to swing automatically, be held open, self-close (sometimes in a fire), and maintain the required tolerance between door and frame (fit tight). They must lock in place not only at the side but also sometimes at the top and bottom, allow or prevent air passage and sound, and still look good after being kicked.

The exterior wood doors of a house are 1-3/4″ thick, while the interior doors are often just 1-3/8″. These doors can be hollow core or solid core, which describes the interior construction of the door. Residential doors are often only 6′-8″ tall, and the widest one is usually the 3′ wide front door, while bedroom doors may be 2′-6″ wide. While 7′ high doors, referred to as "seven 0 doors", are not found much in houses, 8′ high doors are sometimes used in large residences.

An estimator must understand the language used in the door and hardware trade. Labor cannot be evaluated without knowledge of the products. Review the terms found below in the sketches of a door and frame, transom, and side-light, and the language in the chapter that follows, including the doors and hardware glossary at the end of the book.

Section 2 Ruling body, the Door and Hardware Institute

The Door and Hardware Institute (DHI) sets the standards for the hardware industry. It is common for architects to specify within Division 8 of the specifications that an architectural hardware consultant (AHC), trained by DHI standards, prepare the hardware submittals for a project after the job is awarded. An AHC knows the requirements for door openings in all types of public, commercial, industrial, and institutional buildings. AHCs coordinate thousands of hardware items and evaluate options to ensure door openings are in compliance with fire, life safety, accessibility, and building code requirements.

The Steel Door Institute sets the standards for steel doors and frames.

The National Fire Protection Association (in the United States) governs fire doors; see section NFPA 80 – Standard for Fire Doors and Other Protectives.

Section 3 Suppliers and distributors

The prime contractor (the CM or GC) has the contract with the owner and buys from suppliers and distributors. The supplier/distributor buys from manufacturers. The contractor is at least two steps from the manufacturer. If the CM/GC places the responsibility for buying the doors with a subcontractor, it places the them at least three tiers from the manufacturer.

The supplier or distributor buying from the manufacturer often plays a key role in the "fabrication" of doors and frames. The legs and heads of door frames, and unprepped door slabs (having no recesses or holes cut for hardware), are often shipped unassembled to the supplier. Before the doors and frames are delivered to the contractor, the supplier puts the frames together and "preps" the doors, performing some fabrication (shop assembly). Door slabs are bored for cylindrical locksets, cutouts are made for mortise locks, and hinge locations are routed out. The supplier also may install spreader bars at the base of the frame (sometimes this is done in the field).

There are doors, there are frames, and there is hardware, sometimes all quoted separately by multiple suppliers on bid day. The author makes it a practice to take all or none, and does not buy doors and frames from one supplier and hardware from another, even if it is cheaper that way. There is too much coordination required. There is going to be enough confusion without getting two suppliers involved in the same openings.

The flow of door products in some ways is like that of structural steel, which comes from a mill and is delivered to a steel subcontractor who partially fabricates it in a shop before the assembly is delivered to the site. Door frames and doors are sent from a manufacturer to a supplier/distributor, where they are prepped for hardware. Prepped is the door industry term for "fabricated". When the partially completed assembly is delivered to the site, the assembly is put together and finally installed in place.

Section 4 Door frames

Hollow metal frames

The jamb "legs" of hollow metal and wood frames usually have a 2″ face, or are said to be 2″ deep. The head frames of hollow metal are usually 2″ or 4″.

Jamb and head details on architectural plans show the "position" of door frames against various wall types. The location of the frame against a jamb can determine the door's "degree of opening", or how far the door will open. The frame can be positioned to "wrap" the jamb (a frame large enough to fit entirely around a block wall jamb with finishes takes longer to install than a narrow one) or align with one side of the wall, or be in the middle of the jamb. This consideration may not matter for the estimator, but being observant will help in picking up wood trim or other materials adjacent to the door. Positioning of the frame to the wall is important to the AHC who prepares the shop drawings, where frame depth and shape is specific.

Consider a high-use facility such as a hospital with a lot of foot traffic and a pair of doors in a corridor with block construction. A frame such as in detail 1/A6.5 of the Dudley Door Plans (see end of chapter) leaves one edge of the wall exposed to wheelchairs and carts and trays being trafficked down the corridor. To prevent the wall edge from being damaged, a frame with a throat wide enough to wrap the entire end of the wall is often used. Another option is an adjustable frame detail that adjusts to fit around a wide wall; see the wraparound frame in Sketch 1.

Review Sketch 1 for door nomenclature. Sketch 2 is of typical frame anchorage.

See the online resources for diagrams 814.1 & 814.2

Residential frames

A "KD", or knocked down, frame has three pieces, two side jambs and a head jamb. They are sent from the manufacturer to the supplier/fabricator unattached to the door slab. When the supplier/distributor attaches and hinges the door slab to the frame, as is often the case with wood doors, it is called a pre-hung door.

Residential doors can be purchased with a wood jamb on three sides called the "frame". Wood trim comes with it that attaches to the frame and is called "casing". Wood casing lies flat against the wall and is perpendicular to the frame. When the frames are wood and already attached to the door with hinges, the doors are called "pre-hung" doors. So the door and three sides of a frame are one unit, and may or may not include a threshold. Sometimes the casing is provided "loose", in other words it is not attached to the frame. A hollow core door only 1-3/8″ thick is light and flimsy and may only have two hinges, while 1-3/4″ thick doors have three.

Section 5 Plans and door schedules

Door schedules

Door and hardware "schedules" contain most of the information the estimator needs to quantify labor and material for this division of work. Door schedule format and organization varies depending on the architect and the project. Regardless, the door schedule serves the purpose of putting all the door, frame, and hardware information in one place for the contractor and supplier to organize in their own fashion.

One purpose of a door schedule is to group doors that are similar and call them a "type". The door type in a schedule defines the "slab" or "leaf" of a door assembly without the frame and hardware. Consider a door with no adornment, a flush slab (flush means smooth or flat in door lingo). It may have an opening in it for a lockset, called a "lock bore", and the door is said to be "bored for a lock". The slab may have small indentations, called mortises, in the edge of the door for hinge locations, but the face of the door slab is flush.

Door schedules can be in the specifications instead of on the plans, such as a listing of the various hardware groups, or sets, as they are called. After the job is awarded and an AHC detailer puts together shop drawings, the information is presented in a more structured way. The estimator views the door schedule on the plans however they are presented, arranging door items on the estimate in *groups of like kind* in an estimating format.

The estimator does not need to study every door on a schedule; the task is to separate frames and doors into like kinds. Some of the doors, especially at exterior locations, are selected by the estimator for viewing on the floor plan(s) and comparing to the door schedule. *Good door schedules reference jamb details for each door.* Plan details should tie every door jamb to a wall type, but this ideal is often not met.

Door numbering

Door numbering should be checked for accuracy, and plans and door schedules compared. Because of the great amount of information needed to define a door opening, errors occur, and the estimator should look back and forth between plan, schedule, and details.

Sometimes the doors are numbered one through one hundred, but it usually isn't that simple. Architects have considerations like multiple floors and major rooms. Door numbering is not the same as room numbering but sometimes takes on similar logic. All the doors on the first floor may start with the number one, second floor number two, and so on. Then, unfortunately, we get to decimals and lettering. A door to a major room, say a conference room, might be numbered 40. If there are five doors inside a conference room, they might be numbered 40.1, 40.2, or maybe 40A, 40B, 40C, and so on. Instead, I'm all for the one to one hundred part.

Door and frame dimensions

For hollow metal architectural door schedules, the dimensions are nominal, not actual. These columns refer to the door slab, or leaf, which is the moving part of a door assembly. While the door may be referred to as 3070, the actual size of the slab may be manufactured 35-3/4″ wide and 81-1/4″ tall. It is the door "frame opening" that measures 3′ × 7′. This paragraph refers to hollow metal doors.

In residential door sizing, the opposite is true. A 3′ × 7′ door is actually 36″ wide by 84″ tall. The frame is oversized to house the door and allow for undercut.

Hardware groups or sets

A hardware "group" or "set" names all the pieces of hardware needed for a door(s) on a door schedule. For example, all the exterior doors of a building might have the same hardware – a keyed lock, a closer, a threshold, weather-stripping, and a rain drip. Call it group 1 and it works for a dozen exterior doors. Grouping hardware into sets means that one hundred doors can be described with five to ten hardware groups.

See the online resources for diagram 815.1

Section 6 Submittals and shop drawings

Doors and hardware are often long lead items, the manufacturer often taking 6–8 weeks (or longer) to deliver the products. This is *after* their receipt of approved submittals. The project manager working for the GC or CM often makes writing the purchase order for doors and hardware one of the very first duties because of the time it takes to get the shop drawings through the submittal process and delivered.

The submittals provided by the contractor to the architect will consist of shop drawings and product data sheets. The shop drawings for doors and hardware are usually done on regular sized paper, not large sized drawing sheets. The product data sheets, sometimes called "catalog cuts", are from the manufacturer and go to the supplier or distributor, then the builder.

Shop drawings require serious study, and they are often prepared by an architectural hardware consultant (AHC). The documents take time to prepare, and the project manager has a detailed task in checking them once received. Door and hardware submittals can be complicated, with a lot of parts, and none of it can be wrong.

The plans should provide the "position" of the frame to the wall, and architectural details should provide how finishes about the door jamb. But the actual "attachment", or "anchorage", of a door frame is usually shown on suppliers' shop drawings, not the architectural plans. The instruction for metal stud anchors, or masonry wire anchors, or existing opening (EO) anchors are shown on shop drawings prepared by the door supplier.

In a renovation project, *if doors are replaced and frames remain*, the hinge locations of the new doors must be matched to the frames. The hinge indentations must be carefully measured as well as the *strike plate location to match the frame*. The *face plate* for the latchbolt or deadbolt of the new door edge must be made to match the strike plate on the frame.

The supplier's shop drawings will stipulate that the contractor provide these onsite measurements. The shop drawings, without these measurements, can be sent to the architect while at the same time another copy goes to the field superintendent for measuring. This will save time in the submittal process. When the approved shop drawing is returned by the architect, the contractor adds the proper hinge locations to the frame elevations and sends them back to the supplier.

Section 7 Egress and fire ratings

Egress

The estimator should be aware of the path of egress and the layout of code-compliant passage to exit a building from anywhere in it. The life safety plan, often one of the first sheets of a set of plans (after the cover page), should be reviewed by the estimator when counting doors or rated walls. This plan contains arrows following the egress path from the furthest point of the interior all the way to the exit door(s).

How to get out of a building (which is what egress is, the path) is an important life and safety consideration, which makes it a major building code topic. A main consideration is that the doors along the *accessible route are possible to open using a single motion, without the use of a key, tool, or special knowledge.* Exit doors should not be equipped with secondary locking devices, such as a deadbolt or slide bolt. This is a legal concern for both the architect *and* builder, because it is the AHC detailers who prepare shop drawings, and they work for contractors.

Fire ratings of walls

The fire rating of a wall dictates the fire rating of doors in the wall. The location of a wall within a building further defines the rating. A few of the common rated walls that contractors often see are rated corridors, mechanical/electrical rooms, and stairwells. In a fire rated corridor, the wall on each side has to extend up to the floor or roof above. If the corridor ceiling (that spans horizontally wall to wall) is rated, the walls can stop at the ceiling height. The estimator knows that the doors in these locations are going to be rated and expects to see them defined on the door schedule as a "labeled" door.

Fire rated doors

Within a rated "opening", all of its components have to be rated. That includes.

Door frame
Door slab or leaf (or leaves)
Hardware
Transoms
Side-lights
Glazing

A door rating is *usually* 75% of the wall rating; doors in a one-hour wall must have a 45-minute rating. Door hardware must be rated the same as the door. *A fire door is required to be* self-closing *and* positive latching. *These two requirements are the most important considerations for fire doors.* The door must self-close, latch tight, and remain closed (shut). It must have a closer or spring hinge that closes the door.

There are five ratings for doors:

DOOR RATINGS	*WALL RATINGS*
20 minute	Some corridors and applications where smoke is a concern.
45 minute	1 hr corridor or rated rooms.
1 hour	1 hr walls between occupancies. (Yes this case does not fit the usual 75% rule)
1-1/2 hour	2 hr walls such as stairwell enclosures and elevator shafts.
3 hour	4 hr walls separating buildings or separating a large building into designated fire areas.

The maximum rating for a metal door is 3 hours (no lite is allowed). Metal transom panels (not wood or glass) can be rated up to 3 hours. The maximum rating for a door with a glass lite is 1-1/2 hours. A rated door frame is a rated frame for all occasions. If a frame is fire rated, it can be used with a one-hour door or a three-hour door. Grout is not required for fire rated frames. Hinges used on fire rated doors must be made of steel.

Positive latching is usually made by a latchbolt closing (thrown) against and being inserted into a "strike plate". If a fire door is held open (perhaps both doors of a pair, for example in a hospital corridor), it must be equipped with a hold open closer or wall/floor magnet that is tied into a smoke detector or fire alarm. If a fire occurs, the smoke detector or fire alarm will release the doors and the closer will self-close the doors.

The size and area of glass is limited in fire rated doors. With the use of fire rated glass and ceramic glass the rules are complicated, and allowable sizes have changed many times in the past. What the estimator must do is ensure that all door glass is found and quoted properly.

Underwriter's Laboratory

In the United States, the testing authority for construction fire ratings is the Underwriter's Laboratory (UL). Their tests meet the requirements of NFPA 80 as well as the International Building Code and the International Fire Code. When products withstand fire tests for a specified period of time, they are "Listed" or "Approved" by the UL for a fire rating. The product is physically "labeled" (on doors the label is on the edge) to show the rating and approval. If the term "label" is used on a door schedule, it refers to a fire rating. NFPA 80 requires once a year inspections on fire rated doors. The label must be present and the condition of the door must be confirmed, including the closer and latch.

Section 8 Lock terminology

The term "lockset" defines a hardware device that requires a key to operate. The term "latchset" describes a hardware device that latches (latchbolt inserts into a strike plate) but does not lock or require a key.

A lockset describes the hardware that makes up the locking or passage device that opens a door. It includes a lever or knob and the latchbolt and strike plate. A standard lockset contains a spring that controls the latchbolt and causes it to extend into a strike and latch the door shut. This is called "positive latching". A dead bolt does not have positive latching because the latch is either "deadlocked" in or out.

Cylinder locks are a type of lockset that require that two connecting and perpendicular holes be drilled into the door, one round opening on the face of the door slab and one on the edge. The hole on the face of the door is called a "bored hole". The other opening extends from the bored hole through the edge of the door (a certain distance away, called the "backset") where the latch bolt and face plate are. (A face plate is on the edge of a door; a strike plate is on the frame).

Mortise locks are a type of lockset. They require that a rectangular "pocket" section at the edge of a door slab be cut out for insertion of the lock assembly. A mortise lock is more complicated than a cylindrical lock and can take longer to prep (by supplier or in the field) and install.

Rim locksets are applied to the face of a door and do not require cutting a hole through it. They are only placed on one side of a door.

Levers are preferred over knobs in public buildings because they are more easily operated by people with a disability, who may open a door with the use of only one hand or arm.

Locksets are grouped into hardware categories called "functions". Most locks in public buildings are defined by only a few functions. However, in commercial and residential buildings, there are many variations.

Classroom function: Key on outside locks or unlocks the latchbolt by turning a full 360 degrees (full turn is only for classroom function). There is no means of locking or unlocking the exterior from inside.

Classroom security or intruder function: Adds key locking to inside while maintaining free egress from inside.

Deadbolts: Deadbolts are often used for added security or as an auxiliary lock. If the door is a means of egress, it must open with one operation, so deadbolts are often used with push/pull hardware. When a thumb turn or key sends the latchbolt into the strike, the latchbolt is "dead" and will not open from the other side. Only the thumb turn or key will open it.

Dummy: Handle only, similar to push/pull.

Entry or Office function: Requires a key to operate one side. The opposite side has a push button, thumb turn, and lever. A key must be used to open the door if the door is locked. There is free egress at all times from the non-keyed side.

Interconnected lock: Two locks, usually a deadbolt and a passage function latchbolt. From inside the room, the lever simultaneously retracts the latchbolt and deadbolt.

Passage function: Latchbolt is operated (can be retracted) by levers on both sides at all times. There is no key or button required to open from either side. Both sides are always unlocked.

Privacy function: Push button or thumb turn on the inside prevents the latchbolt from retracting. Typically used as a bathroom lock. If not depressed, levers from either side will open the door. There is usually an emergency release on the entry side that can be used with a tool to open the door if needed.

Storeroom function: Key on entry side keeps the door locked at all times, and there is no key, button, or thumbturn on the other (inside). The lever on the inside always retracts the latchbolt and the inside always remains unlocked. When the key is removed from the entry side, the door will stay locked. Used on storage and mechanical/electrical rooms.

Section 9 Construction techniques

Left and right handing

To determine the "handing" of a door, stand on the key side facing a door, and:

If the door opens in and is hinged on the left, it is a left-hand door.
If the door opens in and is hinged on the right, it is a right-hand door.
If the door opens out and is hinged on the left, it is a left-hand reverse.
If the door opens out and is hinged on the right, it is a right-hand reverse.

That's all there is to it.

Handling and protection

Material handling and protection of expensive doors can require a lot of labor, and the care of doors is often addressed in scope packages by CMs. Doors are tall products that must be carefully stored and protected and carried to various floors and areas of a building. Considerable labor can be expended before a frame is stood. Sometimes doors are protected while in place with cardboard or other material until the end of the job, when it is removed.

Installing doors and frames

The GC may elect to install doors, frames, and hardware with in-house carpenters or may look to outside help for their installation. There are subcontractors, even one-person entities, skilled and with their own workers' compensation and liability insurance. These two insurance policies qualify the entity as a subcontractor, which means the GC can hire them and pay for their services in lump sum, as opposed to by the hour. Paying lump sum means there are, for the GC, no payroll taxes, withholding, or workers' compensation. A guaranteed price for the work can be agreed to up front. This makes the estimator happy!

For hollow metal frames occurring in block walls, construction managers and general contractors generally place the responsibility for setting door frames with the mason. If the hollow metal frame is located within steel stud walls, this responsibility is usually given to the metal stud and drywall subcontractor. Whoever builds the wall sets the frame. The prime contractor is responsible for coordinating dimensions and schedules back and forth between the trades. The superintendent must get a wall built around a door and is concerned with the overall width and height of a wall opening, that is, the door plus the frame size, which is called the "rough opening" (R.O.) in the wall. There are different trades involved and the superintendent must provide coordination between them.

There is a great amount of labor involved in standing, plumbing, and leveling a steel door frame. In block construction, door frames get "stood" before the block wall is built. Frames are installed and set in place rigidly with bracing as shown in Sketch 3. A "spreader bar" is fastened to the slab between the two legs of the door that keeps the frame at the correct width so that no amount of jostling will disturb its location. A good practice is for a second spreader bar to be placed halfway up the door. The frames must rigidly stand in place while walls are built adjacent to them.

With block construction, a metal anchor is placed into the door frame throat and then placed horizontally in block coursing, held rigid by mortar and the concrete used to fill the vertical cell. Anchors should be installed at hinge levels and directly opposite the strike jamb.

The most common door width of 3′ has an R.O. of 3′-4″ when the face frame legs are 2″ wide. Since 3′-4″ fits blockwork (being divisible by 8″), everybody is happy. With metal or wood stud construction, the R.O. is just as important to coordinate, but various widths (and heights) can be more easily accommodated.

Concerning 6′-8″ high doors, if a 2″ head frame is used, a height of 6′-10″ is achieved. Typical 8″ block coursing is at 6′-8″ and 7′-4″. A 4″ frame head does not solve the block coursing problem with 6′-8″ doors, since this only totals 7′ in height (which is not blockwork). There is trouble with 6′-8″ doors in block construction, and you can keep from being confused further if you don't build houses. There is a good discussion of this in the chapter on masonry, as well as a sketch of a "door header", which is often used when 6′-8″ doors are scheduled in block walls. When this happens, the masonry trade accommodates the door industry by making a precast lintel that is not 8″ high as they usually are, but 6″ high where it crosses the door with 8″ legs on each side. The bottom of the precast door header is at 6′-10″ and extends to 7′-4″ where the next block coursing is.

With 7′-0″ high doors, the standard height in public buildings, frames with a 4″ head are used in block walls. This matches the 8″ coursing of blockwork, which is 7′-4″ (eleven blocks high). Frame F-2 of the Dudley Plan is noted to have a 4″ head because it is in a masonry wall. See the note in the comment column and the head jamb detail 3/A6.5.

The vertical block cells adjacent to door and window openings are filled with concrete to make the opening strong and to hold anchorage devices. To fix the frame in place, hollow metal frames in block construction are usually filled with grout by the mason while the wall is being laid up adjacent to the frame. Grout is used to cement an anchor inside a frame to make a sturdier installation and provide some sound deadening.

A wood frame or metal stud wall will have vertical double studs on each side of the door and a single or double stud above the head jamb. Metal studs often have 2×4 wood (short blocks or continuous wood) inside the hollow metal stud to provide solid material for the door frame to be fastened to (the metal stud is thin). The SDI recommends *against* grouting frames installed in drywall walls.

Frames in block walls are installed *before* the adjacent wall block are laid. Frames in stud walls are usually installed *as* the wall is built; these frames are sometimes called drywall welded frames.

See the online resources for diagram 819.1

When a door frame is installed in an existing wall, it is "punched and dimpled" in three or four different elevations (heights) on the frame. A hole is drilled into the frame and splayed inward so that the head of the bolt (masonry application) or lag screw (frame) can be recessed (countersunk). The door industry calls these fasteners *existing wall anchors* (EWA) or *existing opening anchors* (EOA) and the wall opening an *existing opening* (EO). Once the door is installed, the fastener is either painted over, leaving it visible, or carpenters can use "Bondo" (the same product used on car bodies) to fill in the dimpled area, creating a smooth surface.

See the online resources for diagram 819.2

Section 10 Dudley job estimate

Dudley job plans

The jamb details of a typical job appear below; see Details 1, 3, and 4, from sheet A6.5. Door and frame elevation drawings are often adjacent to schedules; see Schedule D. The dimensions on architectural door schedules, frame elevations, and plan sections are "nominal", not exact. The precise measurement of hinge locations, the anchorage of frames to walls, and other detailed information is completed on shop drawings after the job is awarded.

On architectural door schedules, some mixing without matching goes on. The estimator must know how to interpret the schedule and not be confused, which will happen enough because of plan conflicts. Note that door 2 is a double door (see door size 6070), but the door type is HM-1. The elevation of HM-1 is for a single door, a flush smooth door(s) with no adornment. *It is the door slab that is being defined in the elevation of HM-1.* Just because there are two of them doesn't matter; this is how a schedule is interpreted.

The estimator is cautious, always looking for "extra" work beyond setting the door and frame. Trim lumber or other work may show up on jamb details. For example, a jamb detail (or note on a door schedule) may reveal chair rail or other wood trim against the frame. The depiction of casing or some kind of wood feature at door jambs may not occur elsewhere on the plans and the estimator must be observant when looking at jamb details. The door supplier will ignore any wood trim so the estimator must include the material cost for it separately, as well as the labor.

See the online resources for diagrams 81 10.1, 81 10.2, & 81 10.3

Interpreting door schedules

The elevation for door 6, type HM-2, indicates door glass for this door, which makes it a different door type, HM-2. Because it has a one-hour fire rating does *not* make it a separate door type. It is the glass that makes it a different type; that is how a schedule is interpreted. Glass is a different trade than doors and hardware and is often furnished by storefront and glass sub-contractors, the trade providing door 7. Door 6 would be made at the factory with an opening in it at the top for the glass, 10″ square as noted in the comment column of Schedule D. The metal frame, or "stop", that houses the glass is called a "lite kit". The doors, frames, and lite kit are furnished by the HM door supplier. But not the glass.

Note the instruction for a 3/4″ undercut for type HM-1 doors on the elevation drawing. What the architect wants is 3/4″ distance from the floor to the bottom of the door. This may be to provide continuous air circulation throughout a building with several hundred doors. What the door supplier/fabricator will draw on a shop drawing is a space of 3/4″ from the bottom of the door leg to the bottom of the frame leg (the location of the floor is out of their control). What the contractor builds is the installation of a frame against a jamb, and the frame leg can be sitting on the slab or can be up the wall a quarter or a half inch (perhaps because the door head has been built a bit high). A precise distance from the bottom of the door to the *bottom of the frame* is what is supplied to the job. The door slab and frame will be fabricated with this precise undercut.

Frame F-1 is for doors 1, 5, 6, and 8. The face frame widths are often noted on plans and included in the frame type description while the depth is not. A detailer (the person creating the shop drawings, working for the door and hardware supplier) reviews the jamb details on the plans to determine what the frame depth should be.

The sketch for F-1, which simply shows that the face frame is 2″ on three sides, is for multiple widths. When the term "As scheduled" is used, a separate sketch or drawing is being referred to (in this case the door schedule).

Hardware groups

The following hardware schedule might be a part of the plan sheets or be contained in the specifications. Hardware submittals list the items *in the same order that they are installed.*

HARDWARE GROUP 1:
1-1/2 pair Hinges
Passage cylindrical lockset
Closer
Wall stop
Silencers

HARDWARE GROUP 2:
3 pair Hinges
2 Panic bars
2 Closers
2 Kick plates 6″ high
Door bottoms

HARDWARE GROUP 3:
1-1/2 pair Hinges
Keyed cylindrical lockset
Closer
Threshold
Door bottom
Silencers
Rain drip
HARDWARE GROUP 4:
2 pair Hinges
Keyed mortise lockset
Closer
Armor guard
Silencers

HARDWARE GROUP 5:
1-1/2 pair Hinges
Passage cylindrical lockset

Note: See the specifications for hardware of the overhead coiling doors and storefront doors.

Unit price sheet

The unit of measure for doors and hardware is each, and no arithmetic is needed to figure out the quantity except addition. There is no waste; everything is an exact count. The estimator's work is in understanding and interpreting the information presented in schedules and jamb details.

The estimator breaks down the work tasks into proper "descriptions of work" for "like kind" quantities. Simply "hanging a door" with a total numerical count is not enough information to place on the estimate. A door opening may take 3 hours of work or 13 hours. It is not unusual for 8 hours to be an average amount of time spent installing a door, frame, and hardware. But simply lumping all the doors into one description and using an average amount of time may well miss the mark. *Frames, doors, and hardware groups should be taken off separately.*

The trade scopes may reveal that several trades are involved in frame and door labor. For illustrative purposes, all the frames from Schedule D are assumed to be in-house work and appear on the following estimate.

The estimator should use the door schedule and elevations to separate frame and door sizes. A 6′ wide frame takes longer to install than a 3′ frame. To gather this information, the frame sizes are deduced from the door sizes shown in Schedule D. The frame width for F-1 is not given; the designer has used the elevation to describe more than one width.

Doors 1, 5, and 8 are 3′ wide and in a frame wall; see Detail 4/A6.5. These doors are like kind (door 7 is also 3′ wide, but it is a storefront door and subcontracted; it does not appear here). Door 3 is in a block wall (see 1/A6.5); it is listed separately and there is only one. Door 6 is in a block wall, too, but it is 4′ wide. To describe and count these four doors and frames, see rows 1–6 of the unit price sheet.

Note the $10 per door in the material column (see row 7), estimated for material to account for some lumber and Visqueen to cover and protect the doors and frames until they are installed.

Glass is made by a manufacturer that the storefront and glass subcontractor works with. The contractor takes the lite kit from the door folks and the glass from the storefront sub and installs them in the door opening. See row 14 of the unit price sheet.

The labor to install hardware groups can vary greatly. Counting hardware groups separately allows labor hours to be apportioned accurately. See rows 9–13 of the unit price sheet.

UNIT PRICE SHEET												81 10.4
job	Dudley Door Job	number								date		
Code	**Description**	**MH unit**	**Mat'l unit**	**Sub unit**	**Qty**	**Unit**	**Hrs**	**$ Hr**	**Labor**	**Mat'l**	**Sub**	**Total**
8100	Set F-1 HM 3070 frames in stud walls											
	Doors 1, 5, 8	1.25	0	0	3	ea	3.75	25	94	0	0	94
8100	Set F-1 HM 4070 frame in blk wall											
	Door 6	2.5	0	0	1	ea	2.5	25	63	0	0	63
8100	Set F-2 HM 3070 frame in blk wall											
	Door 3	2	0	0	1	ea	2	25	50	0	0	50
8100	Set F-3 HM 6070 frame in stud wall											
	Door 2	2.5	0	0	1	ea	2.5	25	63	0	0	63
8100	Hang HM-1 3070 doors											
	Doors 1, 2, 3, 5, 8	1	0	0	4	ea	4	25	100	0	0	100
8100	Hang HM-1 4070 door											
	Door 6	1	0	0	1	ea	1	25	25	0	0	25
8100	Material handling all HM	0.3	10	0	11	ea	3.3	25	83	110	0	193
8200	Hang wood 3070 prepped											
	Door 7	1.5	0	0	1	ea	1.5	25	38	0	0	38
8300	Install hardware group 1	2.5	0	0	2	ea	5	25	125	0	0	125
8300	Install hardware group 2	6	0	0	1	ea	6	25	150	0	0	150
8300	Install hardware group 3	4	0	0	1	ea	4	25	100	0	0	100
8300	Install hardware group 4	4	0	0	1	ea	4	25	100	0	0	100
8300	Install hardware group 5	1.5	0	0	1	ea	1.5	25	38	0	0	38
8500	Install lit kit and glass	1	0	0	1	ea	1	25	25	0	0	25
							42.05		1051	110	0	1161
												1161

P/S Sheet

There are three lump sum items placed here, one material quote and two subcontract prices. The material is for hollow frames, hollow metal doors, the one wood door, and hardware. The aluminum storefront door is installed by one subcontractor (who also furnishes the door glass) and the coiling door is another subcontractor. See these costs listed on the P/S sheet below.

P/S SHEET (products and subs)							81 10.5
NAME	Dudley Door Job		NUMBER			date:	
CODE	DESCRIPTION	QTY	HRS	LABOR	MATL including tax	SUB	TOTAL
81-8300	All doors and hardware	LS	0	0	15,000	0	
8600	Alum door 7 3070, and glass for door 6	1 ea	0	0	0	1,500	
8700	Overhead coiling door 15 x 12	1 ea	0	0	0	4,000	
			0	0	15,000	5,500	

Although this chapter is only about doors, to provide a complete estimate some prototype job overhead costs are shown below, and an estimate summary sheet follows it. These four sheets are the entire estimate. The bid price appears at the bottom of the estimate summary.

Job overhead sheet

JOB OVERHEAD								81 10.6
job	Dudley Door Job	number				date		
code	**Description**	**Qty**	**Hrs**	**$ Hr**	**Labor**	**Mat'l incl tax**	**Sub**	**Total**
1100	Permit	LS	0	0	0	2500	0	2500
1100	Supervision	6 mo	1040	30	31200	0	0	31200
1210	Rent temporary toilet	6 mo	0	0	0	600	0	600
1220	Gas/Travel time	none	0	0	0	0	0	0
1230	Truck allowance	none	0	0	0	0	0	0
1310	Test concrete	4 ea	0	0	0	0	800	800
1310	Test dirt	6 ea	0	0	0	0	600	600
1320	Insurance	none	0	0	0	0	0	0
1330	Tools and expendables	LS	0	0	0	1500	0	1500
1340	Sign/Asbuilts	LS	8	25	200	500	0	700
1500	Scaffolding erect and remove	none	0	0	0	0	0	0
1600	Trailer/Phone/Mob-Demob	none	0	0	0	0	0	0
1700	Utilities	owner	0	0	0	0	0	0
2260	Dumpsters	6 pulls	0	0	0	1950	0	1950
2700	Cleanup	26 wks	260	20	5200	0	0	5200
			1308	75	36600	7050	1400	45050
								45050

Estimate summary

ESTIMATE SUMMARY							81 10.7
DUDLEY DOOR JOB		NUMBER			date:		
DESCRIPTION	QTY	HRS	LABOR	MATL	SUB	TOTAL	
Unit Price Sheet		41	1,051	110	105		
Sales Tax on takeoff items	0.06	0	0	7	0		
P/S Sheet includes tax		0	0	15,000	5,500		
Subtotal		41	1,051	15,117	5,605		
Job Overhead Sheet		1,308	36,600	7,050	1,400		
Sub Total		1,349	37,651	22,167	7,005		
Labor Burden	0.25	0	9,413	0	0		
Total Cost		1,349	47,064	22,167	7,005	76,236	
Markup	15%					11,435	
TOTALS						87,671	
bond	1%	(total x .01 then times 1.01)				885	
TOTAL WITH BOND						**88,556**	

PART 9

Finishes

1

METAL STUDS

Section 1 Introduction

Cold-formed studs and track are now used to build most of the interior non-load-bearing walls in new buildings. The use of metal for stud wall construction has surged since the 1970s and 1980s. Wood studs were predominant before that even in commercial construction.

The term "non-load-bearing" or "non-structural" means that while the studs can carry the dead load of materials used to build walls, they cannot support the weight of floors and roofs. The wall components of studs and track are estimated the same way that wood walls were counted in Part 6. With the addition of other materials, either used in the wall cavity or clad to the sides, non-load-bearing steel stud walls can be made sound resistant, smoke resistant, and fire rated.

There are two ways that a steel stud of the same size (depth and width) can increase in strength. One is to use a thicker steel, measured in mils, or 1/1000 inch. As seen in the following charts, a stud of a certain size is often made in several thicknesses, thereby resulting in a sturdier product, one that will flex less and can be used to build a taller wall. Another way to strengthen a given metal framing product is to increase the strength of the steel used to make it. Two yield stresses are used to manufacture the sheet steel that metal framing products are made from – 33 ksi (kip per square inch) and 50 ksi. By upgrading a product to 50 ksi from 33 ksi, a stronger product will result.

The thickness of metal framing products is measured in mils. The important metric is the "design thickness", the manufactured thickness of the steel, which is made in flat sheets, and then "cold rolled" (bent) into shape. The use of the term "gauge" confuses the issue, since historically various manufacturers have had a different idea of how thick a gauge was. Nevertheless, the word "gauge" finds its way into many sets of plans and specifications and seems embedded in the industry lingo. Many contractors can remember that a 25-gauge stud is used for nonbearing walls, but fewer can identify its design thickness, which is 18 mils, short for 18/1000th of an inch, or to use the exact thickness – 0.0179″ (minimum), or 0.0188″ (design). The difference between minimum and design is explained in Section 3.

Non-structural steel studs (and track) are manufactured in conformance with ASTM C645. Structural steel studding (and track) are manufactured according to ASTM C955.

This chapter concerns nonbearing studs, which are found in Division 9 of the specifications. Bearing studs, or structural studs and joists, are found within Division 5. Structural stud thicknesses are designed by structural engineers, while architects often specify thicknesses for nonbearing studs.

Section 2 Ruling bodies, SSMA and SFIA

The steel stud industry is served by two national trade organizations: the Steel Stud Manufacturer's Association (SSMA) and the Steel Framing Industry Association (SFIA). Both organizations publish technical guides, product information, nomenclature, and load tables, which are the source for much of the technical information in this chapter.

The publications of both associations are based on steel that meets the requirements of the AISI S100–12 (American Iron and Steel Institute), "North American Specification for the Design of Cold-Formed Steel Structural Members".

Section 3 Designations

Product code nomenclature

Designation, nomenclature, code name – this is about naming a product. *All metal stud product names have four parts.* For example, the designation of 362S125–18 names a product, reading left to right: it has a depth of 3-5/8″, the S means it is a stud, and 125 (125 times 1/100) is equivalent to 1-1/4″. The product is roughly the size of a 2 × 4 piece of wood and has a steel thickness of 18 mils.

See the online resources for diagram 913.1

To fully designate a steel framing product, the yield point of the steel must be defined. It is placed in parentheses after the 4-part ID code, and the product in Sketch 1 above becomes *600S162–18 (33 ksi).* The crew building a wall out of these studs would call this product a 6″ stud.

Refer to Table 1. The reason that both a "Minimum" and "Design" thickness is used has to do with manufacturing. When steel is manufactured with the intent of achieving an exact design thickness, the finished product will be slightly different simply because the process of making sheet steel isn't perfect. The industry uses a 5% error factor, and the minimum thickness is 95% of the design thickness, which is what the inspectors will expect when they do their quality checking.

The estimator might find either or both of these terms when reading steel stud specifications, so an understanding of the difference is needed. But, there is no problem as long as the specifications use mils as the thickness.

See the online resources for diagram 913.2

Mils/gauges and EQ studs

Another measurement besides mils is often used that does not conform to industry designations, and that is the use of the term "gauge", or "gage". However, the now-used designations are recent, and architects are still using gauges in their specifications. The following are still apt to be seen – 25, 22, 20, 18, and 16 gauge. The steel stud industry does not use these because they are inexact and have no standardization; there was never a uniform thickness for a gauge. Nevertheless, they were used for many years, and they persist in specifications.

Since about the year 2000, high yield stress steel has been used to produce new products. The most widely used of these products are equivalent gauge (EQ) non-load-bearing studs. This stronger steel can be used in a thinner mil thickness to make them equivalent to traditional older studs yet carry the same loads. The thinner EQ studs will flex no more than the old thicker studs, allowing walls to be built to the same height using less steel.

Section 4 Products

Product list for this chapter

"Cold-formed steel" is a term used to describe steel products made from previously produced *sheet steel.* The sheet steel is rolled or bent and pressed into shape by machine at room temperature. The use of the word "cold" differentiates it from the hot temperatures sometimes used to make structural steel (not all structural steel is made "hot" because some steel shapes are assembled from already produced products. See Part 5).

The products discussed in this chapter are:

1 Metal studs – S Profile designation.
2 Metal track – T Profile designation.
3 U channels – U Profile designation.
4 Clip angles.
5 Furring channels – F Profile designation.

Non-structural metal studs

Non-structural metal studs are made out of 33 ksi steel.
Non-structural metal studs are manufactured in these sizes:
1-5/8″, 2-1/2″, 3-1/2″, 3-5/8″, 4″, 5-1/2″, 6″, and 8″.

Metal studs are built with open slots in them to accommodate the horizontal passage of wiring, piping, and channel bracing. Vertical studs are typically spaced at either 16″ on center or 24″ on center, with 24″ being the maximum.

See the online resources for diagram 914.1

Non-structural metal studs can be made to length, up to 24′ or so! They are made in eight sizes, or depths. Regardless of their proper industry four-part designations, when contractors refer to metal studs, they are called six- and eight-inch studs, naming them by their depth. Non-structural metal studs are manufactured in the following depths and thicknesses.

See the online resources for diagram 914.2

There are a total of 50 standard nonbearing metal stud products manufactured in the United States, and they range in weight from 0.27 lbs to 2.51 lbs per linear foot.

The standard width of a wood stud has changed over the years from 4″ to 3-5/8″ and then to the current width of 3-1/2″. The *metal stud industry makes all of these but calls them* depths. *The short dimension is called a* flange.

Metal track

This product is made in many lengths starting with 10′.
Non-structural metal track is made in the following sizes, which are depths.
1-5/8″, 2-1/2″, 3-1/2″, 3-5/8″, 4″, 5-1/2″, 6″, 8″.

See the online resources for diagram 914.3

U channels

All U sections are made with 50 ksi steel.
U channels are manufactured in these sizes, or depths:
3/4″, 1-1/2″, 2″, 2-1/2″.

There are only four products, and their ID codes are:

Depths ¾", 1-1/2", 2", 2-1/2".
Profile U
Width ½" flange
Thicknss 54 mills

75U050-54
150U050-54
200U050-54
250U050-54

All U sections have a flange width of 1/2″.
All U sections are made with a thickness of 54 mils.
These products are called cold rolled channels (CRC) and are typically made in lengths of 10′ and 12′.

See the online resources for diagram 914.4

One use for channels is for the horizontal brace running through studs as shown in Sketch 4. Another is the use, often in combination with hat channel (with the F profile designation), to support gypsum ceilings. See Sections B and C in the office building second floor plans below.

Clip angles

Clip angles are used for various purposes in metal framing construction. One is the attachment of the horizontal channel to each stud as the channel continues through a wall. See Sketch 4 below. Another is the attachment to a stud brace, see Sketch 5 top of brace to the bottom side of a bar joist, and Sketch 6 top of brace to the bottom of the deck above.

See the online resources for diagram 914.5

F-sections furring channels

All (hat) furring channels are made with 33 ksi steel.
There are two depths of F-sections made, 7/8″ and 1-1/2″.
All flanges for F-sections are 1-1/4″.
Both the 7/8″ and 1-1/2″ sizes are made in five thicknesses; there are a total of ten F-sections manufactured:

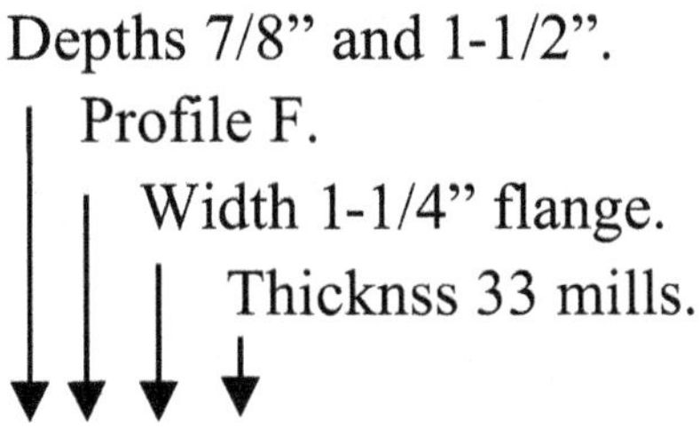

087F125–18 (33 ksi)
087F125–27 (33 ksi)
087F125–30 (33 ksi)
087F125–33 (33 ksi)
087F125–43 (33 ksi)
150F125–18 (33 ksi)
150F125–27 (33 ksi)
150F125–30 (33 ksi)
150F125–33 (33 ksi)
150F125–43 (33 ksi)

The "hat channel" is an "F-section" product as defined by SSMA and SFIA. An "F" in the description of a metal stud product denotes that it is a hat channel. It is only made in two sizes, the sizes being defined by the depth of the furring, and is either 7/8″ or 1-1/2″ deep.

A furring channel "furs" out a wall, it widens it – see Sketch 5, diagram 914.6.

On ceilings, the furring channel (the terms furring and hat channel are both customarily used) is used for gypsum board to be fastened to. See Sections B and C, diagrams 915.3 and 915.4.

See the online resources for diagram 914.6

Section 5 Estimating

Determine wall heights

Consider the estimator's plight when there is a floor-to-floor difference in elevation of 15′ for ten floors, the ceilings jump up and down, and the wall heights are governed by the note, "Build walls 1′ above ceilings". The task is to count the studs, track, and overhead bracing. The takeoff begins by separating the wall heights and arriving at a linear footage for each one per floor. The wall heights are usually not going to be conveniently provided on various wall sections – often it is the finish schedule or reflected ceiling plan that the estimator turns to. The ceiling heights, not the wall heights, will be found there.

Office building second floor plans

Refer to the second floor plan and Sections A, B, and C. Note the use of wall furring channels in Section A and the same product used in Section B to support gypsum board ceilings.

See the online resources for diagrams 915.1, 915.2, 915.3 & 915.4

Sections B and C state that walls must be 1′ higher than any adjacent ceiling. What does this mean, studs and gypsum board or just studs? The answer is only the studs, the elevations show the gypsum board stopping short of the top of the wall in several conditions. This has an important impact on the takeoffs for these two materials.

Finish schedule

In this example, the finish schedule below is where ceiling heights are found. Having the ceiling heights shown on the finish schedule is fine, but wherever they are found, the estimator can bounce all around the plans working on the takeoff or come up with a sketch collecting some of this information. This can often be done by simply making some notes on the floor plan.

See the online resources for diagram 915.5

Construction techniques

Many plans contain instructions to build the interior stud walls to 12″ above ceilings, similar to the notes in Sections B and C. This is often standard procedure because of overhead bracing requirements.

In Section C, the height of the wallboard is different on each side. This height difference is an important consideration for both the superintendent and the estimator. See the next chapter for a takeoff of wallboard.

A non-structural metal stud wall is going to need bracing at the top of the wall. An abutting acoustical ceiling, or a gypsum board ceiling, is not enough to hold a wall in place beyond four to six feet long. Where a wall turns a corner, or it runs into an exterior wall, there is rigid support for the full height of the wall. However, that corridor wall than runs for fifty feet without a turn is going to need a lot of overhead bracing. This is typical construction for metal stud walls. The solution is for 45-degree braces, called kickers, to extend from the top of the wall, alternately left and right, to overhead support, which is usually either a bar joist or the bottom of the floor or roof structure above.

Whether the bracing occurs every four feet, which is more typical, or six feet, and whether it occurs on both sides of the wall or alternately, is up to either what's on the architect's plans, or if no instructions are there, the steel stud manufacturer's instructions. This is not a contractor's decision.

The overhead bracing may or may not be shown on the plans. It is up to the contractor to know it is needed.

If the bar joists are perpendicular to the wall, as shown in Section B, the top of the bracing can connect to the bottom of the bar joist. This is done with the use of a two-sided clip; see the list of products above. Note that the 45-degree brace, which is a stud, is fastened at its bottom to the side of the vertical wall stud in this detail; this is a better construction detail than attaching it to the top track.

The situation changes when the bar joists are parallel to the wall, as shown in Section C. The angled brace extends through the webbing of the bar joist and on up to the floor or roof above. A clip is used to connect the brace.

It would seem that wall sections on the plans would be where the height of walls is found, and sometimes the estimator can determine wall heights there, or find some of them. However, if there are various wall heights due to different ceiling elevations, it may be the reflected ceiling plans and finish schedules that provide this information.

Metal stud takeoff

The gypsum board takeoff can reuse wall lengths from the metal stud takeoff. While the length can be the same, the heights differ. Recall the many different ways that wall heights can be determined – from sections, acoustical ceiling plans, finish schedules, etc.

Counting studs per linear foot of wall is the same as for carpentry. In Part 6, Chapter 3, Section 2, three kinds of walls were studied in detail to see how many studs occurred in a wall 100 LF long. There was a straight wall with no outs, and a factor of 0.79 per linear foot obtained 79 each studs. This is 75 studs at 16″ o.c. plus two extra at end for a corner. This is the least number of studs that can be in a wall 100 feet long with the studs spaced 16″ o.c.

An "average" wall 100′ long was studied, with an average number of doors, windows, and wall intersections. This study revealed that a factor of 1.09, multiplied by the length of the wall, would arrive at the quantity of studs in an average wall. This unit of measure derived from using this factor is "each". This is the quantity of studs needed in the wall, which can be multiplied times the height of the stud to get linear feet.

Then a busy wall was studied. This wall had an abundance of doors, windows, and wall intersections. It took a factor of 1.4, multiplied by the length of the wall, to arrive at enough wall studs to build the wall. Looking over these factors, *an average factor of 1.15 studs per linear foot is used in this textbook* for typical walls.

For the takeoff example here, see the following plan, as it collects wall heights and lengths in one place. These notes about length and height might be written by the estimator on the plans; a sketch and information like this would not be anywhere on the plans. Since the ceiling heights are shown in a finish schedule, the wall lengths are on the floor plan, and the wall details are shown in sections. Here, the walls are numbered so that they can be referenced on a takeoff.

View office 1 with a ceiling height of 10′-8″ and consider how tall the wall studding must be in wall number 1. The corridor is the only room adjacent to office 1, having a ceiling height of 8′. Since the higher of the two ceilings adjacent to wall 1 is 10′-8″, and a foot higher than this is 11′-8″, this is the minimum height the studs can be. However, the estimator would bump the stud height up to 12′ because studs are purchased in increments of 2′, the same as wood studs. However, metal studs can be purchased at any length by arrangement with the manufacturer.

Wall 1 is scaled to be 10′ long and is 12′ high. What the estimator handwrites on the plans is shown in italics. The same analysis is used for all of the walls – determine which side of the wall that the ceiling is highest, add a foot so that the wall is 12″ higher than the ceiling per Sections B and C, and then if an odd number is arrived at, round it off to an even stud length.

See the online resources for diagram 915.6

"Like kind walls" of the same construction *and* height are added together for one length. On the following takeoff, rows 1–5 arrive at a total linear footage for 9′ high walls. Once this length of 21′ is determined on row 5, the track is counted on row 6 with the product of 2 tracks (top and bottom), the length of 21, a waste factor of 10% (use 1.1 as a multiplier), for a total of 46 LF.

Although these walls only have to be built 9′ high and will be built to that height in the field, *the estimator knows that 10′ high studs will be purchased to build the 9′ high walls because they are usually made in 2′ increments, like lumber is.* The typical choice is 8′, 10′, 12′, 14′, and 16′ for the height of studs, unless a special order is made.

See row 7 for the count of 10′ high studs for walls that are only 9′ high.

This same situation occurs with the 13′ high walls; it is 14′ long studs that have to be purchased and are shown in the estimate. See row 27.

See the following takeoff. Read it top down, left to right, and follow the logic. After beginning with 9′ high walls, the next group is 10′ high, then 12′, and ending with 13′ high. Note that 10′ high studs are used for the first group of 9′ high walls, and that 14′ tall studs are used for the 13′ high walls.

The 9′ high walls and the 10′ high walls could have been combined initially instead of showing them separately! Since the studs are going to be 10′ long for the 9′ high walls, the linear footage for these walls could have been combined initially – in other words, rows 1 through 7 and rows 8 through 13 could be shown as like kind. The reason that an estimator may choose to keep them separate at this point is for a later consideration, that of the gypsum board takeoff.

When the gypsum board is counted, the board will not, in some conditions, be as high as the studs. So, by going ahead and counting the length of the 9′ walls here in the metal stud takeoff, the arithmetic is already done for the gypsum board takeoff. Of course, this is an arbitrary decision by the estimator. As is the case in some takeoffs, there are different ways to go about figuring a takeoff.

The column headings for metal studs and gypsum board can vary somewhat, and there are different ways to figure them. Here, near the end of the divisions of takeoffs, there is more creativity and flexibility in how the numbers can be arranged. The metrics of counting concrete, masonry, and carpentry are best achieved with a repeating format such as given in those chapters. However, here in the metal stud and gypsum chapters, some variances in how these divisions can be counted will be pointed out.

METAL STUDS TAKEOFF											915.7
job: 2nd Floor Office		Job number			date			LF	STUDS		
No.	Description	Det.	Each	Pcs.	L	Ht.	%	Track	10'	12'	14'
1	Like kind walls 9' high:	A									
2	Wall 3 length	A			11						
3	Wall 4 length	A			5						
4	Wall 9 length	A			5						

METAL STUDS TAKEOFF											915.7
job: 2nd Floor Office		Job number			date			LF	STUDS		
No.	Description	Det.	Each	Pcs.	L	Ht.	%	Track	10'	12'	14'
5	**Total length 9' high walls**	A			**21**						
6	Track for 9' high walls	A		2	21		1.1	46			
7	**10' high studs** for 9' walls	A	24			10			240		
8	Total length 10' high walls:	A									
9	Wall 5 length	A			10						
10	Wall 7 length	A			9						
11	**Total length 10' high walls**	A			**19**						
12	Track for 10' high walls	A		2	19		1.1	42			
13	**10' high studs** for 10' h walls	A	22			10			220		
14	Total length 12' high walls:	A									
15	Wall 1 length	A			10						
16	Wall 11 length	A			9						
17	**Total length 12' high walls**	A			**19**						
18	Track for 12' high walls	A		2	19		1.1	42			
19	**12' high studs** for 12' walls	A	22			12				264	
20	Total length 13' high walls:										
21	Wall 2 length				13						
22	Wall 6 length				13						
23	Wall 8 length				11						
24	Wall 10 length				11						
25	**Total length 13' high walls**				**48**						
26	Track for 13' high walls			2	48		1.1	106			
27	**14' studs** for 13' high walls		55			14					770
								235	**460**	**264**	**770**

The count continues with bracing next.

In order to study the quantity of braces and their lengths, the wall lengths are first totaled, such as found on rows 5, 11, 17, and 25 in the takeoff above. Next, the length of the braces are scaled from the plans, using the wall sections and sketching out some 45-degree lines. If there are no convenient sections to use on the plans, a drawing similar to Sketch 8 can be used.

Given that the floor and deck heights are known and the height of the walls are determined, the length of the 45-degree bracing is simply geometry or scaling a sketch. Refer to the following sketch for the length of the braces.

With bar joists occurring both parallel to walls and perpendicular to them, as shown in Sections B and C, some braces may only extend to the bottom of bar joists similar to Section B. If the superintendent can get this to work out in some locations, fine. However, for estimating purposes, it's best to assume bracing will extend to the bottom of the roof deck, and "get the quantity covered".

If the instruction is to place bracing on alternate sides of the wall every 4′, the total wall length divided by 4′ will not provide enough braces. That's because many walls, being only 10′ or 15′ long, may require three braces for 10′ of wall, and four braces for 15′ of wall. That's seven braces for 25 LF of walls, so a higher figure than a 10% increase should be used to arrive at the quantity of braces. The number used here to account for the increase (over a straight up 20′ divided by 4′ equals 5 each) is 20%.

See the online resources for diagram 915.8

The following takeoff is only for bracing. Read it left to right.

Start with the total linear feet of walls, which of course have turns in them; they are not straight runs. See column A, where the wall linear footage is divided by 4′. The answer to this division is in column B, and since the bracing occurs on both sides of the wall there are two sides, shown in column C. Column D is where the 20% multiplier is used.

Column E provides the answer to the multiplication preceding it, and in row 2 the answer is 17.4 braces. This is rounded up to 18 each, an even number, and then multiplied times 10′, which is the length of braces for the 10′ walls. The total linear feet for bracing is shown in column H.

METAL STUD BRACING TAKEOFF										915.9
job: 2nd Floor										
No.	Description	Length of wall divided by 4' oc	Each	Sides	20%	Total braces	Each	L	LF
1	Bracing at 10' high walls:	A	B	C	D	E	F	G	H
2	Total length of walls divide by 4' oc	29 / 4 =	7.25	2	1.2	17.4			
3	Qty of braces for walls 3, 4, 5, 7, 9 -					18	ea	**10**	**180**
4	round up from 17.4 to 18 ea								
5									
6	Bracing at 12' high walls:								
7	Total length of walls divide by 4' oc	19 / 4	4.75	2	1.2	11.4			
8	Qty of braces for walls 1, 11 - round					12	ea	**8**	**96**
9	up from 11.4 to 12 ea								
10									
11	Bracing at 14' high walls:								
12	Total length of walls divide by 4' oc	48 / 4	12	2	1.2	28.8			
13	Qty of braces for walls 2, 6, 8, 10 -					30	ea	**4**	**120**
14	round up from 28.8 to 30 ea								
15									

Hat channel takeoff

The takeoff of the wall furring and ceiling furring is next. The wall furring is figured with this logic:

1. Use the wall length and then divide by 16″ to get how many furring pieces will be needed. Since the walls are straight runs, the exact number of pieces can be determined; one extra is added for the distance left over at the end of a wall where there will not be exactly a 16″ width.
2. The furring is taken off in differences of 2′. The ending figures of 39 each tens, 38 twelves, and 19 each fourteen-foot lengths are simply the total number of *each* length. Since the estimate would probably need linear footage, instead of each, these takeoff quantities would probably be multiplied times their length for the quantity used on the estimate.

Furring takeoffs depict how the format used in takeoffs can vary, especially in the later divisions of work. Imagine a straight wall 100′ long and 10′ high, 1,000 sf total, having vertical furring strips on it 16″ o.c. It makes little difference if the length of the wall or the total square footage is used to determine the count of furring strips. For example:

1. Linear foot method: Divide 100′ by 16″ o.c. to arrive at 75, and adding a couple to account for the end conditions, use 77 each ten-foot-long furring strips, which is 770 linear feet of furring.
2. Square foot method: Start with the 1,000 sf of total wall. Since each linear foot of furring will cover 1.33 sf of wall area, divide 1,000 sf by 1.33 to arrive at 750 linear feet, add a couple extra, and the total of 770 linear feet is arrived at.

WALL FURRING TAKEOFF												915.10
job: 2nd Floor Office		Job number			date:							
No.	Description	Det.	Wall L	o.c.	pcs	add	total pcs	%	10' each	12' each	14' each
	Wall Furring Strips										
1	Like kind walls 10' furring:										
2	Wall 14	A	11	1.33	8.271	1	10				
3	Wall 15	A	10	1.33	7.519	1	10				
4	Wall 16	A	10	1.33	7.519	1	10				

WALL FURRING TAKEOFF											915.10
job: 2nd Floor Office		Job number			date:						
5	Wall 19	A	7	1.33	5.263	1	7				
6							37	1.05	**39**		
7											
8	Like kind walls 12' furring:										
9	Wall 12	A	10	1.33	7.519	1	9				
10	Wall 13	A	10	1.33	7.519	1	9				
11	Wall 17	A	9	1.33	6.767	1	8				
12	Wall 18	A	12	1.33	9.023	1	10				
13							36	1.05		**38**	
14											
15	Like kind walls 14' furring:										
16	Wall 20	A	12'	1.33	9.023	1	10				
17	Wall 21	A	9'	1.33	6.767	1	8				
18							18	1.05			**19**
19											

The U channel takeoff further illustrates how a takeoff sheet can be modified to fit the occasion. The columns at the top are just length and width; to arrive at a square footage, and since one linear foot of U channel covers four square feet (in a large room), the sf is divided by four. However, in the example here, the rooms are small and there is more waste, and a 25% factor is used.

U CHANNEL TAKEOFF								915.11
job: 2nd Floor Office		Job number			date:			
No.	Description	Det.	L	W	SF	divide by 4	%	LF
1	Electric room ceiling	B	4	5	20			
2	HVAC ceiling	B	5	7	35			
3	Storage ceiling	B	9	11	99			
4					154	38.5	1.25	**48**
5								

Section 6 Wood blocking

Wood blocking within metal stud walls is a critical issue for the contractor. Compared to what may be a small amount of lumber and material expense, there can be a lot of labor and preplanning involved. With today's practice of dividing up the work, a CM must ensure that someone includes it, as neither the metal stud folks nor the carpentry trade relish this task.

The metal stud industry does have a method for the attachment of some products, but it is not a good one for the pull of the 250 pounds that a grab bar must withstand. Flat "sheet steel" is the metal product used for blocking and sometimes found in the specifications, but this thin metal is ineffective for anything beyond small or lightweight products such as soap dishes.

Sometimes a wall-hung sink is attached to a wall with the use of "flat plate" steel, thicker than the sheet steel used to make metal stud products. Other than this, the preferred method for the blocking within metal stud walls is to use wood. This is best for most of the Division 10 products, such as baby changing stations, shower seats, recessed waste receptacles, and fire extinguishers.

Installing blocking within metal stud walls is slow work and requires preparation. One cannot simply tell a carpenter to grab some lumber and install the blocking. There is a lot of layout involved accompanied by the review of submittals. The placement of wood blocking for toilet partitions requires the review of dimensioned and approved shop drawings. The placement of wood blocking surrounding recessed products requires exacting placement, such as a waste receptacle, which requires wood blocking on three sides in a specific place on a bathroom wall. A fire extinguisher cabinet might be 16-1/2″ wide, as shown on the submittal provided by the Division 10 contractor, which the CM gives to the carpentry subcontractor. Then, there's the height above the floor, which may or may not be on the plans.

There is a lot of preplanning that accompanies wood blocking. The subcontractor given this task might be the metal stud installers, the carpentry trade, or the Division 10 contractor. This is where the saying about project management is true, that the job is completely built on paper before it is in the field.

2

GYPSUM BOARD

Section 1 Introduction

Gypsum panels, commonly known as drywall, describe a family of products consisting primarily of a noncombustible gypsum core with paper backing.

Gypsum contains chemically combined water, which means water exists in the hardened gypsum. When heated to a high temperature, the water evaporates as the heat energy converting the water to steam is used up, thus keeping the other side (assuming the fire is on one side of the wall) of the gypsum board cool until evaporation takes place. Therefore, gypsum board panels are inherently a good material to be used for fire resistance.

Gypsum boards are the primary product used today as a wall and ceiling surface in both residential and commercial construction. Gypsum panel products are flexible but will crack or break if bent beyond their stress limits. The edges and ends of the panels are susceptible to damage when dropped or stacked on edge. The typical thickness used for the majority of construction is 1/2″ and 5/8″. The typical size for gypsum board panels are 4′ × 8′, 4′ × 10′, and 4′ × 12′. Some panels are made in a large 4′-6″ width, while tile backer boards, used in tub and shower areas, are made in 48″, 36″, and 32″ widths.

The most common finish placed on gypsum boards to hide the joints is tape and joint compound. The levels of quality, strictly defined by the industry, are covered here. However, the many plaster operations, which are trowel or sprayed on, are not.

Gypsum board panels can be attached with nails, screws, staples, or adhesives. The length of nails and screws, the spacing of fasteners, and types of adhesives are technical subjects not covered in this chapter.

Building codes throughout the U.S. and Canada require gypsum board to be manufactured according to *ASTM C1396 Standard Specification for Gypsum Board.*

Section 2 Ruling body, the Gypsum Association

The *Gypsum Association (GA)* was formally established in 1930 and represents the interests of all gypsum board manufacturers based in the United States and Canada. The material within this chapter was enabled by researching some of the many technical publications written by the GA. These include:

GA-600, *The Fire Resistance Design Manual.*
GA-214, *Recommended Levels of Finish for Gypsum Board, Glass Mat and Fiber-Reinforced Gypsum Panels.*
GA-214ORG, *Quick Reference Guide to GA-214.*
GA-216, *Application and Finishing of Gypsum Panel Products.*
GA-223, *Gypsum Panel Product Types, Uses, Sizes and Standards.*
GA-234, *Control Joints for Fire-Resistance Rated Systems.*
GA-618, *Building and Inspecting Smoke Barriers.*
GA-801, *Handling and Storage of Gypsum Panel Products: A Guide for Distributors, Retailers, and Contractors.*
GA-1000, *Identification of Gypsum Board.*

Section 3 Gypsum products

Introduction

The manufacturers of gypsum products naturally use different names for their products. For example, a "regular" piece of gypsum board, as defined by the Gypsum Association, is called a *Sheetrock Brand* panel by USG. Georgia-Pacific, touting the 45-minute rating of its 1/2″ regular board, calls theirs *Fireguard 45*, and their whole line of gypsum boards *Toughrock.*

This chapter uses generic names for gypsum products. The most often used boards, the kind that estimators see on job after job, are as follows.

Regular gypsum board

Gypsum boards are encased with strong paper with a smooth finish on the face side, which folds over the long edges lengthwise, but not the ends. The long edges are made with a variety of designs; a commonly used one being "tapered", which is thinner along the long joint, so that it can receive joint compound and end up being a flat surface. *The ends of the boards are square cut and not wrapped in paper.*

Type × gypsum board

ASTM C396 defines type × as gypsum board that provides not less than one-hour fire resistance for boards 5/8″ thick, or not less than 3/4 hour for 1/2″ boards, when attached to 2 × 4 wood studs spaced 16″ o.c. Fire resistant ratings are shown in UL for up to 4 hours for partitions, 3 hours for floors and ceilings, and 4 hours for column fire protection assemblies.

Note that there is no *smoke barrier* board. However, regular and Type × boards can effectively control smoke passage through walls and floors. The most common route for smoke migration is through door openings, vents, shafts, chutes, mechanical air-handling systems, expansion joints, and service penetrations (plumbing, telephone, and electrical lines). Carefully constructed, correctly sealed, and properly maintained wall and ceiling systems built using gypsum panels serve as highly effective smoke barriers. The International Building Code (IBC) typically requires a one-hour fire resistance rating for smoke barriers.

Water-resistant gypsum board

Water resistant does not mean waterproof! These panels are for use in commercial and residential interior areas not subject to excessive or continuous moisture. It is a base for tile applied with adhesive (not mortar) and plastic-faced wall panels. Although these boards have a water-resistant face paper, core, and back paper, they are not to be used in continuously wet areas like gang showers.

For the next step in water protection, see cement boards below. Water-resistant gypsum board is a product for kitchens and bathrooms that are subject to occasional moisture and are a better choice for these locations than regular gypsum board. Think of your kitchen and bathroom at home or apartment.

Water-resistant board is commonly called "greenboard" because of the color of the paper used to wrap it.

Tile backer board

Gypsum "cement boards" are used in wet areas. Called tile backer board, a typical application is behind wall tile in tub and shower areas. These boards are made with Portland cement and are moisture and mold resistant. For ease of use in small rooms, some manufacturers make this product in 32″ and 36″ widths, as well as 48″. As with all gypsum products, the various manufacturers have different names for their cement boards.

While the 1/2″ and 5/8″ versions of these types make up the greatest quantity of boards used in commercial and residential construction, there are others, including 1/4″, 3/8″, 3/4″, and 1″.

Other gypsum products

Gypsum sheathing: Designed for use as a "substrate" that is covered by an exterior wall cladding. After installation, this sheathing should only be exposed to weather for a limited length of time before being covered with a water-resistive barrier. Made in a type × version. To be covered with wood or vinyl siding, metal lath and plaster, masonry or brick veneer, and exterior insulation and finish systems (EIFS).

Abuse resistant gypsum board: In recent years, advances in technology have allowed the manufacture of an extra durable product which resists three levels of damage:

A Surface damage from abrasion and/or indentation.
B Penetration through to the wall cavity from sharp or blunt impact.
C Security breach through the entire assembly from ballistics or forced entry.

Liner panels: A group of panels in commercial construction used for "shaft wall" construction. These are often 1″ thick, may have two or more layers, and are fire rated. They might be used to separate tenant areas where higher fire ratings are needed or used in chases that extend through multiple floors. Combined with a fire rating, these thicker boards are used to wrap and fire protect steel columns and other structural elements.

Section 4 Gypsum identification

ASTM 1396 Standard Specifications for Gypsum Board mandates that boards are to be manufactured in accordance with *ASTM C1264, Standard Specification for Sampling, Inspection, Rejection, Certification, Packaging, Marking, Shipping, Handling, and Storage of Gypsum Panel Products.*

ASTM C1264 requires the back of each gypsum panel to have the following identification:

1 Name of manufacturer or a code that identifies the manufacturer.
2 A code labeling the manufacturing facility.
3 A code identifying the date and time of manufacture.
4 The country of manufacture.
5 The product thickness.

Section 5 Handling and storage

Delivery of gypsum board

Superintendents must carefully plan when gypsum deliveries are made and where it is to be placed, because of the following considerations.

The delivery of board should be near the time of installation. Moisture exposure must be prevented, as well as direct sunlight. Sketch 4 shows risers that are easily constructed of strips of gypsum panels. If a floor is subject to dampness, these risers should be constructed of wood or plastic. Using 3″ square risers by 4′ long, the six risers below would leave 24″–25″ between two risers. Perhaps the biggest consideration with the delivery of gypsum board is its weight.

Stacking of gypsum board

Stacks of gypsum board are large and heavy, and they are temperature and water sensitive. Gypsum board is a sensitive material and if exposed to weather, installed against crooked framing members, or the building shifts, ridging and cracking will occur at panel joints. Before its installation, the loading of the roof and other considerations need to be considered because of the possible expansion and contraction of the building.

Especially in frame construction, large loads need to be in place so that much of the building movement has already occurred. Roofing materials, a heavy load, should be in place not only for loading purposes but to prevent moisture in the building through leakage. Large plumbing items such as bathtubs should be in place, and heavy air conditioning equipment should be set in place.

Gypsum board should be stacked horizontally. Because of its weight, placement near a wall is better than in the middle of the floor. "Risers", or "spacers", are used at intervals so that forklifts or other machinery can lift portions of a stack. A 4 × 12 × 1/2″ gypsum panel can weigh over 80 lbs, and a stack of 25 weighs over one ton!

See the online resources for diagram 925.1

Section 6 Smoke barriers

Gypsum board construction creates very effective smoke barriers. The most common passageway for smoke is through corridors and wall and ceiling openings. Doors, vents, shafts, and chutes can all interrupt a continuous surface of gypsum board. The IBC typically requires a one-hour fire resistance rating for smoke barriers.

It is not important to do nice and neat joint compound work on fire and smoke walls that are above ceilings. These walls often extend "to the deck", meaning they are built from the floor all the way to the structure above, whether that be a floor or roof structure. Smoke and fire walls seal off adjoining spaces. Regardless of the finish below the ceiling, the finish above the ceiling, in the attic or plenum, needs to be functional but not necessarily paint ready. Only one coat of tape and compound needs to be used, and the compound is often applied thick just to ensure the joint is completely covered. It is better to use extra material than risk a thin coat that may require a second coat and additional labor.

Acoustical sealant is a product used to help obtain smoke control. This is a "caulking compound" especially made to reduce sound. The first good place to use it in wall construction is under the wood base plate or metal track. Another obvious place to use it is at the 3/8″ gap between the floor and the bottom of the gypsum board.

Section 7 Fire resistance

Fire codes

Fire resistance ratings are derived first from building codes. As explained in Part 1, Chapter 2, Section 5 Architectural Plan Sets, concerning architectural cover sheets, local building departments often reference a widely recognized building code such as the International Building Code. This code separates all buildings into classifications by occupancy, such as residential, educational, and business. The occupancy, together with the type of construction, is matched to specific fire resistance ratings, such as one hour and two hour, up to four hours.

The Gypsum Association's book on fire resistance is *GA-600 Fire Resistance Design Manual*. This manual defines many ways to achieve fire ratings with various combinations of gypsum products and thicknesses.

However, the definitive source for all fire ratings is UL, Underwriters Laboratories. This textbook uses the 2016 Volume of the UL Fire Resistance Directories. There are three volumes of UL fire code data that mainly effects contractors. They are:

Volume 1 Fire Resistance Directory with hourly ratings for beams, floors, roofs, columns, walls, and partitions.
Volume 2A and 2B Fire Resistance Directory with hourly ratings for joint systems, through-penetration firestop system, and electrical circuit protective systems and duct assemblies.
Volume 3 Fire Resistance Directory with hourly ratings for dampers, fire doors, glazing materials, and related equipment.

From Volume 1, there are drawing sections that detail how to construct a wall, floor, or roof such that it will have a certain rating. UL has conducted fire tests on wall, floor, and roof "assemblies" and have given these assemblies a fire rating. An assembly simply means all of the components, for example an interior wall assembly, would include the gypsum board and its size, stud size and material, insulation, and anything else in the wall construction.

Architects and builders refer to these UL volumes for construction details. A set of plans may contain in meticulous detail the replication of a certain *UL Design Number*, or a note may simply say, "Build wall in accordance to *UL Design Number U305*". The following drawing and notes are an abbreviated version of what's in UL Volume 1. The notes that follow UL drawings are often several pages in length.

Refer to the bottom of note 3 where one manufacturer, starting with the letter A, Acadia Drywall, is listed. In UL Volume 1, over a dozen more manufacturers are listed in alphabetical order together with their qualifying (allowable) products. The following are two short versions of UL designs and notes.

UL Design Number U30, wood studs

UL Design No. 305 is for a wood stud wall.

See the online resources for diagram 927.1

UL Design Number U305, metal studs

UL Design No. U419, shown below, is for a metal stud wall.

See the online resources for diagram 927.2

Section 8 Techniques

Typical installation

Sketch 2, diagram 928.1, shows a typical wall layout of vertical studs (16″ or 24″ o.c.) and gypsum board installed with its tapered long edge perpendicular to the studs.

See the online resources for diagram 928.1, 928.2, & 928.3

Good construction practices

Gypsum panel products should be installed on ceilings first and typically extend horizontally to the wall, where wallboard is installed after ceilings. Wall joints should be located so that none occur within 12″ of the corner of a door or window opening unless control joints are used at these locations, otherwise these joints are prone to cracking. Joints on opposite sides of a wall should not occur on the same stud. Outlet boxes on opposite sides of a wall should be offset so that they do not occur within the same stud cavity.

Where ceiling framing is parallel to a wall, ceiling board should not "cantilever" more than 2″ unsupported before touching the wall, i.e. the ceiling joist, roof rafter, or truss, and should be no more than 2″ from the wall. Since walls often do not occur within 2″ of the overhead parallel framing, a common framing technique is to use "gypsum board nailers", which is wood framing whose only purpose is to provide something rigid to fasten the edge of gypsum boards to. A "gypsum board nailer" can be placed at the ceiling level beside a top plate. Or, the nailers can run perpendicular to a wall, such as 2 × 4s running ladder fashion between trusses and across the top of an interior wall, providing a gypsum nailer on each side of a wall.

For more information about gypsum board wood nailers, see Part 6, Chapter 1, Section 2 Wood Components Not Shown on Plans.

Gaps between gypsum panels shall not be greater than 1/4″.

Gypsum panel products should be installed a minimum of 3/8″ off the floor. This is to keep moisture away from the board and to allow for proper alignment of joints. This is why 8′ high walls are actually built to a height of just over 8′-1″ high.

Ridging and cracking can result from gypsum panels being forced together due to wood framing members shrinking. The same can happen if a stud is slightly turned or bowed.

Steel studs, compared to wood studs, are generally straighter and more dimensionally stable. That is why vertical (or parallel) placement of gypsum boards to steel studs is permissible, but is not recommended against wood studs.

Framing lumber is often installed having a moisture content of 15%–19%. After installation, lumber will often dry, sometimes after many months, to about a 10% moisture content. This causes the lumber to shrink, and its length, width, and height shrink differently.

In frame construction of ceilings and trusses, the use of resilient channels helps prevent movement and minimizes ridging and cracking. The channel should be installed perpendicular to the framing, and the gypsum board applied perpendicular to the channels. The channels provide separation between the structure and the gypsum board, allowing the structure and board to move independently.

Control joints

Because of the expansion and contraction of gypsum board and the structure that it is fastened to, a control joint is used at specific locations. This can be a space or gap between one board and another, or a space left between ceiling boards and wall boards. These joints are used to limit cracking. If the joint occurs along the flat expanse of a wall or ceiling, it is called a *control joint*. If it occurs at the perimeter or end of a wall or ceiling expanse, it is called a *perimeter relief joint* or a *slip joint*.

Good construction practices include the following:

1 Use resilient channels or control joints when ceiling framing spans more than 15 linear feet. See Sketch 6.
2 Use resilient channels or control joints when ceiling framing changes direction.
3 Use resilient channels or control joints when adjacent framing within the same ceiling area has different spans.
4 Use resilient channels or control joints in areas that have variation in temperature and humidity such as garages, or areas that may stay unconditioned over long periods of time.
5 Install resilient channels 16″ o.c.
6 Use a control joint where the gypsum traverses a construction or expansion joint in the structure.
7 Use a control joint when a wall is longer than 30 LF or contains an area greater than 900 sf. See Sketch 3.
8 On ceilings that have a joint (expansion or control of some kind) at the perimeter, termed "perimeter relief", control joints should be installed every 50 feet or less in either direction and the total area between joints should be less than 2500 sf. See Sketches 5 and 6.
9 On ceilings that do not have perimeter relief, where the wall and ceiling gypsum surface is in contact, control joints should be installed every 30 feet or less and the total area between joints should be less than 900 sf. See Sketches 5 and 6.

Where control joints are needed, it is important that not only there be a break in the continuous gypsum board, but also the surface it is being nailed to is separated. If a wall is built with studs, for example, back-to-back studs are used, spaced close together but having a gap between them for separation at the same location as the gypsum control joint. See Sketch 3.

See the online resources for diagram 928.4

This is also the case for ceiling construction; if a control joint is needed in the gypsum board because it is in a large ceiling area, one method to accomplish this is to locate a framing member each side of the joint. Two ceiling joists can be placed side by side, for example. See Sketch 4. Another method is to use furring channels; see Sketch 6.

See the online resources for diagram 928.5

Sketch 5 illustrates what the term "perimeter relief" means.

See the online resources for diagram 928.6

See Sketches 5 and 6. They illustrate good practices numbers 8 and 9.

Sketch 6 shows two ways for stress to be relieved in a gypsum board ceiling. First, resilient ceiling channels are installed perpendicular to the wood joists; the gypsum board is not directly applied to the structure and differential movement of materials can be accomplished. Second, a control joint is installed in the gypsum ceiling.

See the online resources for diagram 928.7

Beads

The following sketches are of two widely used kinds of beads, which are a gypsum board "accessory". A vertical metal corner bead in the first sketch provides a hard edge that would otherwise become worn with use. The J trim in the second sketch allows a uniform gap between two dissimilar materials that are going to expand and contract differently. It's best to place metal trim at the edge of the gypsum board and hold it off the wall a short distance and use sealant to fill the gap. The sealant will move slightly without cracking and provides a good visual surface for paint.

See the online resources for diagrams 928.8 & 928.9

Section 9 Joint compound

The most common method of finishing gypsum board

Joint compound and tape should conform to *ASTM C475 Standard Specifications for Joint Treatment Materials for Gypsum Wallboard Construction.* Joint compound and tape is the most common method of finishing. The many different kinds of plastering operations are not covered here.

The "finishing" of gypsum board consists of concealing tapered or beveled long edges and the end butt joints between adjacent boards. It also conceals the small dimples in the surface of the boards as a result of screws, nails, or staples, as well as accessories. By varying the application and number of coats of joint compound, various "levels" of smoothness can be achieved where joint compound is used. There are five specifically defined levels of finish, each one associated with architectural and other considerations.

If a room is going to be brightly lit and the designer wants the glow and sheen of gloss paint, every little imperfection will show up. A level 5 finish will be specified. Many commercial buildings are built using a level four finish.

See the online resources for diagram 929.1

See Table 1 for a description of all five levels of finish from the Gypsum Association.

Five levels of finish

See the online resources for diagram 929.2

Section 10 Estimating

Wall types and lengths

There are different ways to count gypsum board. One is to use the same wall counts from wall carpentry or metal stud wall takeoffs. This works if the walls were taken off by wall type. Many plan sets have the various wall types lined up side by side in the beginning few sheets of the plans. If the estimator counts the linear feet of these, the studs and gypsum board can be readily counted given that the wall heights are known. Even if the wall heights are different on each side of the wall, the wall length can simply be used twice in figuring quantities.

A typical example of wall types shown on a drawing is below:

See the online resources for diagram 92 10.1

When the wall types are defined as in Sketch 11, the floor plan will typically have the wall numbers shown (referenced) on a floor plan such as Sketch 12.

See the online resources for diagram 92 10.2

Wall heights

It is not enough to count the linear feet of each wall type. The other dimension needed is the height of the wall, and despite the architect's definition of wall types, the estimator has more of them! If one wall type has four different heights, then the length of each one must be determined, and there will be more rows on the takeoff.

On some plans, the wall heights might be determined from interpreting various wall sections, or sections through ceiling soffits. Or, the height of walls might be found on life safety plans with a note such as, "Build shaded walls to deck using 5/8″ type × gypsum board". And then sometimes the wall heights are found on reflected ceiling plans. The estimator may have to search all through the plans to find all of the wall heights.

Gypsum board takeoff

The takeoff below is for the same example used in the metal stud chapter. This is the second floor plan, and Sections A, B, and C, and the finish schedule, where the ceiling heights are found.

The takeoff is aided by Sketch 7 from the metal stud chapter; it shows the wall lengths and heights of the *wall*, which means studs, and in reusing such notes the stud wall *lengths* are ok to use for gypsum board lengths. However, for the *heights* of gypsum board, the estimator must stay alert to where gypsum board stops short of the top of the studs or the count will be too high.

When gypsum board is the same height on both sides of a wall, the counting is simplified. The wall lengths of the *like kind wall types* are simply listed, multiplied by two to account for both sides, and multiplied times the height, with the resulting square footage shown in the right-hand column.

The takeoff method here is to list room numbers and count each of the four walls in the room. This will work for any job. Use room numbers on the left and place north/east/south/west across the top. Fill in the lengths and heights of each wall.

Gypsum Board Takeoff														92 10.3
job: Office Bldg 2nd Fl.			number									date		
			North			East			West			South		
No.	Description	Det.	L	Ht	SF	L	H	SF	L	Ht.	SF	L	Ht	SF
	WALLS													
1	Office 1		9	**12**	108	10	**12**	120	10	**12**	120	9	**12**	108
2	Office 2		11	13	143	12	13	156	13	13	169	9	13	117
3	HVAC		12	**8**	96	9	**8**	72	5	**8**	40	12	**8**	96
4	Electric Room		4	**8**	32	5	**8**	40	5	**8**	40	4	**8**	32
5	Director		9	13	117	12	13	156	13	13	169	11	13	143
6	Bookkeeping		9	10	90	10	10	100	10	10	100	9	10	90
7	Storage Room		9	**8**	72	11	**8**	88	11	**8**	88	9	**8**	72
8	Corridor		4	9	36	30	9	270	30	9	270	4	9	36
					694			1002			996			694
														3386
	CEILINGS		L	W	SF	%	Total SF							
1	Electric Room		5	4	20									
2	HVAC Room		7	5	35									
3	Storage Room		11	9	99									
					154	1.2	**185**							
	OR													
1	Electric Room		6	4	24									
2	HVAC Room		8	6	48									
3	Storage Room		12	10	120									
					192	1	**192**							

The height of the wallboard in the rooms with gypsum board ceilings is 8′; see Sections B and C.

The height of the wallboard in the rooms with acoustical ceilings are 4″ to a foot above the ceiling; see Sections B and C.

Where board heights are about 11′, they are listed as 12′ because a whole board will be used. See all of the walls in office 1.

The gypsum board used to build walls is mostly in 4′ widths, so to build the 9′ high walls and the 13′ high walls, there is the potential for a lot of waste. For walls that are 10′ and 14′ high, there is less waste because a full one-half of a board is used, and there is a better chance for the piece left over to be reused.

Examine the different measurements used for the L and W in the second example of the ceiling count. These even numbers account for the fact that gypsum board comes in sheets of 4′ × 8′ and 4′ × 12′. The first ceiling count above multiplies the ceiling lengths and widths using the nearest whole numbers. In the second example, the estimator uses even numbers, and this is a more realistic way of looking at small ceilings and walls. If a wall or ceiling is 10′-8″ long, a 12′ board is going to be completely used up, so if the estimator uses 12′ for the takeoff, it is just a realistic way of looking at the numbers. A much smaller waste factor can then be used than if exact lengths and heights are used.

Generally, the gypsum board count does not deduct for doors and windows except for large openings. Typical doors and windows only leave leftover oddball lengths and widths of gypsum board that are mostly unusable.

Waste factors vary between estimators.

PART 10
Specialties

Section 1 Introduction

The industry term for this collection of products is "building specialties". They are sold to contractors by suppliers and distributors. Some of these products are expensive, usually much more than the labor to install them. However, there can be handling and blocking requirements that increase the labor for them several fold.

Some of these products are "off the shelf" and the supplier can readily price them for bidders. Others are more complicated and the plans and specifications are sent to suppliers who pass them on to manufacturers. Questions about the documents may result in the contractor sending RFIs to the architect. The estimator needs to allow the time it takes for all this to occur before the bid.

Division 10 products are placed in buildings at very specific locations. As with all dimensioning, the contractor has the responsibility to interpret the architectural drawings and provide exact measurements, which will vary somewhat from architectural dimensions. Wall-to-wall finish dimensions and floor-to-ceiling heights sometimes must be determined well in advance of wallboard and floor tile installation. These dimensions are placed by the GC directly on the shop drawings provided by the manufacturer. These shop drawings may come to the GC through a supplier, but the manufacturer will have created them.

General contractors often install many Division 10 products themselves. Construction managers subcontract them to a growing body of trade contractors. Sometimes the CM includes this work in a subcontract package called "general trades", which might include other work such as carpentry or doors.

Some of the products mentioned in this chapter include:

- Toilet partitions
- Marker boards
- Fire extinguishers
- Corner guards
- Cubicle curtain track
- Signage and lettering
- Mop and broom holder
- Bath accessories or Toilet accessories:
  - Grab bars
  - Handicap mirrors
  - Soap dispensers
  - Waste receptacles
  - Toilet tissue dispensers
  - Paper towel dispensers
  - Baby changing station

Section 2 Suppliers and distributors

Manufacturers make things, and that is what they spend their time doing. They do not sell directly to contractors. Sales are left up to suppliers and distributors who deal with pesky contractors, who always need a quote for materials or advice about installation. The plans and specifications are always being passed back and forth between contractor and supplier, and the business of bidding a job is conducted.

A good distributor that has experienced sales help with knowledge of their product line is valuable to the estimator. They know that marker boards are delivered in pieces and that the tray at the bottom is separate and must be attached to the board after delivery, which affects field labor. They also can help the contractor, who, having installed "wall hung" baby changing stations before, reads a specification referring to "recessed" baby changing stations. This prompts a phone call to the supplier, who furnishes the contractor with a product data sheet from a manufacturer. The data sheet is typically a small drawing on notebook-sized paper with installation instructions. The recessed baby changing station ends up costing more and taking longer to install than the ones the contractor is used to.

Sometimes a manufacturer, maybe a small one or the maker of a singular product, does not have a supplier network. Perhaps the maker of large "letters" for placement on a building may have the admiration of architects in the area but no established distributors. In this case, the manufacturer may have a sales person and sell directly to contractors.

The supplier or distributor of Division 10 products is not involved in fabrication like the door or steel suppliers are. There is no tinkering with the manufactured product. The contractor receives the product as it left the factory.

Bath accessories are sometimes furnished by suppliers straight to the owner (especially for large buildings) without contractor involvement. If a product will need replenished supplies, such as a paper towel holder, a supplier may offer the holder free and then arrange to keep the paper holder full of paper towels later when the toilet room is in operation. The manufacturer in this case is probably in the paper business, and may or may not make the metal holder. The contractor's role is usually to install the paper towel holder but not to buy it. The designation on the plans is OFCI, which means "owner furnish, contractor install." Other acronyms that are used include OFOI and CFCI.

Section 3 Submittals

Product data sheets

Contractors "build the job on paper" by submitting documents to the architect and engineer for approval. The submittal for a product includes a *product data sheet* (the most numerous pieces of information of all the submittals) from the manufacturer, which is used to prove compliance with the technical specifications. Product data gives the product size(s), material(s) it is made of, material gauge or thickness, weight, and other physical characteristics.

Standard signage such as room numbers can be shown in the submittal process with product data sheets and samples. Large lettering and signage is shown on shop drawings.

Samples

On a major project, the specifications may require that a *sample* (which is a submittal) of each bathroom accessory product be furnished to the architect. Upon approval, the products are returned for use in the job. The submittal process also includes *shop drawings* (a submittal) with horizontal and vertical dimensions of the product to walls and floors placing the product at a specific location.

Shop drawings

Toilet partitions, marker boards, and cubicle curtain track are all assembled in the field. Unlike the door and steel trades, where the suppliers get involved in preparing shop drawings, the manufacturer prepares them on these Division 10 products. This is a big distinction for the project manager. The manufacturer prepares the shop drawing per *their* format with a place to select the colors and another place on *their* form for the contractor to insert field dimensions. The project manager, upon receipt of the shop drawings, sends them to the job superintendent, who often must predetermine the exact finished dimensions before wallboard and tile is installed. For example, the plan drawing for the side-by-side placement of toilet partitions will require the distance (length) between the walls that are on either end of the partitions. The elevation of the partitions may require the floor-to-ceiling finished height.

The submittals for products hanging from ceilings take special consideration. In a hospital, heavy curtains hang from cubicle curtain tracks at the ceiling. The track must withstand the possible occurrence of someone accidently grabbing the curtain for support. Blocking must be provided above the ceiling for the cubicle track to fasten into; the track is fastened through the ceiling and into wood or metal support. This work will require shop drawings.

The acoustical ceiling layout needs to be shown on the shop drawing for ceiling-mounted cubicle curtain track. Within this grid is an assortment of air conditioning grilles (supplies and return), lighting all over the place, and TVs. To properly draw the track configuration at the hospital bed areas (and prevent conflicts upon installation), the architectural reflected ceiling plan must be coordinated with the mechanical air conditioning plan and the electrical plans. All of this must be coordinated by the GC, and a separate drawing is made to show the placement of wood blocking for the track and TVs. This is a lot of shop drawing work for the contractors.

Section 4 Wood blocking

Division 10 products are not hard to install, but they must be placed in very specific locations and there must be blocking in the wall (or ceiling) in exact locations for them to be attached to. In a public building, these products get a lot of use and must be securely installed. Ensuring that grab bars support people without fail (testing weight is often 250 lbs, or one big architect) is a life and safety issue, which makes this rise to the highest level of concern for the contractor and the building department. Getting four out of five right doesn't work! Careful layout is needed and exacting measurements are required before product placement. *More time can be spent in preparation than in installing the product.*

Continuous blocking (see Sketch 1) is used for everything from grab bars to wall rails that run the length of several corridors. Continuous blocking is installed above the floor at the level of the product without the bother of placing it on center in small pieces to match the fasteners. This may cost more in material, but the labor, which costs more than the material, is cheaper. Isolated products like toilet tissue holders simply need a linear foot or two of wood or metal blocking.

Some products, like waste receptacles, are installed within the wall. Blocking for recessed products takes longer than installing continuous blocking. *Recessed blocking* must closely surround a recessed waste receptacle or fire extinguisher cabinet. A "box" must be built within the wall on three or four sides of the recessed unit so that product and blocking can be fastened together. This blocking must be very carefully measured and placed in advance of wallboard and product installation. Recessed products take extra time for both blocking and product installation.

The blocking for a recessed product interrupts the spacing of wall studs and necessitates extra studding (beside the product) and some horizontal framing (above and below the product) to provide the opening. If the opening is in a block wall, the mason spends more time building the wall than if it were solid. And, after an opening is created in either a steel stud or masonry wall, wood or other blocking (the box) must be installed. The layout and placement of blocking for recessed waste receptacles and fire extinguisher cabinets takes far longer (per linear foot or square foot) than it does for surface-mounted products.

The support above a ceiling for TVs, cubicle curtain track, cameras, etc., can be substantial. The amount of labor and material it takes depends on the structure above the ceiling (whether bar joists or concrete floor, etc.) and how far this structure is above the ceiling. The support will not be shown on the architectural plans. The estimator must account for it at bid time prior to having shop drawings to go by. The labor for this can be far greater than that of installing the hanging product.

Manufacturers will often provide fasteners for their product. However, it is not unusual for estimators to include some miscellaneous expense in the estimate to cover fasteners, handling, and temporary protection.

See the online resources for diagram 10 4.1

Section 5 Schoolhouse plans, specifications, and legends

Life safety plans and fire extinguishers

The architectural plans will include, per a building code requirement, a sheet called the life safety plan. There may be several life safety sheets for many areas and several floors. These define the path of egress (how to get out of the building) from the farthest reaches of the building to the exit door(s). This path is drawn with a line or dashed line with arrows and traces the footpath down the middle of corridors.

These corridor plans are the perfect place for the architect to show the placement of fire extinguishers (FE) and fire extinguishers in cabinets (FEC). The estimator knows to look on life safety plans for them, as well as on the floor plans of storage, electrical, and mechanical rooms.

Specifications

Sometimes the specifications contain information about quantities. It is common for room signage (including handicap signs and symbols) to be covered in the specifications and not be mentioned on the plans. (Parking signage is a different product and is found in Division 2). Read the following short examples of language from Division 10 specs:

> "Provide one mop and broom holder in each janitor room."
>
> "Provide room numbers for all rooms except corridors. Signage to be placed on the wall adjacent to the lock side of doors sixty inches above finished floor."
>
> "Provide one soap dish for every sink location, or provide one toilet tissue dispenser for each toilet."

There is a symbol in the janitor room of the schoolhouse plan that appears to represent a mop and broom holder, but it does not show up on the legend. However, instructions for it could appear in the specifications. Information about Division 10 products can be found either on the plans or in the specifications, and it is the estimator's task to study them both.

Schoolhouse floor plan and marker boards

The following schoolhouse plan illustrates one way that marker boards, FE, and corner guards can be shown on the architectural plans. Also, refer to the legend following the plan.

See the online resources for diagrams 10 5.1 & 10 5.2

Women's restroom plan and bath accessories

The following women's restroom plan illustrates how bathroom accessories are often shown on the architectural plans. The legend that follows explains the symbols. This plan would be accompanied by wall elevations giving the elevations of the products above the floor.

The specifications may require that all the bath accessories be manufactured by one company. Accessories are common enough that the supplier may know the prices for them without having to inquire. However, toilet partitions are a custom assembly, and the manufacturer that builds them must see the plans to quote the job.

Division 10 products are often found on easy-to-read "enlarged" floor plans because the scale of typical floor plans, 1/8″ and 1/4″, is too small to clearly show all the items. This is the case for bathroom accessories, because public restrooms are crammed full of products. Large-scale elevations of the restroom walls are shown. Another feature of the plans regarding Division 10 products is the often used "legend". A legend is the "key" that links a *plan symbol* to a *product.*

See the online resources for diagram 10 5.3

Section 6 Schoolhouse estimate

Material and labor pricing of the schoolhouse

A unit price sheet and a P/S sheet are shown for the schoolhouse project. A job overhead and summary sheet are not shown.

Bathroom accessories are not hard to count. Since the unit of measure is simply "each", there is no involvement with L/W/Ht or even SF. In a large building with many restrooms, the key is to format a takeoff that documents the count for each one so that an accurate summary can be made. The estimator can use the QS as a collection sheet to separate floors, buildings, etc., which is a good practice, as this organization helps prevent addition mistakes. On small jobs, quantities can be observed on the plan and placed directly on the estimate.

To estimate labor, all unlike items must be taken off separately. For example, there may be three kinds of mirrors, all taking a different amount of time to install. This can include handicap tilt mirrors, full-length mirrors, and small mirrors with a stainless-steel edge. To facilitate the ordering of materials (if the job is awarded) and the estimation of labor, it is best to separate these products in the takeoff count from the beginning.

The total *material* cost for marker boards, toilet partitions, bath accessories, and any other Division 10 categories should each be totaled including sales tax and delivery. The following two pricing sheets are examples of how contractors estimate this labor and material. This is not an example of an entire estimate; there is no job overhead sheet or summary page. These quantities are from the two schoolhouse floor plans.

If wood blocking is continuous, the takeoff is a simple linear foot count. The blocking is typically a 2 × piece of lumber, perhaps a 2 × 6 or 2 × 8. If the toilet accessory product is recessed, the linear feet of blocking that it takes to surround the product is assessed and then multiplied by the product quantity.

Unit price sheet

UNIT PRICE SHEET												10 6.1
name	Schoolhouse and Womens Restroom					JOB #:				date:		
code	DESCRIPTION	Man	Mat'l	Sub	QTY	unit	Hrs	$	Labor	Mat'l	Sub	
		hr unit	unit	unit				rate				Row
10100	Waste receptacle, recess	1.25	0	0	3	ea	4	30	113	0	0	1
10100	Paper towel dispenser	1	0	0	2	ea	2	30	60	0	0	2
10100	San napkin disposal	1	0	0	8	ea	8	30	240	0	0	3
10100	Toilet tissue holder	1	0	0	8	ea	8	30	240	0	0	4
10100	36" grab bar	1.5	0	0	1	ea	2	30	45	0	0	5
10100	42" grab bar	1.75	0	0	1	ea	2	30	53	0	0	6
10100	Soap dispenser	0.7	0	0	8	ea	6	30	168	0	0	7
10100	Framed miror/shelf combo	1.5	0	0	8	ea	12	30	360	0	0	8
10100	Baby changing station	2.5	0	0	1	ea	3	30	75	0	0	9
10200	Markerboard 1 4' x 4'	2	0	0	12	ea	24	30	720	0	0	10
10200	Markerboard 2 8' x 4'	3	0	0	6	ea	18	30	540	0	0	11
10600	FE	0.7	0	0	2	ea	1	30	42	0	0	12
10600	FEC	1.5	0	0	2	ea	3	30	90	0	0	13
							92		2,745	0	0	
1	2	3	4	5	6	7	8	9	10	11	12	13
						3 x 6 = 8			8 x 9	4 x 6	5 x 6	

These descriptions and quantities are placed directly on the estimate (not backed up on a takeoff). All of these products require various work-hour units to install so the unit price sheet is where the labor for them is figured.

Product and subcontractor P/S sheet

P/S SHEET (major products and sub prices)								10 6.2
name	Schoolhouse and Women's Restroom				Job no:		date:	
code	DESCRIPTION	QTY	HRS	$ Rate	LABOR	Mat'l include tax	SUB	TOTAL
10100	Bath accessories	LS	0	0	0	4,000	0	
10200	Markerboards	LS	0	0	0	5,000	0	
10530	Corner guards	8 ea	8	30	240	500	0	
10600	FE and FEC	LS	0	0	0	600	0	
					240	10,100	0	
1	2	3	4	5	6	7	8	9
					4 x 5			

Example of labor appearing on the P/S sheet - single quantity of like kind products appearing on one row.

Lump sum product quotes incl. tax and freight

PART 11

Construction documents

1

WHAT IS (AND ISN'T) A CONTRACT DOCUMENT

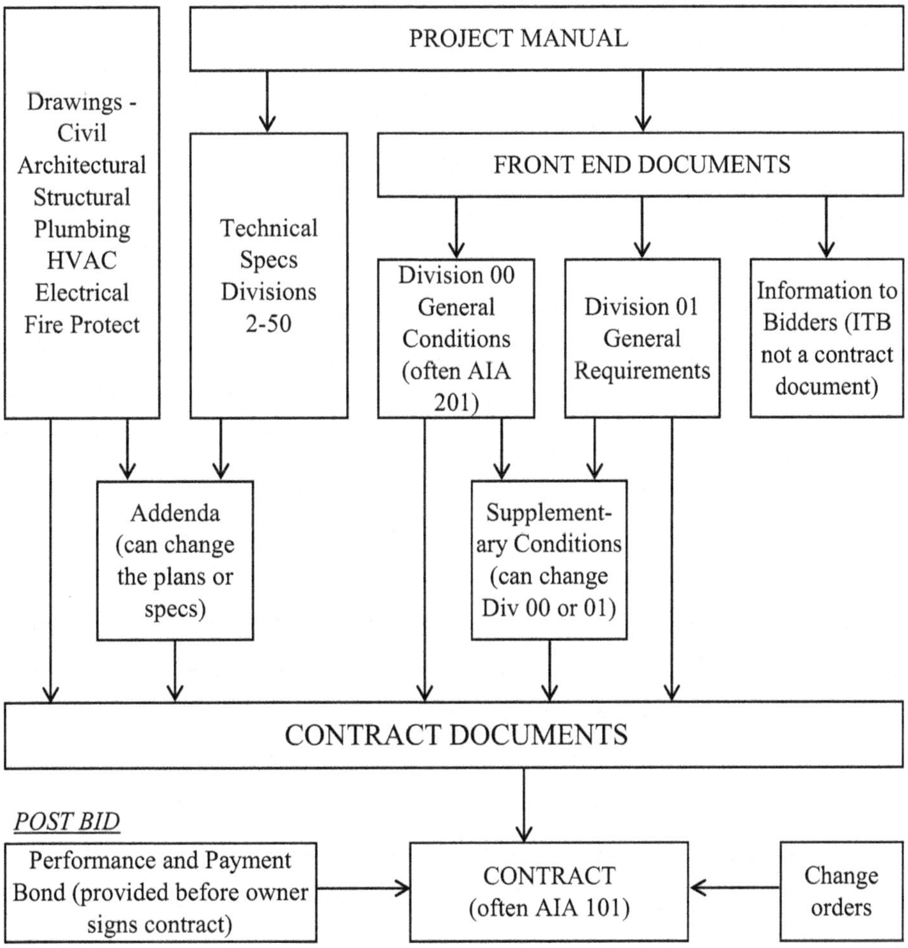

Chart 1 ORGANIZATION of CONSTRUCTION DOCUMENTS

Section 1 Construction documents
Section 2 The project manual
Section 3 The drawings and technical specifications
Section 4 Addenda

Section 5 The generals and their conditions

General conditions
General requirements
Supplementary conditions

Section 6 The contract between owner and contractor, *articles 5.3.1, 1.1.1, 1.1.2*

Introduction
Flow-down contract language, article 5.3.1
Summary of the contract documents

Section 1 Construction documents

The term "construction documents" does not mean or identify any particular set of drawings, specifications, the contract, the submittals, or anything. However, it is such an obvious choice of language it must be clarified here in order to avoid confusion.

Construction documents are not mentioned in AIA 201. However, since this Part 11 concerns all of the documents concerned with contracting, *this term is used here as a blanket description within which all of the others reside*. Hence, the name of the above chart, "Organization of Construction Documents".

The term "construction documents" has no legal relevance or definition. It could be used to describe a purchase order, some delivery instructions, or practically anything to do with a project.

What is and isn't a "contract document" *is* relevant and important to understand.

Section 2 The project manual

There are two nontechnical divisions of the specifications that are placed in the project manual, Division 00 General Conditions and Division 01 General Requirements. The technical specs, from Division 2 to 50, are also placed there. All of these divisions are contract documents and available to the contractor pre-bid.

Sometimes the owner's request for proposal (RFP), or a document called the invitation to bid (ITB), is placed at the front of the project manual. It precedes the nontechnical and technical specifications. RFPs and ITBs contain a short summary of the project and the bid forms. On public bids, the contractor is not allowed to alter this document, or the bid will be non-responsive. Qualifications and notes written on the bid form are not allowed. These "bidding invitations" are *not* contract documents.

The contract documents do not include bidding requirements. Bidding instructions are separated from the contract documents and include the invitation to bid (ITB), portions of addenda relating to bidding requirements, requests for proposals (RFP), and the contractor's bid (however, the bid form itself is usually in the ITB). The Invite To Bid is sometimes called the Advertisement For Bids. The ITB or Advertisement often has a section within it called Instruction to Bidders (which is not the ITB!).

The entire *front* portion of the project manual, consisting of the ITB and the two divisions of nontechnical specifications, 00 and 01, are often referred to as "front-end documents". The nontechnical Divisions 00 and 01 are contract documents; the invites to bid are not.

The documents used to amend or clarify the nontechnical specifications (Divisions 00 and 01), are called supplementary conditions, and these supplementary requirements are contract documents.

Section 3 The drawings and technical specifications

Plan reading is covered in a separate chapter, which also contains a section about the specifications. This chapter concerns legal aspects of the plans and *technical* specifications.

The plans are contract documents, as is a document called an addendum that is sometimes used to amend or clarify the plans and technical specifications (Divisions 2–50) before a bid.

The entire set of specifications is arranged in divisions, from 0 to 50. The first two are nontechnical; see Section 4 below. Technical Divisions 2–50, or 02–50, begin with 02 earthwork, are followed by 03 concrete and 04 masonry, and are meant to define one trade at a time. All of the technical specifications are organized in a three-part format – General, Products, and Execution (see Part 1, Chapter 3, Section 3 Plan Reading).

The contractor must understand that the plans and specifications contain so many instructions that some of it will be inaccurate. The preparation of them is a voluminous undertaking, and the architect only does it once for a project. It can be obvious to a contractor that a note on sheet 3 is in direct conflict with another note on sheet 9, but these occurrences are going to happen. The contractor's duty is to report it upon discovering the inconsistency.

The set of drawings, which are a contract document, often contain a survey concerning physical characteristics of the site. Surveys are covered in Part 1 and are often found within the civil portion of the plans. In Article 2.2.3 of AIA 201, the language notes that surveys are "owner provided", as opposed to being provided by the architect. This is important – the architect is providing distance between him/her and the survey.

A very important statement in the General Conditions is that the contractor can "rely" on the survey. This might seem unnecessary to say – of course the contractor is going to rely on it. However, this statement, given the fact that field occurrences are sometimes found to be at variance with the plans, is helpful to the contractor. The contractor can trust the plans and move forward until they are proved wrong. The contractor can take what is given and run with it. This is good; the contractor is not put into the position of having to distrust the plans.

It is the obligation of the contractor to thoroughly study the plans and all of the contract documents, which are complementary. The contractor is to use them for preplanning and preparation, and for the purpose of facilitating the work, *not* for the purpose of discovering errors and omissions. This can have important consequences when dealing with concealed conditions. The contractor should study and compare the various plans that affect each portion of the work, including the location of items on a survey by field measuring. That's what contractors do, preplan.

After a problem surfaces, the owner or architect may think the contractor should have seen it coming, and perhaps should have figured out that a contractors claim of an unforeseen condition was knowable and should have been predicted. However, contractors can rely on the information given and are not expected to figure out a puzzle or have hindsight beyond what is in the documents.

However, while the contractor does not have to suspect that there is a mistake in the plans, there is an obligation to report any inconsistencies or errors.

Section 4 Addenda

During the bidding period, the contractor has the ITB (invitation to bid), and the plans and specifications. There is a bid date soon and the bidding race is on. A pre-bid conference was yesterday, or tomorrow, and the bid opening is in seven to fourteen days. The contractors are asking questions, in the form of RFIs, to the architect. The architect responds with answers to these questions (to all of the bidders) in the form of addendums issued before the bid, which become contract documents. There may be several addenda (plural). An early addendum (singular) might be to change the date of the pre-bid conference or bid date, with subsequent addendums answering technical questions.

The architect or owner will announce, sometimes in the ITB or RFP, sometimes at the pre-bid conference, that there is an ending date for asking questions. What the architect and owner need is a cut-off date of about four to seven days before the bid in order to answer RFI questions and issue a final addendum.

Section 5 The generals and their conditions

General conditions

Division 00 General Conditions and Division 01 General Requirements are both known as nontechnical specifications.

The General Conditions usually consist of AIA Document 201 and are located inside the project manual. Because of their importance, a whole chapter is devoted to them; see Chapter 2. The federal government has its own form, but AIA 201 is used by other government agencies and many private owners.

Division 00 General Conditions is a contract document.

General requirements

The second use of the word "General" is to define various owner requirements. They are found inside the project manual and is called Division 01. Because of its importance, a whole chapter is devoted to it; see Chapter 3.

Division 01 General Requirements is a contract document.

Supplementary conditions

A confusing part of construction documents is the second use of the term "conditions". This is a sometimes-used document named "supplementary conditions". The sole purpose of it is to change, add, or delete language in the two nontechnical divisions – Division 00 General Conditions and Division 01 General Requirements. The owner is usually the party making these changes, and a city or county may have some policies and procedures that are different from the standard General Conditions, or may amend their standardized General Requirements for a particular project.

Supplementary conditions are usually contained in the project manual and follow Division 00 and Division 01. They can also be issued through an addendum if not initially provided for in the project manual.

The supplementary conditions are contract documents..

Section 6 The contract between owner and contractor, *articles 5.3.1, 1.1.1, 1.1.2*

Introduction

For the benefit of bidding contractors, an unsigned contract is often a part of the front-end documents. When a bid is awarded, the contractor signs this contract, which *references* all of the other contract documents.

The contract documents, which form the contract for construction, in no way forms a relationship between anyone except the owner and contractor, such as the architect and contractor, or the owner and a subcontractor. The architect shall, however, be entitled to performance and enforcement of obligations under the contract intended to facilitate performance of the architect's duties.

The contract, after the award, may be amended or modified only by a modification, of which there are three kinds: *(1) a change order with a description, cost, and possible time extension, and is agreed to by the parties, (2) a construction change directive, for when the parties do not agree and the architect instructs the contractor to proceed anyway, or (3) a written order for a minor change in the work issued by the architect.*

Any of the three ways to effect a modification is referred to as a "change order".

Flow-down contract language, article 5.3.1

The contract language of AIA 201is required to be included in all subcontracts so that all the contractors on a construction project operate with the same obligations. The prime bidder has the responsibility to insert language in subcontracts that links the General Conditions to everyone on the job. The language of AIA 201 must "flow down" to the subcontractors, through the GC or CM, so that subcontractors are bound by the same contract documents that the prime contractor is bound to. This can be done by referencing AIA 201 in the contractor/subcontractor agreement.

Summary of the contract documents

The contract documents, forming the entire contract between owner and contractor, are the following:

1 Owner/contractor agreement, the contract.
2 Division 00 General Conditions.
3 Division 01 General Requirements.
4 The entire set of drawings.
5 Technical specifications, Division 02 up to 50. These divisions are divided into three parts –

Part 1 General, Part 2 Products, and Part 3 Execution.

6 Addenda, which can change the drawings or technical specifications.
7 Supplementary conditions, which can change the General Conditions or General Requirements.
8 Modifications (three kinds of change orders)

2

DIVISION 00 GENERAL CONDITIONS

Section 1 Introduction

1997 version of AIA 201
Getting things done
Design intent or concept 1.2.1, 3.2.1, 4.2.7
Unforeseen conditions

Section 2 Owners and users of facilities

Owner's representative 2.1.1, 4.2.1
Owner can carry on other work 6.1.1
Owner's authority to dismiss the contractor 2.3.1, 2.4.1, 14.2.1, 14.3.1, 14.4.1

Section 3 The architect's authority

Inspections by the architect 4.2.1, 4.2.2, 12.1.1
Architect and owner can object to contractor's subcontractor 5.2.1, 5.2.2, 5.2.3
Architect can reject the work 4.2.6, 13.5.1, 13.5.2, 13.5.3

Section 4 The contractor's responsibility

Bidding due diligence 1.5.2
Contractor's authority 1.2.2, 3.2.1, 3.2.2
Scheduling 3.10.1
Supervision 3.3.1, 3.9.1, 10.2.6
Substantial and final completion 4.2.9, 8.2.3, 9.8.1, 9.8.2, 9.8.3, 9.8.4

Section 5 The contractor's submittals

Submittals defined 3.12.1, 3.12.2, 3.12.3, 3.12.4
Timely, correct, and full submittals precede professional field production
Product data sheets
Shop drawings
Schedules
Samples and mockups
Test reports
Concrete mix designs
Certificates
Contractor's approval of submittals 3.12.5, 3.12.6, 3.12.7
Architect's approval of submittals 3.12.8
Substitutions and/or equal
Submittal log

Section 6 Change orders

Changes in the work, general 7.1.1, 7.1.2, 7.2.1
Minor changes in the work 7.4.1
Construction change directives 7.3.4, 7.3.5
Change orders, summary

Section 7 Claims

Claims defined 4.3.1
Claims time limit 4.3.2
Unforeseen conditions 4.3.4
Giving notice for a claim 3.2.3, 4.3.5, 4.3.7
Architect's role in reviewing claims 4.4.1, 4.4.2, 4.4.4, 4.4.5
Mediation 4.5.1, 4.5.2, 4.5.3
Arbitration 4.4.6, 4.6.1, 4.6.2, 4.6.6

Section 8 Delays

Delays and causes for delay 8.3.1
Hazardous materials 10.3.1
Emergencies 10.6.1
Delays as cause for contractor to terminate the work 14.1.1, 14.1.2, 14.1.3, 14.1.4

Section 9 Payment

Schedule of values 9.2.1
Draw requests 4.2.5, 9.4.1, 9.4.2, 9.5.1, 9.7.1, 9.10.2
Stored material 9.3.2
Final payment 9.10.1

Section 10 Closeout

Closeout introduction
Record drawings and contract documents 3.11.1
Demonstration and training, operation and maintenance manuals
Warranties and guarantees 3.5.1, 12.2.1
Consent of surety

Section 11 Unforeseen fire line case
Section 12 Gooseneck faucet case

Section 1 Introduction

1997 version of AIA 201

This chapter is about the document AIA 201, created by the American Institute of Architects, first used in 1911. It has fourteen "articles" which are known as "General Conditions". The AIA revises this document every five or ten years, but the 2007 version has not been endorsed by the Associated General Contractors of America (AGC), and the 1997 version of AIA 201 is, as of this writing, still being used.

General Conditions address the legal framework of the contract between owner and contractor. Its paragraphs of construction law are arranged into articles. This makes it the only division in the project manual not formatted using the three topics of General, Products, and Execution.

This document is included as a part of many city, county, and state construction projects across the country. It becomes part of the owner/contractor contract. It may not appear in full written form, it may simply be "referenced", but in either case it is a part of Division 00 of the specifications. It is used in many private construction contracts as well, but not federal work.

The fourteen articles of AIA 201are named:

1 General provisions
2 Owner
3 Contractor
4 Administration of the contract
5 Subcontractors
6 Construction by owner or by separate contractors
7 Changes in the work
8 Time
9 Payments and completion
10 Protection of persons and property
11 Insurance and bonds
12 Uncovering and correction of work
13 Miscellaneous provisions
14 Termination or suspension of the contract

These articles are long steeped in legal history, meaning that they have been used in court many times and precedent has been set. Attorneys can look up the outcomes of cases similar to the one they face.

AIA 201 has 14 articles and an index noting where subjects appear throughout the document. *This textbook does not follow the same order and format as AIA 201.* This would at first seem to introduce confusion, but in defense of the organization of this section, the following is noted.

This textbook is for construction students. AIA is a document made by attorneys working for architects. The most important subjects for contractors – the subjects of time and duration, payment, performance, design concept, and submittals, are scattered throughout the articles in AIA 201. The reason this textbook follows a different format, rather than simply explaining each article one by one, is that topics that are priorities for contractors are explained more or less in one place.

Getting things done

The responsibility for moving a project forward is given to the prime contractor, by the owner and architect, through the General Conditions as defined by AIA 201. The job of getting all things done, even getting the architect to do his/her job on time, is the sole responsibility of the contractor. The energy to move everything forward comes from the contractor, and to do this the contractor must have a staff of estimators, project managers, superintendents, bookkeepers, and support staff. Once the project starts and the clock is ticking, experienced project managers and superintendents are trained not to let anything get in their way. The construction superintendent will vocalize to *anybody* (architects, owners, the governor) their displeasure with the lack of progress or a possible future delay. And, the project manager is adept at documenting subcontractor absences and shortages of labor, hurrying the architect to approve submittals, tracking weather delays, and always documenting.

The contractor is the entity responsible for *time*, with making sure that *all* parties do their job within the scheduled completion date. At the end of a project not completed on time are liquidated damages and worse. After the job starts, the time restrictions on the owner and architect are light compared to the load put on the contractor.

An everyday example of the detailed management that is required by the contractor, through the General Conditions, is the requirement of a submittal log. The submittal process is the contractor's (and subcontractors') voluminous submissions of paperwork describing the materials, products, and equipment that they select to be used in the project. Completing submittals on a job is a long and drawn-out process that requires constant attention by the project manager and interaction with subcontractors, fabricators, suppliers, and the architect and engineer. The contractor, not the architect, tracks all of this activity, updates it, and presents it to the architect upon request for review.

The General Conditions give the contractor the responsibility to watch the clock, keep the submittals and as-builts organized and up to date, and revise the schedule. Trained in estimating, scheduling, and construction techniques, the contractor is, as soon as the job starts, in charge of everything and in a hurry. There are problems dealt with every day, and the contractor must have a production attitude that doesn't allow delay and that gets things done. The schedule must be met.

The contractor is fully responsible for "means and methods", including safety. How things are done, and in what sequence, and by whom, is all up to the prime contractor. The architect and owner put this is the contractor's hands, and make it clear, over and over in AIA 201, that they have nothing to do with means, methods, and safety.

Getting things done is all up to the contractor, as this chapter will explain.

Design intent or concept 1.2.1, 3.2.1, 4.2.7

This concept is so important it is addressed in advance of the other articles of the General Conditions because it can affect the interpretation of the entire plans and specifications. It is linked to, or associated with, two other ideas – that the plans are complementary, and that plans should be read as a whole.

The design concept is a way of "interpreting" the documents, and this is how the architectural profession views them. The construction student can best comprehend the individual articles of AIA 201 if the definition of "design concept" is tackled up front. Understanding this architectural approach to interpreting plans and specifications is paramount in a correct reading of the General Conditions.

The term "design concept" phrases a particular view of the overall purpose of the plans and specs that is somewhat different from a literal translation. This term is linked, in AIA 201, to two more – that plans should be read "as a whole", and that the plans are "complementary". These three terms appear over and over in Division 00 General Conditions. Grasping what these terms mean is a key part of understanding the entire General Conditions. The architect figures there are occasions when it is obvious what is meant in the documents, despite minor conflicts and confusing instructions.

The design concept is important to understand in the architect's approval of submittals. AIA 201 paints a line down the middle, with architects and contract documents on one side and contractors and submittals on the other, and attempts to limit the architect's liability to what the architect has drawn and specified, even after stepping across that line and approving a submittal. *Approval of a submittal is limited to the extent that the product "Conforms to design intent".* This refers to the totality of the architect's drawings and specifications, and the concept that *meaning can be given to the* design *beyond what is specifically drawn.* This is a clear concept for designers, but a fuzzy one for builders. The intent of the General Conditions is to make all of the contract documents work together to create one "whole" project. Similar language is repeated several times in AIA 201.

A correct use of this concept is when the contractor's submittal is unclear or not complete. The architect may have approved the submittal, but later information shows there to be a problem. The design concept language can be used as basis for the architect to change direction and disavow the approval. Perhaps some pertinent data was left out of the submittal, or some options were not included and later became known.

An incorrect use of the design concept is for it to be used to defend someone (owner or architect) for changing their mind.

Unforeseen conditions

Some construction legal problems are studied here, including the common situation of encountering a concealed condition. Unforeseen conditions are a huge concern for contractors because they often meet resistance when it comes to approving change orders resulting from them. After all, they are often an embarrassment to the A/E and certainly inconvenient for the owner. These parties are too often of the opinion that the contractor needs to deal with it without a change order.

Unforeseen conditions can be very problematic for the contractor, and this subject is explored in detail in this chapter. The "fire line" case near the end of this chapter studies an unforeseen condition.

Section 2 Owners and users of facilities

Owner's representative 2.1.1, 4.2.1

The owner has two representatives. One legally acts for the owner and is employed by the owner, and the owner will tell the contractor who this person is in writing. The other is an agent, the architect, who also "represents" the owner. The General Conditions refer to the architect as an owner's representative. This representation occurs during construction and until final payment is due. It can continue during the one-year warranty period if correction of the work is needed.

The owner's representative, if a government employee, will sometimes be the owner's representative on several projects at the same time. They are entrusted to sign change orders. An owner's rep will often know construction law, especially if they are a rep for the federal government, or a state or county, while the owner's rep for a small town that builds one project a year will not be well versed.

For the contractor, there is a big distinction between the owner's representative and a user of the facility. The principal at a school, who is a user of the project, has no authority concerning the contract and cannot effect change orders with the contractor. The fire chief, though held in great regard, is a user of the facility and not the owner, and does not direct the contractor. The architect does not make decisions for the owner either, except in the course of administrating the contract through inspections and pay applications.

The users of the project are the people utilizing the space after completion. Sometimes there is confusion from this quarter, and contractors deal with upset staff. If the user gets too involved in a project, the contractor needs to proceed with caution. Construction law, complementary plans, and interpretation of the documents as a whole are not understood by laypeople. Sometimes one or two lines on a set of plans can be (erroneously) used to stake a harsh opinion, while the contractor reads the plans as a whole and follows the construction law of AIA 201. *See the case of the unforeseen fire line and the devastating effect that one note, note 4 from the plans, would have had if the contractor had not also had the General Conditions to rely on.*

It isn't the contractor's job to educate the end user. The architect and owner representative are the go-to people, and if the contractor has a PR problem with the user, go through the former.

Owner can carry on other work 6.1.1

During the course of the project, the owner can place their own employees or other contractors on the site and have work done concurrently. It is the owner's site, and the owner has the right to perform other work.

Owner's authority to dismiss the contractor 2.3.1, 2.4.1, 14.2.1, 14.3.1, 14.4.1

If the contractor is *not* doing a good job, several severe General Conditions can apply. The one at the end of the line is "termination for cause", which means the contractor is being fired for good reasons, and the bonding company is called in to take over using another contractor, in which case it is the end of the first contractor, who will likely never get a bond again.

The reasons for outright termination include bad workmanship and improper materials, or not paying for them. If the contractor has constant battles with the local building department, or the fire inspector who has jurisdiction over the project, the owner has grounds for termination. The owner may terminate the contract if the contractor:

1. persistently or repeatedly refuses or fails to supply enough properly skilled workers or proper materials;
2. fails to make payment to subcontractors for materials or labor in accordance with the respective agreements between the contractor and the subcontractors;
3. persistently disregards laws, ordinances, or rules, regulations, or orders of a public authority having jurisdiction; or
4. otherwise is guilty of substantial breach of a provision of the contract documents.

However, before outright termination, the owner can step in and "carry out the work", which is the term used in 2.3.1 to describe what can happen if the contractor isn't completing the work or has deficiencies. This is quite different from the subject of carrying on "other work", meaning other construction at the same time, which is the subject of 6.1.1. above. Article 2.3.1 concerns work that is the contractor's obligation. This is a remedy the owner can accomplish short of kicking the contractor off the job. It involves a specific "default" or "negligence". After seven days of receiving notice from the owner, and then a second notice of three days, the owner can step in and correct the deficiency. The cost of this will be charged to the contractor through a deductive change order.

Or, exercising stern action, the owner can "stop" the work if the contractor has failed to correct. This is not a good situation for the contractor to be in, but it is another remedy the owner has short of kicking the contractor off the job. It involves work that is not in conformance with the contract documents. The owner can order that the contractor stop work in whole or in part until the "cause for such order has been eliminated".

There may be a time that, through no fault or cause by the contractor, and while the contractor is executing good work on time, that the owner needs to *suspend* or *terminate* the work. Some dire development may occur, perhaps a natural disaster hits the city and leaves the local government with a couple dozen buildings to rebuild. Contractors build for the owner, owners are the client, they have the money, and if they decide to stop the project, they can cancel the contract. It's in the General Conditions. The contractor, of course, does not have this option, even if the budget is found to have a few holes in it.

The terms for partial and full contract cancellation by the owner are "*suspension* by the owner for convenience" and "*termination* by the owner for convenience". If the owner has to shut the job down, the contractor will be paid to date.

Oh, well. The owner has the money and calls the shots, and the contractor is lower down the food chain.

Section 3 The architect's authority

Inspections by the architect 4.2.1, 4.2.2, 12.1.1

The architect is the representative of the owner concerning inspecting the work for conformance to the plans and specifications and keeping the owner informed about the progress and quality of the portion of the work completed. Contract language is guarded about inspections – it states that its purpose is to determine "in general" if the work will conform to the documents, and that inspections will not be "exhaustive". And once again, as stated several times in the General Conditions, the architect is not responsible for means and methods.

The architect is the interpreter of the contract documents concerning "performance" and can be asked by owner *or* contractor for judgment.

The contractor must ensure that the work is inspected before it is covered up. Shining a light on all aspects of the work is the best policy for the contractor. Otherwise, the architect can insist that the contractor revel the concealed work (in floors, walls, ceilings, and underground) for inspection at the contractor's expense and delay.

Architect and owner can object to contractor's subcontractor 5.2.1, 5.2.2, 5.2.3

There is a process for the owner and architect to reject a subcontractor that the contractor proposes to use. Sometimes, the bid form requires a list of subs, or at least some of them, and the subs are known from the bid form. On other occasions, the names of all the subs are later furnished to the owner and architect per the General Conditions. Sometimes, a major owner, or the architect, may have used the subcontractor or fabricator before, or have knowledge of their reputation, and may make an objection to the contractor, which must be "reasonable".

The contractor is also not required to contract with anyone to whom the contractor has made reasonable objection. But if the owner/architect does make a reasonable objection, the contractor shall propose another to whom the owner or architect has no reasonable objection.

And, if the new subcontract price is higher, the contractor is entitled to an increase in cost, if the original subcontractor was credible and judged "reasonably capable" to have been able to perform the work. If the proposed but rejected subcontractor was reasonably capable of performing the work, the contract sum and contract time shall be increased or decreased by the difference, if any, occasioned by such change, and an appropriate change order shall be issued before commencement of the substitute subcontractor's work.

Architect can reject the work 4.2.6, 13.5.1, 13.5.2, 13.5.3

Testing plays an important part in the rejection of work. First, there is the required testing called for in the contract documents, building departments, and other public agencies. The contractor gets the tests done and pays for them; however, sometimes the owner is charged with this expense so there is no conflict of interest.

Additional testing, beyond those required by the documents, is sometimes needed. An owner, a public agency such as the health or fire department, or the architect can instruct the contractor to conduct additional tests at the owner's expense.

If the testing or retesting reveals failures, then all costs revert to the contractor. The architect is an inspector and interpreter of the documents and has the authority to approve or reject work done at the site.

Section 4 The contractor's responsibility

Bidding due diligence 1.5.2

The estimator has to study the site, and site visits are mandatory. By executing the contract, the contractor represents that the site has been seen and that the contractor is generally familiar with local conditions under which the work is to be performed.

Contractor's authority 1.2.2, 3.2.1, 3.2.2

The contractor has the duty to study and compare the "complementary" drawings and to report any errors that are found. However, the contractor, in the review of the plans, is not an architect or engineer and therefore is not expected to be as knowledgeable about laws, statutes, codes, and regulations. Furthermore, contractor preplanning is for the purpose of facilitating the work, and is not for the purpose of discovering errors or omissions. The contractor is not required to ascertain that the contract documents are in accordance with applicable laws, statutes, ordinances, building codes, and rules and regulations. That is the architect's job.

The contractor controls how to distribute the work of the various divisions, regardless of the organization of the plans and specs. When it comes to splitting up the work into trade components, the contractor is in full control; if one specification division of work is divided between the contractor's own forces, a subcontractor, and to a sub to a sub, that is up to the contractor. If the mother-in-law's decorating company is hired to paint the lobby, and the remainder of the painting is divided between two cousins and a brother, so be it. Just make sure they have insurance, give them a contract with flow-down provisions, and make sure they do a good job on time.

Scheduling 3.10.1

The contractor, promptly after being awarded the contract, must prepare and submit a construction schedule. The schedule has to be revised at appropriate intervals as required by the conditions of the work and project. Often, the contractor works with two or more schedules at the same time. A beginning-to-end formal construction schedule is made by project managers or a specialized scheduler and turned into the owner at the beginning of the job. This schedule is the main one and is updated maybe a few times during a major job.

The "two- or three-week look ahead schedule" is what the superintendent and project manager work on inside the construction trailers onsite. They need more detail than is provided by the overall schedule, which is only updated every once in a while. Showing daily activities with specific descriptions, the look ahead schedule is accurate and depicts daily tasks for three weeks or so. These short-term schedules can be prepared on Excel spreadsheets with Monday–Friday across the top and activities filled in the landscape portion of the sheet. Or, as has been seen in the construction trailers of more than one large CM, sticky notes are precisely arranged across a wall on marker boards!

Reviewing a detailed schedule for the next two or three weeks or so seems to be standard operating procedure for some owners. A beginning schedule, a good overall CPM (critical path method) forecast of the work, was recently prepared by the author for the FDOT (Florida Department of Transportation). This schedule, once the job got underway and weekly meetings between the owner and contractor began, was hardly ever referred to. It was a major requirement for the contractor to complete, had to show the completion date within the contract time, but was simply updated once and kept on file. However, the superintendent's "three-week look ahead schedule" was huddled over in detail by all concerned at weekly meetings.

An "involved" owner needs to know what's going to happen tomorrow and the next day. Are the kids in the classroom next door going to be disturbed when the interior stud walls are shot to the floor on Tuesday?

Supervision 3.3.1, 3.9.1, 10.2.6

Here is where the contractor is in complete control, and if the contractor wants to build the building upside down and then flip it over, that falls within means and methods and have at it. How to build, how to install, and the sequence are totally up to the contractor. If the documents do happen to slide into the area of means and methods, such as some specific installation instructions, it is still the contractor's job to evaluate and be responsible for safety and techniques, sequencing, and procedures.

The contractor shall employ a competent superintendent and necessary assistants who shall be in attendance at the project site during performance of the work. The superintendent shall represent the contractor, and communications given to the superintendent shall be as binding as if given to the contractor. Important communications shall be confirmed in writing. Other communications shall be similarly confirmed on written request in each case.

The contractor is completely responsible for onsite safety. Superintendents often obtain "certificates" for attending OSHA and workers' compensation related classroom instruction. On large jobs, there may be a full-time safety officer, while often the person designated for the prevention of accidents is the superintendent.

Substantial and final completion 4.2.9, 8.2.3, 9.8.1, 9.8.2, 9.8.3, 9.8.4

A completion time of 180 days means until substantial completion. It is the contractor's job to have enough forces to accomplish the work and get the work done on time. Typically, thirty more days are given to reach final completion. A construction schedule usually shows substantial completion as the end of the job. To reach substantial completion, the owner has to be able to move in and use the space (or the designated portion of space – sometimes areas of the project are finished at different times). If the air conditioning doesn't work or the punchlist is simply too long, the architect can determine that the project is not ready for use.

When the substantial completion inspection is asked for, the contractor is supposed to make a list of uncompleted items and provide it to the architect preceding the inspection by the architect. Then the architect (and engineers) make their inspection, provide their own punchlist, and if they agree that the project is ready, issue the official documentation in writing that the project is substantially complete.

At the time of substantial completion, a punchlist should be comprised of just fixing some door hardware, installing the vinyl base in some rooms, and having the painter around for a few days doing touchup – minor things, even if there are a lot of them.

The substantial completion date is formalized in a "certificate of substantial completion" that the architect prepares. Warranties required by the contract documents shall commence on the date of substantial completion or designated portion thereof unless otherwise provided in the certificate of substantial completion.

The architect will conduct a final inspection, the date of final completion will be forwarded to the owner, written warranties and related documents assembled by the contractor will be forwarded to the owner, and the architect will issue a final certificate for payment upon compliance with the requirements of the contract documents.

Section 5 The contractor's submittals

Submittals defined 3.12.1, 3.12.2, 3.12.3, 3.12.4

The mounds of paperwork required to do government contracting stem a great deal from the requirement of submittals. The government, an experienced owner, knows what it wants and insists on a good submittal process. Submittal requirements on private work run the gamut; large owners like phone and insurance companies that have been around for a long time and have buildings across the country require thorough submittals, while those who do not build frequently may not even understand the process.

The best way for the student to understand a submittal is to consider that every piece of material in a building, every product and fastener (down to the last nail size), and every piece of manufactured mechanical and electrical equipment is going to go through a paperwork (or digital) approval process. All of the specific products used throughout the 50 divisions have to be determined. The choice of products must be narrowed down to a single manufacturer and a single model number. Additionally, some "means and methods" are reduced to shop drawings, and this too becomes a submittal. It is all a big job; the project manager has a lot to do. Consider it preplanning and worth the effort because such thoroughness helps eliminate mistakes.

Although everything given to the architect, such as a pay application or the warranties, can be said to be "submitted", the focus is narrowed here to make the current topic understandable. The "submittal process" defines the physical pieces and parts, colors, model numbers, and dimensions of the building. A partial list of required submittals is sometimes contained within the specifications. But it is up to the contractor to provide complete and full submittals.

The reason that the contractor goes through a submittal process is that the architectural plans are not precise enough and the specifications are not specific enough for constructability. Plan dimensions are rounded off and are not exact. The structural plans, with their dimensional specificity, are more exacting, but they, too, require extensive shop drawings. Shop drawings are required to provide finished sizes of everything from structural steel members to cabinets to rough openings for windows and storefronts. The choice of products must be narrowed down to a single manufacturer and a single model number.

There may be unresolved building code issues going back to the bid pointed out by the hardware supplier. These kind of details must be fixed in the submittal process, sometimes resulting in change orders. In addition, there is a lot of "coordination between the trades" that must be worked out. The electrician may need to consult with several other trades concerning door alarms and mechanical equipment. The push button device that automatically opens a door for people with disabilities involves two trades, the door supplier and the electrician. The size of roof hatches provided by the roofing contractor must be transmitted to the steel subcontractor so that bar joists and angles provide support on all four sides of the hatch.

Through the submittal process, a contractor "builds the job on paper" by documenting, through various kinds of submittals, the materials and products and equipment in the contract documents. The plans and specifications take the project to a point, but there is a "handoff" here, where the contractor takes over and provides dimensions, specific products, means and methods, and other details. The submittal process should be viewed as good preplanning, and it must be completely thorough.

Timely, correct, and full submittals precede professional field production

The experienced estimator will often recognize the source of some of the language presented in the specifications. Language from major manufacturers becomes instantly recognizable with experience. This helps speed along the review of specifications and submittal requirements. The estimator, charged with complying with a book-length set of specifications, does not have the time to read them like a novel. It is important to recognize what it *different*; what is familiar can be dealt with quickly. It is what is unusual that must be noted and checked out.

The General Conditions do not try to define the various items that can be a submittal. The term "shop drawings" is used at times to define not only drawings, but also schedules and other "data", apparently referring to the entire collection of submittals. Actually, shop drawings are a small part of the submittal process, and it is confusing to combine the definition of shop drawings with other submittal types. To a carpet installer, carpet samples are a submittal. To a bricklayer, a submittal is a mockup of a brick wall. For most items, the submittal is a product data sheet.

To induce a supplier or subcontractor to prepare submittals, which is work and an expense, the contractor will issue a "letter of intent", purchase order, or subcontract. This is a commitment to the vendor that they "have the job". The language of the purchase will require that submittal approval is required before the goods are "released" for delivery and incorporation in the project.

It would seem to most contractors that after a submittal is approved by both contractor and architect, the submittal would "replace" the document that came before it, i.e. that portion of the plans and specifications that the submittal deals with. Since the plans and specifications are a part of the contract between owner and contractor, an approved submittal (with dimensions, clarity, specificity, and detail) would seem to become part of the contract, but this is not so according to the General Conditions. *Shop drawings, product data, samples, and similar submittals are not contract documents.* The basis for not recognizing submittals as a contract document lies with understanding "design concept and intent".

Also within the limiting nature of the architect's approval of submittals is language about the contractor being responsible for dimensions and quantities. The architect will not check dimensions and is not responsible for this kind of coordination. General condition language also is clear that the contractor is responsible for means and methods and safety.

The industry standard terms that describe many recurring submittals include (this list is not from AIA 201):

Product data sheets
Shop drawings
Schedules
Samples and mockups
Test reports
Concrete mix designs
Certificates

Product data sheets

These are by far the most voluminous submittal type; they are the manufacturer's description of a product or piece of equipment, providing its uses – for example, "This product is a vapor barrier. It is used in wall construction to help prevent moisture" – and going on to provide product limitations, size(s), weight, strength, gauges, colors, and installation instructions.

A manufactured product means "standard units mass produced." The manufacturer will provide tests that the product meets, such as those by the ASTM (American Society of Testing Materials), and codes such as the fire ratings by UL (Underwriters Laboratories). Product data sheets are handled by the thousands by subcontractors and project managers. Look on the back of a gallon of paint for its chemical and other characteristics to get an idea of the information on a product data sheet.

The following is a partial list of some common items typically submitted using product data sheets *written by manufacturers:*

Concrete accessories:

Metal keyways, expansion joints, curing compounds, vapor barriers, form ties.

Masonry products:

Brick ties, mortar mix, horizontal joint reinforcing, keyways.

Carpentry:

Fire treated wood, chemical treatments, fasteners, metal straps, glu-lam beams.

Thermal and moisture:

Shingles, felt, insulation, joint sealers, roofing products, curbs for A/C equipment.

Doors, windows, glass:

Wood and hollow metal doors, hardware, glass, skylights.

Metal studs and gypsum wallboard:

Gauges of metal studs, fire rated and moisture-resistant board, joint compound, tape.

Paint:

Paints and stains for various surfaces, caulking and sealants.

Acoustical ceilings:

Grid, tile, wire, access doors (hatches).

Bath accessories:

Grab bars, tissue holders, soap dishes, waste receptacles, mirrors.

Plumbing:

Piping (various kinds), pipe insulation, exhaust fans, fixtures.

Air conditioning:

Equipment, thermostats, duct insulation, exhaust fans.

Electrical:

Conduit, wire, devices, fixtures.

The following product data sheet is formatted for an imaginary "caulking compound", or "sealant".

See the online resources for diagram 11 25.1

Shop drawings

Shop drawings are made by various trades to amplify, detail, and dimension parts of the work that may only be vaguely defined on the plans. The architectural and structural plans may be said to be "diagrammatic", while the shop drawings are specific. The dimensioning on shop drawings is extremely important, has to be right, and is one of the central purposes of submittals. Small shop drawings often accompany product data sheets, but shop drawings are commonly plan sheet size and done by competent drafters for a specific job.

Many products require several submittal types to define them. Consider lockers, which require shop drawings with an elevation view, and wall dimensioning. In addition to a floor plan and elevation(s), which are shop drawings, a locker submittal includes product data sheets to explain the kind of material the lockers are made of, for example 18 gauge steel, their weight, and other characteristics. A third type of submittal needed to complete the locker submittal is a color chart.

Examples of items shown on shop drawings (all of these would usually be accompanied by other submittal types):

A Rebar placement within concrete.
B Precast wall panels showing sizes, joints, cutouts.
C Structural steel including connections.
D Steel joists and decking.
E Miscellaneous steel, handrails, shelf angles, ladders, anchorage.
F Cabinets, drawers, glides.
G Wood and steel trusses. Glu-lam beams and arches.
H Tapered roof insulation showing thicknesses and slopes to drains.
I Windows and storefronts, component detail and finish dimensions, fastening and caulking.
J Toilet partitions.
K Lockers.
L HVAC duct layouts, sizing of ducts, location of supplies and returns.
M Fire protection, sprinkler layout, pipe sizing.

The GC is heavily involved in the shop drawing process because of the coordination required between the trades. Products provided by one trade are openings for another, and information is traded back and forth through the leadership of the GC. Rooms with busy acoustical ceilings include light fixtures supplied by the electrical trade, air conditioning supplies and returns, and cubicle curtain track, all of these being different trades, and all of it requiring coordination. Requests for information from one trade go through the GC with questions for another. Manufacturers supplying shop drawings, for example toilet partitions, request site dimensions from the GC. The project manager is a busy person here and requests information needed from the field to the superintendent.

There are many versions of shop drawings, and some are used by contractors only. This is so in the steel trade, where there can be erection shop drawings, used only by the contractors.

The word "fabrication" is closely linked to shop drawings. As opposed to manufacturing, which involves mass production, fabrication is the process of taking materials or manufactured products and, in a shop prior to job delivery, assembling them to meet "individual design requirements". Besides structural steel, which is cut to length and bent in a shop before job delivery according to very precise shop drawings, other fabricated items include rebar, wood cabinets, and trusses. Fabricated items require shop drawings, and when put together are an assembly.

In the case of steel reinforcement (rebar), it is usually cut and bent in a "plant" that assembles, according to their own shop drawings. They deliver to the site, where the reinforcement is installed by a subcontractor employing "rodbusters" with spools of tie wire on their belts. Structural steel is different; the manufacturer of standard shapes delivers steel to a fabricator, who is usually the steel subcontractor that has the onsite job working for the prime. The fabricator does the shop drawings, partially assembles the steel, and then delivers the steel to the site where the pieces are put together.

There are only a few examples of manufacturers that retain control over the shop drawing process compared to the many divisions of work that utilize suppliers or distributors to prepare shop drawings. Manufacturers of lockers and toilet partitions do their own shop drawings. These industries do not typically use the supplier/fabricator model of shop drawings, even when their products are being sold through a distributor to the contractor. Wall-to-wall and floor-to-ceiling measurements are provided by the contractor by writing directly on the manufacturers' shop drawings. The drawings are then relayed through the distributor and sent back to the manufacturer of lockers and toilet partitions.

Unknown dimensions should not slow down the submittal process. The architect is not responsible for dimensions anyway, and they will approve shop drawings with unknown dimensions yet to be determined. Because it might take six or eight weeks for toilet partitions to be manufactured, the contractor may need to determine tile-to-tile dimensions before the tile (and maybe the gypsum board, too!) is installed. It makes sense to place the dimensions on the shop drawing at the last minute, after architect approval, when the contractor is ready to return the shop drawing to the distributor.

The GC might submit a shop drawing for a "heavy gauge wire fence" (the type separating chemicals from students in a chemistry lab, including a wire fence door), with question marks on it from the manufacturer of the wire fence for wall-to-wall and floor-to-ceiling dimensions. If the room isn't fully built yet, that shouldn't keep the shop drawing (and accompanying product data sheets) from being submitted to the architect. The GC can go ahead and get the wire fence drawing approved, the door location noted, the gauge of wire approved, and determine fastening requirements while the room is constructed. The architect will approve/disprove the submittal with the dimension question marks remaining on the drawing. As soon as the finish dimensions are determined, the GC places them on the approved shop drawing and returns them to the manufacturer of the fence. Time saved.

Now that the contract documents, issued from the architect and owner, and the submittals, issued by the contractor, have been defined, and noting that the submittals are specific, it would seem that they would become a contract document after approval by the architect. This is not the case.

Schedules

Schedules are those submittals that resemble Excel spreadsheets that list items like room or door numbers on one side and related information across the page. Schedules are usually prepared by suppliers or manufacturers. Often accompanied by samples and color charts, they can contain a mass of information. Especially in the case of doors and hardware, they require a lot of checking and verification by the GC. Door and hardware submittals are sometimes called "catalog cuts" because the schedules often contain small drawings of locks, closers, and other hardware. Perhaps long ago they came from catalogs, because this word is still used.

Refer to the chapter on doors for an example of a door schedule.

Products on schedules are often long lead items, and it is important to deal with them early. Loose ends can easily ensue, especially on renovation work when existing site conditions have to be verified. If there are a hundred doors on a project and there are all kinds of wall types and thicknesses, there can be a lot of information to assemble, taking much time and effort. In addition, door and window sizes are sometimes changed by the architect at the submittal stage, and getting these submittals approved can be a huge task. After approval, it can take six or eight weeks or longer to get doors and hardware to the job, hence their long lead status.

Samples and mockups

These are actual materials or products furnished as a submittal. Brick, paneling, shingles, and acoustical ceiling panels and grid are submitted as samples. Complete site mockups of short brick walls are sometimes built to serve as a representative example of the mortar jointing and overall quality where workmanship is part of the submittal. Mockups are usually built onsite and approved by the architect, and the actual work must match the sample.

Small sections of miscellaneous steel are often made as samples by fabricators, such as handrails or ladders, as well as windows, skylights, and fencing supplied by manufacturers. Samples of carpet and vinyl flooring, ceramic tile flooring and walls, and floor transition strips and thresholds can all be supplied for approval. Samples for plastic laminate, a covering for countertops, are on small "chips".

Paint colors are provided on "color decks". It is not the contractor's job to make *any* decisions concerning finishes and colors. On a private job, if asked to "match the existing color" or "paint it white", the project manager should insist that the owner or architect make the decision. The only color the contractor should match is one on a manufacturer's color deck. On most projects, the architect will insist on making all of the color selections at one time. In this case, paint, flooring, and other finishes, all from various sources, must be provided early in order to prevent delay.

Test reports

For mass-produced goods, test reports of manufactured products can be found on product data sheets. These tests are done by major testing laboratories that follow elaborate protocols. Those doors and windows for that project near the coast? Since the wind load for the coastal zone might be 130 mph, testing must prove that this pressure can be withstood. The actual test makes for interesting reading, "seven-foot-long 2 × 4′s were hurled against the exterior side of the window at a force of 150 mph", etc.

Two large organizations that provide construction testing are Underwriters Laboratories (UL) and the American Society of Testing Materials (ASTM).

Test reports are also needed on some local materials, and these test reports are different. Local testing is done on basic materials like sand, earth, concrete, and concrete blocks. Since sand is obtained from the same source for long periods of time, the testing is done periodically, as is the crushing strength of a concrete block from a block plant. Local labs provide letters certifying characteristics and strengths of materials. The test report for sand states that it passed through a sieve of a certain size grid and provides its weight and moisture content. These tests may be on file at the plant when the project manager asks for the testing verification required in the specifications. Sand does not have to be constantly tested for every single job.

Other tests are made in the field at the time of the work or after. Concrete cylinders are made the day of the pour, taken to a testing facility, and tested at 7, 14, and 28 days. The contractor receives the results and transmits them to the architect.

The specifications often stipulate that the owner pay for and provide testing, in order to avoid a conflict of interest.

Concrete mix designs

A concrete design mix describes the quantity of sand, cement, and water used to create concrete for a specific pour(s). The engineer might define it in the specifications or require the concrete plant to submit a mix design for approval that meets a certain strength.

Certificates

These are a unique type of submittal that may involve life and safety issues such as medical gas testing. In hospital rooms containing oxygen, nitrous oxide, nitrogen, carbon dioxide, etc., an independent firm is hired to verify that these lines are properly labeled and perform to standard. They issue a certificate stating that the lines work properly before they are used on patients. After all, a patient needing laughing gas, lying in the bed in a hospital, shouldn't accidently get hooked up to a vacuum hose.

Sometimes a certificate is a letter signed by the GC or manufacturer, or both, stating that the submitted and approved products were in fact used in the project. This may seem redundant, and it is, but it is to help prevent fraudulent substitutions. The government realizes many different people from various companies handle submittals and installation and wants a single person or company to be responsible and to hold at fault in case of an accident.

Contractor's approval of submittals 3.12.5, 3.12.6, 3.12.7

The contractor represents that the submitted product is going to fit, that it is compatible with the surrounding construction and "requirements" of the contract documents. In other words, the contractor takes full responsibility for the product fitting in with the design concept, and if anything goes wrong, it's the contractor's fault!

The contractor should not incorporate any materials or products into the work without having an approved submittal. The project manager should begin the submittal process immediately upon receiving a contract, and work on submittals until they are completed, often midway through the job!

The contractor must place and sign an approval stamp on the submittal. The architect does the same, finding some room on an often-crowded sheet to place a stamp and make a few notes. The submittal can get busy, and sometimes they are redone just to get them more organized and less sloppy.

The following format is an example of a separate sheet being used for the contractors' approval or disapproval of a submittal. It can be used to forward to the architect or send back to a supplier/subcontractor. Note that the submittal is given a submittal number, and current owners are increasingly providing (to the contractor) their own numbering requirements and protocol. Another important number is the specification division that the product is a part of, the specification number.

If a rubber stamp is used, the submittal is often accompanied by a transmittal sheet stating what is being provided; the following cover page serves as both an approval stamp and a transmittal.

See the online resources for diagram 11 25.2

Architect's approval of submittals 3.12.8

The architect's approval of a submittal contains a big qualification, which is "approval is only to the extent that this product/ material/equipment meets the design concept". This approval is tentative. It is not the architect's job to flesh out all of the information and see to it that the contractor's paperwork is full of the appropriate data. There are numerous details that can be contained in a submittal – parts and pieces and sizes and colors. What if the contractor's submittal is found, after approval, to be lacking in some manner?

Even when a submittal is approved, the contractor is not "relieved of responsibility for deviations from the contract documents". This can lead to what the unfortunate contractor considers a circular statement – "The approval of submittals is not an excuse for failure to comply with the plans and specs!"

The following are two typical examples of stamps that architects use:

See the online resources for diagram 11 25.3

Note the words used in the following stamp that are the equivalent of an approval. The term is "No exceptions taken". (See the approval code "NET".) This lacks an affirmative response. Reminds one of the recent usage of the phrase, "not a problem!"

See the online resources for diagram 11 25.4

The design concept was introduced earlier in this chapter, which defines a perception of interpreting the drawings by the architectural profession. One more viewpoint is now made here concerning the interpretation of the drawings and specifications and of the approval of submittals, and that is of the law profession.

Attorneys, of course, look at all of the documents. They concentrate on what is specific, not on general statements. Great importance is given to what is large and what is small, what is clear and what is not. These are drawings, notes, and specifications. The lawyers look at what is shown in greater detail. Clarity and intelligibility on the plans and within specifications are paramount and govern over concepts. Weight is placed on length and specificity of notes, and specification language versus shorter passages and vague language.

Substitutions and/or equal

The term "substitution" is related to the term "or equal". The specifications may name three or four possible manufacturers for a piece of equipment. This is to allow for competitive bidding. Or, one manufacturer's name might be accompanied by the words "or equal". A term used to define this occurrence is an "open specification". Contractors (through their suppliers, who may have worked on the project at the bid stage) often make submittals concerning products not in the specifications, but the burden is on the submittal to provide enough information for the architect to determine that the product is equal to the one specified. As in any submittal, it should be full of supporting data and contain warranty information, manufacturer's test results, installation instructions, etc.

Sometimes there is a "sole-sourced spec" requirement, meaning that only one product and manufacturer can be used. The owner may want to keep the same manufacturer of equipment because their vendor or employees know how to work on it. Or, an architect may have made a thorough search for a product that best meets a special need of the client. The specifications may state, "No substitutions, use product XYZ." This is known as a "closed spec".

Sometimes there is a requirement that any substitutions must be made at the time of the bid. This is especially hard to do on an ordinary job, but if a project has some custom needs, the product specifications may be written very narrowly. Such narrow language is also known as a "restrictive specification".

The GC must be especially prudent when submitting "or equal" products or equipment. Suppliers and manufacturers can sometimes be biased, trying to make a sale. Their claim for "equal" status should always be followed up with a full and complete submittal. And, to consider a related issue, manufacturers are always replacing one product for another, and of course the new one may be a little bit different from the old one. The specifications often contain outdated information, and the contractor must make sure to note when a "substitution" is being made, and why.

Submittal log

The submittal process can be a huge paperwork exercise for the project manager. A job of just a few million dollars can require hundreds of product data sheets, a dozen or more shop drawings, as well as tests and samples. Submittals are processed at different times, and some of them start over and go back through again. Without a good tracking checklist, it can be an organizational mess. Tracking of submittals should be broken down by division and by product. After submittal approval, the contractor can transmit instructions to the fabricator, vendor, or subcontractor to proceed with the production of fabricated materials (shown on shop drawings), the delivery of standard materials and products (described on product data sheets), or the ordering of manufactured equipment (described on product data sheets).

Submittals, as they wind their way up and down the food chain, are often commented on, approved as noted, approved, or rejected. Resubmittals are common, even if just to "clean up" the submittal by ridding it of cluttered notes. It is a process; they move along from place to place. With dozens of submittals, all at various stages of approval, the contractor needs a good tracking system. Indeed, the General Conditions make it a requirement.

The "tracking" of each submittal, of where it is along the approval process, is the contractor's job per the General Conditions. Every submittal goes through a repetitive process that includes requesting data from subs/manufacturers/fabricators, approving it and transmitting to the architect, receiving it back sometimes approved, but often with comments to amend and resubmit, or having it rejected. These steps, in order to keep the entire collection of submittals organized, are "tracked" on a submittal log where the dates for each step are documented. The architect does not keep track of any of this. Instead, the contractor, charged with all things involving time, is required by the General Conditions to keep such a tracking list, called a submittal log, and to show it to the architect for his/her review as requested!

The contractor keeps a spreadsheet to track the progress of each submittal. Of course, the contractor is charged with making sure the whole submittal process proceeds in tandem with the timeline schedule.

An example of a submittal log is as follows:

See the online resources for diagram 11 25.5

Section 6 Change orders

Changes in the work, general 7.1.1, 7.1.2, 7.2.1

Although the owner and contractor already have a contract for a set amount of work, they agree, through the General Conditions of the contract, that it can be changed. These changes are called modifications. A modification (commonly called a change order) is composed (potentially) of three things – a description, a cost, and time. A description is the only one of the three that is always part of a change order, which may not have a cost or take any extra time.

General condition language differentiates three kinds of modifications, although they are all commonly called change orders – *minor changes* issued by the architect that don't need contractor approval, a *change order* agreed to and signed by the parties, and a *construction change directive* where the parties may be in disagreement, but the architect issues written instructions to the contractor anyway.

If the architect walks the job and tells the construction superintendent to move a cabinet from one wall to another, this is a *minor change* that can be effected in the field but should be followed up in writing by the architect. Or, it may not be documented until the contractor completes as-built drawings showing the location change. A *change order* is based upon agreement among the owner, contractor, and architect. A *construction change directive* requires agreement by the owner and architect and may or may not be agreed to by the contractor.

Minor changes in the work 7.4.1

The architect has the authority to make minor changes to the work that do not involve cost or time. For example, "put the shelves over here instead of where they are shown on the plans". Given there are no extra shelves and the instruction is made in advance of the work, the contractor obliges, and the change is made through a minor change order.

Move a mirror from one wall to another, eliminate a ceiling return air and place it at the top of a wall . . . easy to do changes that don't involve (at least not much) cost can be done by an architect on a site visit and followed up with something in writing later. Even changes that involve a few dollars are done by contractors willing to trade things out here and there.

Construction change directives 7.3.4, 7.3.5

The usual purpose of a construction change directive is for when the architect disagrees with the contractor about a change order amount. At an impasse, the architect can issue a construction change directive, which directs the contractor to move ahead with the construction of the change; the arguing can continue, but the work proceeds! The change directive can be used in a contentious situation, which is illustrated in the following fire line case. A directive can be used to instruct the contractor to proceed with the work when the facts are disputed. This is often better for everyone, as no one wants the entire project to languish because of a dispute that can be carried on during construction. Of course, the poor contractor is going to pay for any extra work as the argument continues, and there may be times when the contractor does not want to proceed. However, the General Conditions are clear – the contractor must proceed with the work when given a construction change directive.

This hybrid of a change order can also be used simply to make a minor change that needs to be made immediately (so as not to impede the work), with the contractor left to come up with a price later.

Signing a construction change directive must be carefully considered by the contractor. The architect or owner may insist on the contractor's signature, even if the price is not agreed to. It is clear that the contractor has to proceed, but signing can indicate approval of a price. Sometimes the proper action for the contractor to take is to sign under protest, "signed under protest, John Doe".

Change orders, summary

Dealing with change orders is often a burden for the contractor. Most of them are small and take a disproportionate amount of time compared to what the contractor is allowed to charge. Project managers can spend hours preparing short detailed estimates that accompany change orders. Documentation has to include work-hours, labor costs, material costs, and there's often a lot of working back and forth with subcontractors' and suppliers' pricing and their breakdowns. Then, down at the bottom where the markup is placed, the owner has limited the poor contractor to a punitive percentage that doesn't come close to allowing breakeven. Then, after a change is approved, the project manager must prepare a cost-coded breakdown and send it to bookkeeping, and the contractor's original budget is revised. More work. Then the superintendent has to deal with the change, which is a hassle because often there are no plans to accompany the change, and it is an interruption to the original planned work.

This contractor is still waiting, after quite some years, to understand the common perception that contractors make a lot of money on change orders, or to find a contractor that bid the job at cost, figuring change orders would "make it up". To make any money on change orders they have to be large, and for every one of these there are a thousand small ones. The author would gladly trade away all of the change orders that he's ever had, which is a substantial number.

Section 7 Claims

Claims defined 4.3.1

A change order, agreed to by the parties, is not a claim. But when the architect makes an interpretation in writing that the contractor considers extra work, and the architect disagrees, they are at odds. The contractor should make a claim for additional cost and time, and if it takes very many days to do the pricing, the contractor should send a notice immediately stating that a claim is *going* to be made.

A claim is a demand by the contractor for payment or time (extension), or both. Claims are made by giving "notice". Notice is a big word in construction law; contractors should not be timid in its use. A claim must be made within twenty-one days of an event causing the claim, and it is important for the contractor to keep working and "proceed diligently".

Claims time limit 4.3.2

When a contractor makes a claim, it must be made within twenty-one days of the event. It is initially referred to the architect for a decision (unless there are hazardous materials), and must be done before a contractor can take the next step. Additional information may be requested, and *the contractor has ten days to respond.* The architect then decides within ten days and states the reasons in writing.

Unforeseen conditions 4.3.4

There are two areas in which many unforeseen conditions are encountered, one in new construction and one in renovations. In new projects with foundations in the ground, a concealed condition may be clay or other unsuitable material not in an owner's geotechnical report, or pipes running all over the place (and in the way of new construction) like spaghetti. In remodeling work, concealed walls and attics can be the location for many an obstruction.

These occurrences are called unforeseen conditions. Usually they result in additional cost to the owner, since the condition was not foreseen by the architects and engineers and therefore not on the plans. Contractors find these problems early in the job. They result in a change order, but one initiated by the contractor and as such are a claim; this distinction places them here in the section about claims instead of change orders, where the architect and owner are the ones usually initiating changes.

Claims for unforeseen conditions are often contentious. Although an issue may seem obvious to a contractor, the owner will obviously wish the matter hadn't come up and figure that the architect or contractor had considered it previously. The architect, when confronted with claims of unforeseen conditions, will exercise great lengths to stretch the meaning of design concept and try to find reasons for the contractor to have the work included in the contract (base bid). In addition, the architect will research the General Conditions (AIA 201), looking for some "catch-all" language that might ensnare the contractor.

The contractor must give written notice within twenty-one days of observing the condition or lose the chance to claim it as a change order. The architect will promptly investigate such conditions and, if they differ materially and cause an increase or decrease in the contractor's cost of, or time required for, performance of any part of the work, will recommend an equitable adjustment in the contract sum or contract time, or both. If the architect disagrees with the contractor and that no change in the terms of the contract is justified, the architect will notify the owner and contractor in writing, stating the reasons.

Giving notice for a claim 3.2.3, 4.3.5, 4.3.7

If the contractor believes cost or time needs to be accounted for, *written notice should be given*. This action should be taken in written communication even if the cost or time is not known yet. The contractor should shine a light on the situation and say there is going to be additional cost, or time, by giving notice.

Notice of additional costs should be given prior to executing the work. An exception to the practice of giving notice is for emergencies. If there is additional time involved, the duration should be stated.

If a delay is ongoing, the contractor does not need to send out a letter every week documenting the occurrence. A delay only has to be documented once.

Architect's role in reviewing claims 4.4.1, 4.4.2, 4.4.4, 4.4.5

When a contractor makes a claim for a change order, the architect makes an initial decision. If there is no disagreement to this decision, the architect's decision is final. If there is disagreement, demand must be made for arbitration within 30 days if the contractor wants to appeal the decision.

If the contractor does request mediation, the architect must respond within ten days and do one of the following:

1 request additional supporting data from the claimant or a response with supporting data from the other party;
2 reject the claim in whole or in part;
3 approve the claim;
4 suggest a compromise; or
5 advise the parties that the architect is unable to resolve the claim if the architect lacks sufficient information to evaluate the merits of the claim or if the architect concludes that, in the architect's sole discretion, it would be inappropriate for the architect to resolve the claim.

If the architect needs more information, the contractor has ten days to answer. The architect will then either reject or approve the claim in whole or in part, making a final stand. The approval or rejection of a claim by the architect shall be final and binding on the parties *but subject to mediation and arbitration.*

Mediation 4.5.1, 4.5.2, 4.5.3

Mediation is the first step for the contractor to take when the architect is disagreed with and the contractor does not want to accept the decision. Or, if the architect ignores a claim made by the contractor, the matter can be headed for mediation after a thirty-day wait. Requests for mediation are "filed" with the *American Arbitration Association*, and a demand for arbitration can be made at the same time. Costs for mediation are shared by the parties, and the results, if agreed to, are enforceable.

Arbitration 4.4.6, 4.6.1, 4.6.2, 4.6.6

If mediation does not solve a dispute, the matter "goes to arbitration", unless the parties agree otherwise.

Contractors are grateful that the General Conditions have an arbitration clause! This is a huge consideration, because of the time and expense of trying a case. Mediation and arbitration are lesser evils.

When an architect makes a decision that the contractor disputes, the matter must be taken up within thirty days or the decision becomes final.

Claims relating to aesthetics are not subject to arbitration; in this area, the architect's decision is final. The award rendered by the arbitrator or arbitrators shall be final, and judgment may be entered upon it in accordance with applicable law in any court having jurisdiction thereof.

Section 8 Delays

Delays and causes for delay 8.3.1

A contractor must be careful when approaching the subject of delays. Owners and architects may only see "partial" work stoppage when the contractor knows valuable time is slipping by and the critical path is being affected. So, it is important for the contractor to work "all the way up to" the delayed condition, do all the associated work around it, and clearly be at a standstill. Delays can be confusing to understand when a building is being constructed and activity is occurring all around.

The General Conditions list the following potential causes for delay:

1 Act or neglect by the owner or architect.
2 Change orders.
3 Labor disputes.
4 Fire.
5 Unusual delay in deliveries.
6 Casualties or causes beyond the contractor's control.
7 Delays authorized by owner pending mediation or arbitration.
8 Hazardous materials.
9 Emergencies.
10 Other delays authorized by the architect.

For the above reasons, the contract time shall be extended by change order for such reasonable time as the architect may determine.

Hazardous materials 10.3.1

If the contractor encounters a hazardous material, written *notice* should be given to the owner and architect. Work should stop in the affected area. This is an "excusable delay", and additional time will be added to the contract.

Emergencies 10.6.1

In case of emergencies affecting safety of persons or property, the contractor's actions during the emergency can be grounds for additional time and money. An emergency is an "excusable delay".

Delays as cause for contractor to terminate the work 14.1.1, 14.1.2, 14.1.3, 14.1.4

The contractor can request verification that the owner has the financial means to complete the job per section. When requested, this evidence must be furnished as a condition precedent to commencement of the work. Lack of this information can cause a delay and is grounds for the contractor to terminate the contract.

Whether it is the fault of the architect or the owner, the contractor can terminate the work if not paid.

The contractor may terminate the contract if the work is stopped for a period of thirty consecutive days because of a court order, or a public authority requires all work to be stopped, or a declaration of national emergency. If the owner causes a delay of sixty days, by failing to carry out owner obligations, the contractor can terminate the contract. When there are repeated delays by the owner, the General Conditions provide a formula for computing the overall delay and give the contractor a basis for terminating the contract.

The contractor gives a *seven-day* notice in advance of terminating the contract. Note that payment is due from the owner, including profit and overhead for work in place.

Section 9 Payment

Schedule of values 9.2.1

The schedule of values is used for the purpose of making applications for monthly payments. The contractor takes their entire estimate and divides it into a page or so of line items, with the painting being worth so much and the roofing a certain amount. Both the painting and roofing may have several components and consist of multiple line items. This schedule of descriptions and amounts is not an exact replication of the contractors' costs for the listed line items because the contractor's job overhead and profit is "spread" over all of the various line items. The job overhead, consisting of supervision and scaffolding and such, may not be spread evenly across all of the line items.

Furthermore, by combining costs into concise breakdowns, some arbitrary decisions are made. The cost for steel handrails may be thrown into "concrete sidewalks", a schedule of values item. The cost of handrails doesn't even show up. On a renovation project, a carpentry knee wall, cabinets, and some gypsum board work may all be combined into a category on the schedule of values named "kitchen cabinets".

For this reason, if the owner deletes cabinets from the contract, and the schedule of values has them pegged at $50,000, the owner cannot use the scheduled value as a basis for a change order credit.

There is another reason for the schedule of value (SOV) items to be skewed. That is the purposeful "unbalancing" of overhead and profit across the line items. Contractors, until receiving final payment, are always behind in being paid. First, ten percent "retainage" is withheld from a "draw" (payment). Second, payment isn't usually made for thirty days or so. The ongoing costs of the job are carried by the contractors. For these reasons, a job can be "in the hole", or have a negative cash flow until final payment is made. To partially counter this, contractors engage in "front-end loading". This is apportioning more profit to activities at the beginning of the job than the ending activities. See the following example of front-end loading:

See the online resources for diagram 11 29.1

Before the first application for payment, the contractor submits to the architect a schedule of values allocated to various portions of the work, supported by such data to substantiate its accuracy as the architect may require. This schedule, unless objected to by the architect, shall be used as a basis for reviewing the contractor's applications for payment.

Draw requests 4.2.5, 9.4.1, 9.4.2, 9.5.1, 9.7.1, 9.10.2

Many architects make it a practice, which is not a requirement of the General Conditions, to review with the contractor the monthly pay application a few days before draw requests are due. They may walk the job with the contractor or superintendent, and let the contractor know if any percentages of completion are disagreed with. It is usually satisfactory to the contractor if an item or two is reduced a few percent or the value of stored material is reduced. The main thing for the contractor is to get a payment application approved, whether it is for $500,000 or $475,000. Turn in an amount that will be processed; maintain the cash flow!

The architect has seven days to process a draw request. If approved, a "certificate for payment" is issued to the owner. If not approved, the architect will notify the contractor and owner in writing "reasons for withholding certification in whole or in part".

There are several qualifications to the architect's approval of payment, including being subject to, guess what, conformance with the contract documents, later inspections and tests, and later correction of minor deviations. The architect does not warrant that an "exhaustive" inspection has been made, or has reviewed means and methods, or studied subcontractor draws and prior payments to them.

The reasons that a draw request can be withheld include:

1. A loss to the owner for which the contractor is responsible.
2. Defective work.
3. Nonpayment to subcontractors and suppliers.
4. Reason to believe that prior payments are greater than actual percentage of completion, i.e. the contractor has been overpaid.
5. Reason to believe that the contractor is not going to be completed on time.
6. Persistent failure to carry out the work in accordance with the contract documents.

If the contractor and architect cannot agree on a draw amount, the architect will issue a certificate for payment for an amount the architect approves.

The contractor has the right to stop work if the architect or owner does not act in processing a certificate for payment. If either the architect doesn't process a certificate of payment within seven days or the owner is over seven days past the contract stipulated date of payment, the contractor can stop work after seven days of written notice.

In addition to completing the punchlist made at substantial completion, as evidenced by the architect, the contractor, before receiving final payment, will provide warranties and guarantees and other close out documents. The General Conditions list these items as precedent for final payment:

1. An affidavit that payrolls, bills for materials and equipment, and other indebtedness connected with the work have been paid or otherwise satisfied.
2. A certificate evidencing that insurance required by the contract documents to remain in force after final payment is currently in effect and will not be canceled or allowed to expire until at least 30 days' prior written notice has been given to the owner.

3 A written statement that the contractor knows of no substantial reason that the insurance will not be renewable to cover the period required by the contract documents.
4 Consent of surety.

An example draw request is shown below. It is for the second pay request on a $1.1 million project. Two change orders have occurred for a total of $3,000; see rows 12 and 13 of page 2 of the draw request, where the schedule of values are shown below.

A draw (certificate for payment) is made by the contractor using two forms. Sheet 2 uses the schedule of values approved at the beginning of the job; see columns B and C. The sum of payments approved in prior months are shown in column D; the current monthly application is in column E. Column H is the percent complete.

See the online resources for diagram 11 29.2

The total completed to date, column G above, is sent to the first page of the draw request; see row 4 below. The other amount sent to Sheet 1 is the total of previous applications, column D, which appears on row 7 below.

See the online resources for diagram 11 29.3

Stored material 9.3.2

Being paid for uninstalled materials delivered and onsite is customary. There is a column on standard pay requests for stored materials. If the scheduled value for gypsum board is $100,000, and the material is delivered but not hung on the walls and ceilings, the material can be submitted for payment on a draw. The contractor might apply for $30,000 of stored material, even though there is no line item for material only. The contractor may be asked to provide invoices to verify the cost of the material.

Being paid for material or equipment *offsite* is different but can be done. Often, the requirement is made that the goods be stored in an "insured warehouse". That means a warehouse suitably secure to allow insurability, with the insurance paid for by the contractor and coverage made payable to the owner.

A fabricator or supplier may have legitimate need for being paid before delivery to the site. This presents a problem to the owner, who would rather have the goods on their own property than in storage somewhere. However, these instances occur, the owner's site is never big enough, and the General Conditions cover this occurrence.

Final payment 9.10.1

When the architect finds the work acceptable and the contract fully performed, the architect will issue a final certificate for payment.

Section 10 Closeout

Closeout introduction

Closeout "tasks" are covered in both of the nontechnical specifications – Division 00 General Conditions and Division 01 General Requirements. The following closeout requirements are covered in the General Conditions, and are a condition precedent for final payment. Waiting until substantial completion is too late for a project manager to start working on "closeouts", as these items are called. The date of substantial completion will need to be known in order for it to be placed on warranties, but there are other tasks, such as the record drawings, that can be done.

It takes time to amass the closeout documents, project management time. Sometimes the project manager has assistants help with this work – assistant project managers, interns, or the bookkeeper. Usually the expense involved is considered home office overhead and not charged to the job. However, many estimators have a line on the job overhead portion of the estimate for "closeout docs". They might place some labor expense in the estimate, plus the material cost of paper and notebooks.

Record drawings and contract documents 3.11.1

The contractor is charged, through the General Conditions, with making sure the architect and owner have complete copies of approved submittals at the end of the job.

The need for as-built drawings (also called project record documents) occurs when something is placed in a different location than is shown on the plans. When piping (underground or above ceiling), the electrical service, or mechanical equipment, because of job conditions, is placed in a different location than that shown on the plans, the contractor must field mark it on a set of job plans. This is a job that the superintendent does.

The superintendent collects this information throughout the job, making notes on a field set of plans. At the end of the job, the field set is sent to the office and perhaps redrawn or placed on AutoCAD. These are called the "record drawings", or simply the as-builts. They are a closeout document, and they are transmitted to the architect.

Although the as-built drawings can be for any division of work, important ones include the drawings locating MEP equipment, piping, ductwork, and site utility locations. It is important that record be made of actual in-place locations of underground sewer and water lines, above-ceiling ductwork, and the electrical service. *After all, the A/E on the next job will not verify onsite locations of these items; they will use the as-builts to produce their plans.*

Demonstration and training, operation and maintenance manuals

Part of every job is instructing owner personnel on how to use equipment and turning over written instructions. These may be simple tasks, like how to operate the air conditioning thermostat, or more complex and numerous learning tasks that might take several days and include multiple vendors. There are elevators to run, air conditioning and thermostats to operate, and fire alarm instructions to understand, and all of these require instruction by the contractor to owner. This is where the contractor shows the owner how to operate those huge garage doors at the fire station or simply change out the air conditioning filters. Onsite meetings are held at the end of jobs to instruct owner personnel how to operate the equipment.

Written operation manuals contain instructions on how to operate the keypad for door locks, the air conditioning equipment, and the smoke detectors. These are consolidated into a notebook, or assembled in an electronic file, and transmitted to the architect and owner at the end of the job.

All of this work is part of the job, and the time it takes should be budgeted for, meaning it should be a line item on the job overhead portion of the estimate.

Warranties and guarantees 3.5.1, 12.2.1

The prime contractor is usually required to furnish a one-year guarantee for labor and materials. This is a general guarantee and is usually supplanted by several others provided by manufacturers and subs. The following is an example of a typical and simple one-year guarantee by a GC:

See the online resources for diagram 11 2 10.1

The substantial completion date, the date that the architect certifies by an inspection and an official document of substantial completion, stops the clock for project duration. The definition of substantial completion is, "The project is ready for the owner to use as intended."

The substantial completion date is the all-important date used as the beginning of the warranty period. All warranties must use this date, despite many a subcontractor's attempt to use their installation date! The hot water heater, air conditioning equipment, and the front door lock, which may all have various warranty periods extending beyond one year, all start with the same *beginning* warranty date of the substantial completion.

The project manager must coordinate the warranty closeout effort with various vendors and ensure the proper date is used.

Warranties on new roofing work are among the most important guarantees that are provided by the contractor to the owner. The manufacturer for a new roof on a school or campus building, which will probably be replaced every 15 years or so, may be the most important decision that the architect and owner make in regard to the overall project. The warranties of several manufacturers may be studied, and manufacturer representatives may visit the architect's office and provide samples of both material and warranties.

The warranties from major manufacturers on roofing can be compared to tire warranties, where the depth of the tread is expected to diminish over the warranty period. The roof is not expected to stay like new over the warranty period; it will have wear and tear but will not leak. While the roofing subcontractor may be expected to provide a *labor* guarantee lasting a year or two, the owner (the government or major owner) doesn't care, in the long term, about a guarantee from local Ronnie's Roofing Company; it wants that warranty from the huge manufacturer (supplying the roofing product) that has been in business for fifty years and will still be in business for a lengthy warranty period.

Mechanical and electrical equipment have extended warranties beyond one year; five-year periods are common. A subcontractor buys the equipment from a major manufacturer, with the required warranty period stated in the specifications. At the end of the job, the substantial completion date is provided by the prime contractor to the subcontractor, who then obtains the warranty from the manufacturer. The warranty is then sent back to the subcontractor and up the chain to the prime contractor and the architect, and then it is finally provided to the owner.

Several years after the project is completed, when the GC is retired, the owner might be dealing with a manufacturer about a warranty issue.

Consent of surety

At the end of a job, the bonding company is asked if they know of any issues that should be dealt with prior to closing out the job. This is done in the form of asking them to provide a "consent of surety". This consent is where they state, "We know of no reason why you shouldn't pay our guy or gal their due. Nobody is making a claim for nonpayment. We know of no problems!"

This is an item covered in Division 01 General Requirements and is a closeout requirement, but it does not have a cost associated with it.

Section 11 Unforeseen fire line case

Perhaps the most problematic situation for contractors is an unforeseen condition. These are usually found very early in the construction schedule, often immediately after starting the job. With a new building, the problem is often piping (water, sewer, fire line) found either to be underneath the building footprint, where it is not supposed to be, or far away from its anticipated location, costing more to connect to. In a renovation project, unforeseen conditions are found early in the job above ceilings and within existing walls.

The author has experienced many unforeseen conditions and was involved in the following two occurrences of unforeseen fire lines being located under new buildings. Both times the owner tried to make a case for its relocation to be a part of the contract documents – in other words, relocated at the contractor's expense.

At a community college, the project was to build a three-wall addition to an existing building. The campus had not kept up with the location of its underground fire lines and, unknown to them, for it wasn't on the plans, a fire line traversed directly under the middle of the proposed addition. Relying on general condition language about the contractor being responsible for a complete job, not specific plan instructions, the architect demanded the fire line be relocated. Arbitration settled the community college matter in the contractor's favor.

The second occurrence of a fire line occurring underneath a proposed new building is covered in detail as follows and concerns many parts of the General Conditions. Ironically, it occurred at a state fire college! The author did not make this up.

A five-story "burn simulator building" was to be constructed amidst a campus of a dozen or so buildings. The steel building had a footprint of only about three thousand feet, was designed to withstand fires, and had gas piping to various locations with furnaces that could be set aflame. No one would want to be in this place except a firefighter clad in asbestos or better. The point was for rookie firefighters to climb a ladder and enter the building through a window, fire hose in hand, and practice putting out fires.

The site was fairly flat and grassy. There were no obstructions above ground. A fire hydrant sat in one corner, and the plans indicated underground fire lines ran to it.

The important aspects of the site are shown in Sketch 1.

See the online resources for diagram 11 2 11.1

The earthwork specifications required that excavation *extend five feet beyond the building footprint and five feet under the concrete footings*. It was to be backfilled with lime rock, the same yellow soft gravel, easily compacted, that is used under asphalt roads. The contractors noted that the plans indicated the fire line to be at the very edge of the overexcavation on the east and north sides.

Here are two notes from the civil plans:

2 The location of all existing utilities shown on the plans has been determined from the best information available. The engineer assumes no responsibility for accuracy. It shall be the contractor's responsibility to notify the various utilities and to make necessary arrangements for any relocation of these utilities.

4 The contractor shall field verify the horizontal and vertical location of all existing utilities within the limits of the project envelope shown prior to commencing work. The contractor shall call all utility companies to have the locations of all utilities field marked prior to commencement of work.

In the days preceding the start of excavation, superintendents for the contractors were onsite. The utility companies "located" water and sewer lines. This was done by spray-painting the ground along the route of the underground pipes. The word "locate" is commonly meant to locate a pipe's path by indicating it on top of the ground. It is not common practice, when being told to locate utilities and fire lines, to dig them up. Utility companies do not do this. Nor does the contractor unless the plans indicate replacement. The water and sewer lines were found to be well outside of the building footprint and presented no concern. The utility companies were not responsible for fire lines, and the superintendents asked the owners (the fire college) maintenance staff to locate the fire line.

Often, an owner will know where fire lines and water and sewer lines are located because of maintenance done over the years. However, the state fire college had no clue about where the underground fire lines were. The aboveground fire hydrants were the only indications of their comings and goings. *What the contractors had, prior to excavation, was the fire line location shown on the plan, which came from a survey and used by the architect on the current building plans.*

The superintendents pre-planned the following:

1 Trucks entering and leaving the site would use the corner labeled number 1 on Sketch 2. This was the "high corner", by less than a foot, but any dirt disturbed by the trucks would wash downhill towards silt fences on the other side.

2 Some of the dirt initially excavated would be piled at area 2 within the rectangle shown at the southwest corner of the site, see Sketch 2, available for backfill later. Most of the dirt would be hauled away in trucks. After excavation reached the bottom elevation, *returning trucks would be filled with lime rock to be used for the five-foot backfill.*

3 Excavation would commence on the "far side" of the building, see note 3, and work in the direction of the entry/exit.

It was a plan. Work commenced on a Tuesday morning. By Thursday afternoon, excavation was completed on the west rectangle of the building, and half was done on the east side. *On the east side, at a depth of about four feet, fire line piping was found under the stairs portion of the building. See Sketch 2.*

Now go back to notes 2 and 4, and particularly the first sentence of note 4. In this chapter's Section 2 Owners and Users of Facilities, mention was made about how a user can erroneously interpret "one or two lines" on a set of plans, without viewing the plans as a whole or understanding the construction law within AIA 201. In this case, note 4 was powerful ammunition in the hands of the state fire marshal. Why should the contractor be paid for relocating the fire line when the contractor should have uncovered it? Surely this language, intended for utilities (water, sewer, electric) extended to fire lines!

Follow the fire line case over the next few pages and see how note 4 has to be put in context with other instructions.

See the online resources for diagram 11 2 11.2

No one knew that a fire line extended under the structure. As soon as an unforeseen condition is found, the contractor is going to need to do all the documentation, field locate and photograph the problems, make sketches, document it all, and do it in a hurry. The documents put all the work on the contractor, who has the duty to do all the investigating and present it all to the owner and architect for their review. When the contractor can offer solutions, it is best to try to push the ball forward with options presented to the owner along with a price. This is the contractor's job, and the clock is running. A notice of delay should be written, and if more time goes by, another notice should be given of a continuing delay.

Work stopped in the stair area on Thursday afternoon. A superintendent spray-painted the location of the piping and took photos with a phone, and called the office. The project manager and other management thought about what to do. Contractors are good at going forward. Stopping work is a big deal and has a lot of ramifications. But this was clearly an unforeseen condition, or at least the contractor thought so. New instructions were needed. The decision was made to backfill the west side of the excavation up to the level of the fire line on the east side. Otherwise, there would be a "hole" in the ground that would collect water. Another reason to fill in this area was the principle of "working all the way up to a delay".

On Friday morning, the project manager emailed the photo of the partially exposed fire line to the architect. Excavation had stopped at the top of the fire line, and the superintendent had traced its path in red paint on the ground. Phone conversations continued throughout the morning and included the site superintendent. The PM developed a sketch, similar to sketch 2, which outlined the building, the extent of excavation, where the fire line was supposed to be, and where it made an unexpected "L" turn under the building.

The contractor did the backfill of the west side on Friday, demobilized on Monday, and began pricing the relocation of the fire line. On Tuesday, the architect sent out a cryptic email explaining that all options were being considered, including moving the building, encasing the "L" portion of the fire line in concrete, and footing "gymnastics". The contractor priced the relocation option and turned it in to the architect, who passed it on to the owner rep and the user.

At the end of the week, the architect paid a visit to the contractor's office and explained that the owner blamed the contractor with part of the problem. The owner offered to pay for materials, but wanted the contractor to pay for the labor. The contractor, the architect explained, per note 4 on the plans, should have located the fire line ahead of time. Also, if the contractor had started excavation at the other end, where the fire line was, the fire line would've been found earlier, leaving the owner more options now, at least in moving the building.

The contractor asked the architect if he agreed with the owner. The answer was yes, and that a "change directive" would be written, instructing the contractor to commence work. The fire line would be relocated but the expense was not agreed to.

The contractor disagreed with the architect, and explained the decision to backfill on Friday. The situation had been handled "by the book", and the change directive would be "signed in protest", and the contractor would take the matter to arbitration. The contractor asked for a face-to-face meeting with the owner.

A change directive was received by the contractor, signed by the architect, but unsigned by the owner, so the contractor let it sit. Various options were discussed, and major and minor movements of the building were contemplated. The contractor said that in his experience, pipes were relocated, not buildings. The architect complained that prices were not being handled fast enough. The contractor sent the architect a letter in defense of the contractor's handling of the problem, which was labeled as an unforeseen condition, and a bold "Notice of Continuing Delay" was at the head of the letter.

After two weeks, the architect announced that a decision had been made about the work, and the fire line was going to be relocated. There had been no reply to the contractor's letter; nothing was in writing stating the owner's view, only the initial conversation with the architect at the contractor's office.

At the meeting, the contractor stated that the "L" portion of the fire line was an unforeseen condition as defined by Article 4.3.4 of AIA 201:

> If conditions are encountered at the site which are (1) subsurface or otherwise concealed physical conditions which differ materially from those indicated in the contract documents or (2) unknown physical conditions of an unusual nature, which differ materially from those ordinarily found to exist and generally recognized as inherent in construction activities of the character provided for in the contract documents, then notice by the observing party shall be given to the other party promptly before conditions are disturbed and in no event later than 21 days after first observance of the conditions.

This paragraph goes on to state that the architect will investigate the conditions and if they differ materially from the contract documents, prepare a change order. If there is disagreement, he or she will state the reasons in writing. Claims by either party must be made within twenty-one days of the architect's decision.

The architect had situated the building so that the excavation extended to beside the fire line on the east side. And, that is where it was found to be. But, no one had suspected that the line would turn where it did, poking its way under the stairs.

The contractor knew that a literal translation of note 4 had to be discredited. What did "Field verify the horizontal and vertical location of all utilities" mean? Could it be the engineer actually meant for the contractor to expose the fire line for its full length, so that one could look into an open trench and follow its full route, in advance of other excavation?

This is Article 2.2.3 of AIA 201:

> The owner shall furnish surveys describing physical characteristics, legal limitations and utility locations for the site of the project. The contractor shall be entitled to rely on the accuracy of information furnished by the owner but shall exercise proper precautions relating to the safe performance of the work.

Since the owner had placed a survey in the plans, the contractor was able to say that it could be relied on and assume that the location of the fire line was as shown on the plans. The contractor can rely on what is given and move forward. And, that's what contractors like to do.

Nor does the contractor have to suspect a problem. See Article 3.2.1:

> Since the contract documents are complementary, before starting each portion of the work, the contractor shall carefully study and compare the various drawings and other contract documents relative to that portion of the work, as well as the information furnished by the owner pursuant to section 2.2.3, shall take field measurements of any exiting conditions related to that portion of the work and shall observe any conditions at the site affecting it. These obligations are for the purpose of facilitating construction by the contractor and are not for the purpose of discovering errors, omissions, or inconsistencies in the contract documents; however, any errors, inconsistencies or omissions discovered by the contractor shall be reported promptly to the architect as a request for information in such form as the architect may require.

By use of the word "complementary", certain clauses of AIA 201 could be used to help explain the full meaning of note 4. And, very importantly, the contractor is given the drawings and surveys for the purpose of "facilitating construction", which means going forward, not looking backward. The contractor is not purposed with discovering errors!

The contractor also made the point that had the owner/architect wanted to influence the construction sequencing, they could have done so, and they had the same information that the contractor did prior to the work. The contractor is in charge of means and methods, "unless the contract documents give other specific instructions concerning these matters." The documents had not shown instructions, and the architect had not given any.

The architect/engineer knew, or should have known, that with the given overexcavation requirements, the location of the fire line might be uncovered. It was going to be close. However, if the owner/architect wanted the contractor to start work at one corner, they had their chance ahead of time, but not with hindsight. However, no such instructions were received, nobody said, "start here", or even brought it up in conversation.

A construction schedule had been prepared by the contractor and provided to the owner and architect. It did not start with an event labeled, "Isolate and review location of fire line." Furthermore, the architect had not voiced any intent ahead of time, or during the startup, of coming to the site to see an exposed fire line. If the architect/engineer had been expecting to see exposed fire lines in advance of the work, they had forgotten to come and look.

The owner and architect were convinced and agreed to pay the contractor. They asked that the old fire hydrant on the north side be replaced as well, and for the entire line along the east side be moved a few feet further from the building.

The government, through the General Conditions of AIA 201, is not out to trap anyone, or to look with hindsight and second guess. There is no could have, would have, should have, or "that unforeseen condition is the contractor's fault". After all, it was the owner's ground and the owner's fire line.

Section 12 Gooseneck faucet case

This actual case, where a subcontractor failed to properly submit, is simple but points out some very important aspects of the submittal process.

Now that the contract documents and submittals are defined, consider what it means to "fully submit" a job. An incomplete submittal can cause all sorts of problems, with the architect later crying "foul" and taking back his/her approval. This means that *all* of the product data sheets about a product be turned in, and to be thorough the project manager should use arrows to point or highlight important parts, because manufacturers often write data sheets for equipment, or windows, that come in various sizes or weights or tons. There is a duty for the contractor to be complete.

Following is the problem that the plumber had with gooseneck faucets. For a science lab, the plumbing engineer employed by the architect had specified the faucets, specified the manufacturer, and even provided the model number. Case closed, there it is. The plumber provided these very same faucets, beautifully displayed in sinks near the end of the job atop rows and rows of base cabinets in a science lab. The plumbing engineer came for the substantial completion and exclaimed the faucets were wrong! No, said the plumber, they are exactly what you specified!

The engineer called in the lab teacher, who had not seen the faucets (this occurred during the summer vacation). Seems the teacher had told the engineer late in the planning process they wanted gooseneck faucets. The engineer had placed the wrong specifications in the project manual, but intended to make a correction when the submittals were made. Only he never did get submittals on the faucets because the plumber figured there was no need, the exact specified faucet was in fact still being made by the manufacturer, the plumber had all the information needed to define the faucet, so the specified faucets were ordered and installed without a submittal.

The plumber had to remove all of the existing faucets and provide the goosenecks. Since no submittal was made, the contractor had no basis for arguing. All because a full and complete submittal was not made. The lesson learned was that the submittal process can be an occasion for an owner/architect to change their mind.

The plumber could have protested that the engineer should have made a correction to the specifications earlier. However, there is no language in the General Conditions to support this. All of the work to uncover problems rests with the contractor by fully submitting the job.

To "fully submit" a job means that every single component of the building, down to the nails and fasteners, is submitted for approval by way of its appropriate documentation – whether that be by product data sheets, a sample, or shop drawings. The quantity of submittal data has increased over the years due to the complexity of building construction and the use of manufactured products. While submitting minor materials may seem burdensome, it is beneficial to preplanning and a key to efficient production free from errors and delays.

3

DIVISION 01 GENERAL REQUIREMENTS

Section 1 Introduction

Division 01 General Requirements are instructions to the contractor from the owner. They are near the beginning, or "front end", of a project manual, right after Division 00 General Conditions. These first two division numbers are part of the old sixteen-division breakdown and remain the same in the new fifty-division format by the CSI, the Construction Specification Institute. General Requirements are sometimes contained within a checklist of standard clauses that the purchasing department of a city or county uses on its construction contracts.

These two divisions are sometimes confusing, first because they both use the word "general" to define them. Second, when Division 01 General Requirements is divided into parts and subparts within project manuals, they can be numbered Division 1 or 01000 or anything in between. Division 00 General Conditions is safe from such a range of numbers, because zeros can't be divided up, although some architects have probably tried it. Additionally, there is no need for architects to provide a breakdown of Division 00 General Conditions in the project manual because it already consists of sequentially numbered articles (if AIA 201 is the source).

A third confusing aspect about these two divisions is that they contain many of the same subjects. Closeout documents, which are a collection of many items, are an example. These include "demonstration and training" and "project record documents". The language about closeout documents in Division 00 General Conditions includes the requirement that they be sent to the architect/owner in satisfactory form before final payment is made and other lawyerly language. In Division 01General Requirements, the language about closeout documents includes that owners need to place all the project record documents, or as-builts, on a flash drive, or on 11 × 17 sized paper, or other such requirement. There is the difference – Division 00 General Conditions is about legalities, Division 01 General Requirements is about owner-customized needs such as formatting and training.

Some General Requirement topics are costs, usually defined as "job overhead", but some are not, such as the preconstruction conference. A job trailer is in the category of "job overhead" and attributable to the job, not "home office overhead". It does not contribute to the physical building structure, but a construction trailer is a necessary thing, and so are temporary toilets, and it is these items that are a part of Division 01 General Requirements and a part of a construction estimate.

Job overhead costs include onsite utilities, building permits, and an item called tools and expendables. Note that while utilities and permits are often discussed in Division 01, tools and expendables (expendables are things like water coolers, drinking cups, parking cones, and wood ramps) are not. However, all of these costs appear in the same part of the contractor's estimate for a job, while home office expenses, such as the boss's salary and office rent, do not. Job overhead expenses are often 5%–10% of total project costs, and sometimes more on small jobs. Supervision, considered job overhead, is required full time on most government or private projects of any size. As job size increases, assistant supers and/or foremen expense is added to job estimates and considered job overhead. It is labor.

Cleanup is a major expense for prime contractors, and the budget for it never seems to be enough. It is a job overhead expense and is labor. Dumpsters, where the cleanup crew deposits debris, is a material cost. Job overhead expenses largely depend on job duration. In fact, in estimating language, the "unit of measure" for job overhead is often figured in weeks or months. Job overhead expenses are best judged near the end of estimate completion because that is when the estimator knows the job well.

Some contractors treat concrete and other testing as a job overhead expense. This is because of the timing of when certain aspects of the estimate are done. It is best to estimate all the concrete work first and then figure the testing after all the pours are identified and yardage counted on the takeoff *and sent to the estimate.* Pricing the concrete testing is a function of total yardage and number of pours, so it should be studied after all the concrete has been taken off. Testing, which is a direct expense related to accounting, is placed into the job overhead part of the estimate because of timing.

The estimator can best judge job duration and complexity after he or she is thoroughly familiar with the job. That means figuring the job overhead costs last! Do the takeoffs, transfer the quantities to the estimate, and start adding up labor and material. Throw in plug numbers (guesstimates) for the major materials and subs. Then, arrive at a rough total for the job. This is the best time to complete the job overhead portion of the estimate – after the other costs are known, or a plug number is used. That means job overhead is often completed just a day or two (or the night!) before the bid.

Many of the same job overhead costs are encountered over and over from job to job. That's why compiling a checklist for them is good practice. Many companies compile a page of job overhead costs and have them preprinted on an otherwise blank estimate form. If it applies, enter the cost, if it doesn't, delete it.

See Part 2, Chapter 2 Pricing for an example of a job overhead sheet.

Section 2 Pre-bid

ITBs and RFPs

When a project initially goes out for bids, the owner publishes a short document called *information to bidders* or *request for proposal*. This is the first thing an estimator will review when considering whether to bid a project. The bid forms, which the bidders must use, are where the base bid and alternate prices are placed. The bonding requirements are stated, and a short job description, called a *scope of work*, is included. ITBs and RFPs, while made available separately from other documents, are often inserted into project manuals at the very beginning of the front-end documents.

A paragraph, or a few paragraphs, would seem too brief to be of much use to the estimator when starting a complex estimate of fifty sheets of plans and several hundreds of pages of specifications. And, a scope of work in either an ITB or RFP may carry little legal weight. However, sometimes there are defining clues in the summary of work that help narrow the scope or point the estimator in the right direction. For example, the estimator might be confused when the third, fourth, and fifth floor plans are included in the plan set when the job was advertised as a fourth floor renovation project.

The answer to why the floor plans for the third and fifth floors are included in the plan set is answered in the summary of work by these words – "there is new construction in the stairwells connecting the third floor to fourth floor, and fourth to fifth. However, there is no other work on the third and fifth floors". This clarification saves the estimator time in searching around the plan set for details and sections that might pertain to the third and fifth floors, when they were only provided for information and clarity. This is an often used plan element – *provided for clarity*. The summary of work can direct the estimator to the limits of the work, of inclusions and exclusions, that are all-important criteria for the estimator.

Scopes of work vary greatly in their length and depth of explanation. Most are just a paragraph or two. An example of a short scope is:

> The improvements under this contract consist of an addition to an existing weight station on highway ____ in ____ County for the State of ____ Department of Transportation. The existing building will be renovated, and an addition will be constructed that ties into existing conditions.

An example of a detailed summary of the work is:

> This project consists of a 20,000 sf addition to the existing City Hall Complex located in ____. The building pad will require the import of clean compacted fill to raise the proposed building slab to align with the floor level of the existing building.
>
> The goal for the work is to leave the existing building functioning until the new addition is near complete and openings are cut into the existing wall in order to minimize disruption in daily operations.
>
> The base bid construction consists of a stem wall foundation, slab on grade, and block walls. The base bid includes re-shingling the existing buildings and shingles on the addition.
>
> There is an additive alternate to place a metal roof on the new addition. The metal roof will be over a 40-mil peel and stick ice and water shield over 5/8″ CDX plywood over pre-engineered wood trusses. Color of roof panels to be selected by architect from manufacturer's full range.
>
> The exterior walls consist generally of 12′ high walls with the top 2′ being a concrete masonry tie beam. The blocks are filled with insulation and the exterior finish is stucco.
>
> Interior walls consist generally of 4″ and 6″ metal stud walls. Some contain sound batt insulation. Some are fire walls; consult the plans for various wall types.

List of subs

The owner often requires that the bidders submit with their bid a list of the subcontractors that they will use on the project.

There are two reasons that an owner wants to know who the prime has teamed up with. One is to evaluate the quality of work that can be expected. The owner and architect are often familiar with the local subcontractor community and have experienced their performance on prior jobs. They will be able to spot the good and mediocre performers. Indeed, there is language in the General Conditions, Division 00, that allows the owner/architect to object to the use of a listed sub.

The other reason that an owner requires a subcontractor list is an ethical consideration, and that is to keep the prime from shopping sub-bidders after a bid. The government and large private owners are invested in the long-term health of the competitive bidding process. Towards this end, requiring the prime to list subcontractors prevents them from being changed later through shopping. The owner is forcing ethical conduct on the prime contractor.

The number of subs that are required to be listed varies. Sometimes there are simply a few empty rows on the bid form with an instruction to list major subs, leaving it up to the bidder to insert trades and subcontractor names. Other times there will be instructions to name the roofing sub and MEP (mechanical electrical plumbing), and that's it. It is not usually the case that all subcontractors have to be listed.

Section 3 The bid

Alternates

An alternate is, for the estimator, a separate bid defined in Division 1, and perhaps in the plans and the technical specifications. A reason for providing a separate price, which is what an alternate is, may be that the owner and architect think the work in the alternate may not be affordable, and if included in the base bid might make the project over budget. Another reason may be that the item is being paid for by a different funding source than the base bid, a different department or agency of the government, or it may be that a grant is paying for something specific and the cost needs to be isolated.

Careful attention is paid to these alternates by the estimator. Long lists of alternates can cause confusion over where certain costs should be allocated, and woe be the estimator that waits too long to start figuring these separate prices.

Typical general requirement language about alternates is as follows:

> An alternate is an amount proposed by the bidder for work that may be added to or deducted from the base bid amount if the owner accepts the alternate. Include as part of each alternate, miscellaneous devices, accessory objects, and similar items incidental to or required for a complete installation whether or not indicated as part of the alternate. Include within the alternative bid prices all costs, including materials, installations, and fees.
>
> The description for each alternate is recognized to be incomplete and abbreviated, but implies that each change must be complete for the scope of work affected. Refer to applicable specification sections and to applicable drawings for specific requirements of the work regardless of whether references are so noted in description surrounding work as required to properly integrate with the work of each alternate. It is recognized that descriptions of alternates are primarily scope definitions, and do not detail full range of materials and processes needed to complete the work as required.

The language of the preceding paragraph, taken from an actual project, is sweeping and far reaching. It says that the description of the alternate may be lacking, but the scope nevertheless is for a fully detailed job. In other words, "We haven't defined it fully, but you (the contractor) must build for us a complete job, including all of the labor and material and miscellaneous devices and accessory objects as needed".

An example of an alternate is:

> *Remove and replace windows in addendum number 1. If the alternate is not accepted, clean existing windows identified in addendum number 1.*

Failure to bid an alternate can be cause for the owner to throw out an entire bid. That's what it will state in many bid documents. However, this is not automatic, and the matter turns on what is in the owner's best interests.

Enter the world of "bid discrepancies" and "minor informalities".

With all the numbers involved that estimators deal with under the pressure of an absolute deadline, and with sub bids and quotes received sometimes at the last minute, plus additive alternates, deductive alternates, lists of subs to fill out down to their address and license numbers, give us a break!

Stuff happens.

A one-million-dollar bid to remodel a Clerk of Court office building had three alternates. They were all deductive, meaning the items were included in the base bid. Alternate 3 was to deduct the cost of an awning on the exterior of the building, a simple alternate. Two of the three bidders had deducts of approximately $25,000. Bidder 2 was also near $25,000 for alternate 3, but it was positive, an add.

The bids were as follows:

	Base Bid	*Alt 1*	*Alt 2*	*Alt 3*
Bidder 1	808,000	−3,500	0	−27,500
Bidder 2	798,890	−2,560	0	+23,232
Bidder 3	779,100	−5,516	−1,400	−25,700

Bidder 2 had sent a "bid runner" to the bid who could offer no explanation (it is the author's opinion that the estimator should always go to bid openings, just in case something comes up) as to why one alternate was a positive number. It didn't really matter; it didn't affect the low bid, which was the author's.

When the bids were published, as public bids are, the awning alternate for bidder 2 had been changed to a negative. The estimator for bidder 2 had called the city purchasing agent and said there had been a mix-up with the bid runner. So, the purchasing department changed it.

This was a bid discrepancy. Government owners have language in their procurement policies that they can "waive minor informalities." Of course, the owner decides what is minor, but a disgruntled bidder can protest. This was an actual case, but changing the circumstances, what would have happened if the low bidder for the base bid had an "add" for the awning alternate and the other two bidders had negative bids?

	Base Bid	*Alt 1*	*Alt 2*	*Alt 3*
Bidder 1	808,000	−3,500	0	−27,500
Bidder 2	798,890	−2,560	0	−23,232
Bidder 3	779,100	−5,516	−1,400	+25,700

In this case, changing the positive alternate to a negative would change the results of the bid (given alternate 3 was accepted), so there would be a major consequence. The owner would probably allow the change because it was simple and obvious. However, bidder 2 might lodge a protest. Just as a public owner has language concerning bid discrepancies, they also have formalized rules about bid protests, at least an initial step or two, such as referring the whole thing to the architect or city attorney for a written opinion. Beyond that – the courts.

To discuss the bid instruction, "Failure to bid an alternate can be cause for the owner to throw out an entire bid", consider the following bid:

	Base Bid	*Alt 1*
Bidder 1	4,000,000	−50,000
Bidder 2	3,750,000	−42,000
Bidder 3	4,250,000	−56,000
Bidder 4	3,500,000	no bid

For whatever reason, bidder 4 did not provide a price, and after a bid opening is prevented from doing so. The owner can still accept the low bid by bidder 4, save 250,000, and consider the lack of an alternate price a minor informality.

For the estimator, each alternate requires a separate estimate. Alternates are usually not as simple as deleting an awning. Often, they require a range of quotes and prices from suppliers and subcontractors. Multiple alternates complicate an already difficult endeavor of an estimating task lasting two or three weeks. Subs and suppliers may not pay as much attention to detail as the prime bidder does, and a bid with alternates requires additional management. Then, on bid day, there is the manipulation of perhaps dozens of prices into the base bid and separate alternates. The potential for error is simply a function of the amount of arithmetic done during a hectic day.

The complexity of a bid increases with the number of alternates, and so does the probability of bidding mistakes.

Allowances

An allowance is an owner-specified cost that is included in a bid. One reason allowances are used is to provide money for an item not designed or defined in detail at bid time. Perhaps the architect and owner cannot provide the detail to define the cabinets but know they will need them. Cabinets are a complex subject for the designer, with upper and lower cabinets and drawers here and shelves there, and cabinetmakers can't settle on a price without all the details.

An allowance might be for a group of appliances, with the manufacturer and model numbers to be provided by the owner later. An allowance can be for most anything. By describing an allowance in Division 1, and possibly repeating it on the plans and in the specifications, the owner has a sum, maybe not an exact amount, but nevertheless has a budget included for the allowanced item in the base bid. This way, a major change order is avoided later. A minor change order will still be executed eventually, to account for the over/under difference between the allowance and the actual cost.

The prime bidder will not be allowed to charge for profit *within* the allowance. If some cabinets are allowanced at $50,000, the prime bidder places this amount in a bid totaling, say, five million. Later, exact drawings are made and the prime bidder has the cabinet work done by a subcontractor for $45,000. The prime is not allowed to charge the owner $49,500 (45k plus 10% profit), only $45,000. Careful attention is made by the estimator for associated costs such as supervision, barricade and protection, etc. While delivery and unloading may be viewed as part of the allowance, supervision, care of the product, and markup would not.

Job overhead

As discussed in the Introduction, many Division 1, or 01, or 01000, are costs and placed within job overhead of the contractors estimate. The company cost code will usually begin with a Division 1 code, and all kinds of "indirect costs" will be placed there. While not all Division 01 items from a project manual will have a cost (there is no cost associated with a preconstruction conference), if it does have a cost it will be placed in the job overhead portion of the estimate.

There are some contractor expenses that are not mentioned in Division 01 of a project manual. These job overhead costs include:

A Travel time.
B Gasoline.
C Truck allowances.
D Tools and expendables.
E Rentals such as scaffolding, cranes, lifts.

If a jobsite is more than twenty or thirty minutes from the home office or the employee's house, the contractor may pay for some of the time that it takes to get there. If the job is an hour and a half away and the contractor plans on sending twenty people from the home office area to the jobsite, travel time can be a considerable expense.

The car and truck mileage for employees to get to a distant site can be considerable. The contractor may use the mileage rate established by the federal government that reimburses an employee for gas plus wear and tear for a vehicle. The wear and tear portion of the reimbursement is usually higher than the cost of the gasoline reimbursement, and this job overhead cost can be quite high.

If the superintendent or some carpenters routinely travel around with company tools and equipment, they are often given a weekly stipend to reimburse them for wear and tear on their vehicle. This is called a truck allowance and should be charged to the job. If the company owns pickups and furnishes them to employees, some or all of this truck expense will be charged to job overhead and be a part of the estimate.

Saws, electrical cords, and all kinds of small tools are provided by either the company or employees or both. Many small items, things like water hoses, will hardly last the length of one job. Taken as a whole, tools and expendables can be a hefty job overhead expense.

Scaffolding and other job logistics can be very expensive. The estimator uses the "rentals" cost code to account for (place on the estimate) temporary toilets, scaffolding, walk-behind compactors, cranes, etc.

Referenced standards

This specification states that there are many other publications that are a part of the contract documents. Just as Division 00 "references" AIA 201, which is a separate document and not placed in the project manual, technical standards are simply listed and by reference become a part of the specifications.

A project manual can't possibly contain all the applicable building codes, tolerances, or good practice standards for each industry, so specifications often outsource instructions by using industry standard publications. There are dozens of trade organizations that conduct research and write specifications for their own industry. For example, the American Concrete Institute issues voluminous data regarding the formwork, reinforcing, and pouring of concrete. Other industry "institutes", such as the American Hardware Institute (AHI) and the Steel Joist Institute (SJI), are "referenced" by construction specifications because these industry trade associations are the root source of specifications and best practices for their industries.

There are many testing organizations that specifications refer to. The American Society of Testing Materials, known as ASTM International, has scores of specifications for the steel industry, such as:

ASTM A328 / A328M Standard Specification for Steel Sheet Piling.
ASTM A709 / A709M Standard Specification for Structural Steel for Bridges.
ASTM A1064 Welded Wire Reinforcement (aka wire fabric, wire mesh).

Another testing organization is Underwriters Laboratories (UL). It is a safety consulting and certification company that has offices worldwide. UL is one of several companies approved to perform safety testing by the Occupational Safety and

Health Administration (OSHA), an agency of the U.S. federal government. The "labels" that are on the side of rated doors are issued by UL.

The National Fire Protection Association (NFPA) is another hugely important group that writes codes governing fire, electrical, and related hazards. They issue several publications, including NFPA 1 *The Fire Code*, NFPA 13 *Standard for the Installation of Sprinkler Systems*, and NFPA 101 *The Life Safety Code*.

Section 4 Preconstruction

Preconstruction conference

The preconstruction conference is an important meeting held prior to the job start between the newly awarded prime bidder, owner, and architect. This is not a social gathering. Lunch will not be served. The preconstruction meeting will be held at the owner's conference room, or at city hall, and may be at the same location as the bid opening. (This concerns public bids with government owners that have formal procedures.)

The owner makes sure the contractor understands what the owner requirements are. Owners and architects are well prepared for these meetings, and come armed with often used requirements – like how to number the submittals, how to send them electronically, and to have the draws to the architect by the 30th, and so forth. User concerns are often discussed (for example, commission meetings are held at 1 pm on Wednesdays – don't work in the adjacent rooms during commission meetings!).

The preconstruction conference usually marches to a predetermined agenda made by the owner's representative, the architect, or both. They will have a set list of issues to discuss. The contractor can ask questions like where the dumpster can be placed, but this is not the time or place to discuss construction techniques or means and methods.

The project manager and superintendent both attend the preconstruction meeting, as do the major subcontractors.

The plans and specifications have already been studied, and RFIs written and answered with addenda; the precon, as it is called, is not about construction. The agenda is a checklist of items created by the government owner and purchasing agent, updated from time to time depending on technology. The checklist is used over and over from job to job and concerns how to make the project run smoothly for the owner.

During the bid, the contact person for bidders is the purchasing agent for the owner. Bidders do not contact the A/E directly. All questions and RFIs are sent to the owner's representative. At the precon, it is usually announced that the contact person is shifted to the architect, at least concerning submittals, draw requests, and everyday running of the job. That's what the owner has hired the architect to do, manage the job. The purpose of the precon is to go over the owner's, and architect's, protocols for the job.

Sometimes the contact person is employed by the public works department of a city or county or school board. A larger or more formal owner, such as a state or a university, will name the architect as the contact person for all correspondence, submittals, and draw requests.

The owner may improvise protocols at the preconstruction meeting. Perhaps the architect has to drive halfway across the state to make a site inspection for a new park. The city commission has to approve each month's draw request because it is a small town and there is no purchasing department. The city commission meets the first Wednesday of every month, so the architect and a city commissioner decide at the precon that the contractor should provide a draw request by the 18th of every month. That gives the architect from the 18th until the 1st Wednesday of the following month to make an inspection and forward the draw to the owner. The contractor is paid the week after commission meetings, about a 30 day cycle.

A university or federal job would be more formalized, with a predetermined date for turning in a draw and very specific requirements demanded of the contractor concerning the transmission of shop drawings and other submittals. For example, specific project management software may be required.

In any event, the preconstruction meeting works out the logistics of the paperwork (or electronic) interaction between the parties.

Photo documentation

In renovation projects, there is often a general requirement for the contractor to make photo documentation at the beginning of the project, which is a good idea for all concerned. It is to the contractor's benefit to have a record of the condition of beginning finishes. Otherwise, later scrutiny of wall paint blemishes and countertop scratches may be perceived the fault of the contractor.

The subject of before and after condition of finishes can become a huge concern at the time of substantial completion. Did the increased traffic of material handling cause the wear and tear and scratches on the floor? When the new door opening was made in the wall, how far down the hall should the contractor be responsible for painting? There are scratches all the way down the corridor...

The superintendent must take extreme care to protect corridors where wall openings are made. And, if the floor needs protection, using either plywood or cardboard or other material, this is cheaper than replacing flooring! Temporary protection is a cost that should be a part of the job overhead portion of the estimate, and so are photos, because of the time it takes to take the pictures and the printing expense.

In addition to before and after "precautionary" photos, the specifications may require "progress photos" be made weekly or monthly. The requirements for progress photos may include a certain quantity, arrangement by date, and other protocols. The project may even require that a professional photographer be used.

Photos are a cost; that's job overhead for the estimator, whether the photos are shot with in-house personnel or professionally.

Construction schedules

The owner's instruction in Division 01 about the schedule might be to:

> Graphically show by bar chart the order and interdependence of all activities necessary to complete the work, and the sequence in which each activity is to be accomplished. Highlight the "critical path" through the schedule to illustrate those interdependent activities that cannot be delayed without impacting the overall completion time.

The expense to prepare a schedule is usually not "charged to the job", in other words not a part of the costs in the estimate. Since a schedule is a part of every project, the in-house work done on them is considered project management, and this time is expensed to home office overhead. Most large construction managers and GCs provide their own in-house schedules. These are periodically updated as the job progresses. However, if special software is required, or an outside firm or consultant has to be hired to complete a schedule, then this cost shows up on the estimate.

Section 5 During construction and the project site

Supervision

The contractor shall provide a qualified, full-time superintendent at the project site throughout construction. The superintendent shall maintain, at the job site, a complete and accessible file containing all submittals, shop drawings, and samples approved by the architect/engineer as well as supplemental erection or installation instructions.

Sometimes there is a further requirement beyond typical supervision for technical observation of a trade to ensure compliance with good practice. An example of this is a Division 1 requirement to:

> Maintain a full-time non-working supervisor on the job site during roofing work in progress. Supervisor shall have five current years minimum documented experience of roofing work similar to scope of specified roofing.

Supervision is a cost directly attributable to a job. The expense is placed directly on an estimate just like the cost of concrete. The only difference is that it might appear on the job overhead portion of the estimate, and if so might be the largest item of job overhead. To an accountant, this is an "indirect cost", because it is not a material or labor item that is part of the structure. But it is an expense the contractor cannot do without – jobs don't run themselves, and a hardworking and demanding superintendent is a fixture on any successfully completed job.

Site

There are many factors about a site that can influence job overhead costs. They may or may not be mentioned in Division 1. A new building on a large open site presents the fewest issues. At the other extreme, an extensive renovation situated within a congested area has a lot of job overhead costs concerning staging, material handling, deliveries, offsite storage, and parking. There are many logistics to deal with at a busy site, and estimators must account for their costs. At a campus, workers sometimes have to park away from the site and the contractor has to arrange bus or van service to take them back and forth.

Utilities

Temporary utilities and outages may be covered in Division 01 General Requirements.

Planned utility outages are occasionally required for repairs, maintenance, or construction. Major owners, such as a university, formalize the process of planning and executing a utility outage. The contractor will provide a written request ahead of time, and the outage will not proceed until authorization is obtained. Notification will be made to affected departments. Simply digging in the ground may require a "dig permit".

On private property, contractors can call city utilities departments for the line locations. Electric, gas, water, and sewer lines will be located by paint sprayed on the ground.

Fire safety

The contractor is required to take responsibility for a buildings' safety during construction.

Some Division 01 language is as follows:

A Prohibit smoking within buildings under construction. Designate area onsite where smoking is permitted. Provide approved ashtrays in designated smoking areas.

B Establish fire watch for cutting and welding and other hazardous operations capable of starting fires. Maintain fire watch before, during, and after hazardous operations until threat of fire does not exist.

C Maintain fire extinguishers, fire blankets, and rag buckets for disposal of rags with combustible liquids. Maintain each as suitable for the type of fire risk in each work area. Ensure that nearby personnel and fire-watch personnel are trained in fire extinguisher and blanket use.

D Sprinklers: When sprinkler protection exists and is functional, maintain it without interruption while operations are being performed. If operations are performed close to sprinklers, shield them temporarily with guards. Remove guards at the end of work shifts, whenever operations are paused, and when nearby work is completed.

E Fire-prevention plan: Prepare a written plan for preventing fires during the work, including placement of fire extinguishers, fire blankets, rag buckets, and other fire-prevention devices during each phase or process. Coordinate plan with owner's fire-protection equipment and requirements. Include fire watch personnel's training, duties, and authority to enforce fire safety.

Maintenance of traffic (MOT)

Construction sites often require temporary barricades and warning signs in the street. An urban setting, for instance a downtown project, needs a full treatment of signage (flashing lights and warning signs) starting several blocks from the site. This can get complicated, with departments of transportation often requiring a "traffic control plan" be drawn, submitted, and approved prior to implementation, so there are companies that specialize in this service. Sign and barricade companies do this; they have people who know what the department of transportation requires and will quote the job as a rental, per week or month. It is up to the prime contractor to multiply the rental times the estimated number of months they will be needed.

The same process might be followed with the campus police for a university project, or with the city public works department, etc. MOT can include lane closings shifting during construction, flashing lights, warning signs, message boards, and other kinds of barricades and "orange cones". It can all shift locations during the work and be a large expense.

Mockups

A mockup is a small piece of the work actually built onsite as a sample of what an assembly of materials or products is going to look like. The specifications may require that a short brick wall be built onsite that is representative of brick and mortar work. Once the mockup is built and observed and the brick coursing and mortar joints are all to the architect's liking, the wall will stand and remain in place, an example or standard to match when the actual walls are constructed.

The cost of building mockups should be included in the estimate.

Temporary or construction facilities

The word "facilities" is used to describe a host of infrastructure subjects from utilities to temporary access roads. It includes temporary toilets, trailers, and signage both in and out of the building.

The format in a project manual of Division 1 General Requirements is the same as for the technical specifications. That is – *General, Products, and Execution.* The following collection of facility items are grouped how they would be seen in Division 01 of a project manual.

GENERAL

Work Included:

1. Temporary utilities such as water, electricity, and telephone.
2. Field offices and sanitary facilities for the builder's personnel.
3. Enclosures such as tarpaulins, barricades, and canopies; traffic control and pedestrian control devices.
4. Erosion control measures.
5. Directional and informational signage.

Submittals:

1. The builder shall present a jobsite management plan in the form of a scaled, marked-up site plan for the owner's review at or prior to the preconstruction conference. This drawing shall identify, at a minimum:
2. Temporary fencing with gates and point(s) of access.
3. Materials delivery and storage areas.
4. Field office or storage trailers.
5. Temporary accessibility features including paved or unpaved roads, sidewalks, bicycle paths, ramps, curb cuts, canopies, barricades, or other means of maintaining safe and ADA-accessible routes through or around the site.
6. Waste collection (dumpster).
7. Signage and striping.
8. Paths for emergency egress.
9. Onsite staff parking.
10. Tree protection.
11. A formal maintenance of traffic (MOT) plan.

PRODUCTS

1. Temporary water including necessary piping.
2. Temporary electricity including necessary wiring.
3. Temporary telephone and internet.
4. Temporary sanitary facilities.
5. Field offices and sheds.
6. Temporary fencing.
7. Temporary signage.

EXECUTION

1. Install signage simultaneously with fencing and temporary roads or paths.
2. MOT plan shall be per department of transportation (DOT) standards.
3. Take precautions to protect roofs or other critical openings in the building.
4. Remove temporary facilities as progress of the work will commit.

Temporary controls

The following can found in Division 01 General Requirement items or on the plans.

The word "controls" as used here should not be confused with some electrical and mechanical devices named controls. A thermostat is used to control air conditioning but is not the subject at hand. These controls are about limiting dust, pollution, and erosion during construction.

Pollution control can be changing air conditioning filters often during construction, or of limiting the use of VOCs (volatile organic compounds). Pollution control can also be the collection of lead paint removed from old walls. Polyethylene or tarps are required to be placed on the ground when lead is removed, so that the pieces of removed paint can be bagged up and properly disposed of.

Erosion control is "on the ground" prevention of soils and grasses to migrate to areas away from the work. When excavating earth, temporary control means keeping soil and grass where it is, and not letting it wash or blow away. Silt fences, which are short barricades of fabric, are a common expense in the use of ground control. The sitework contractor will often have this in their scope of work.

Testing

Testing is covered both in Division 01 General Requirements and within the technical part of the specifications. Division 1 will contain *general contract language*, and more specific testing requirements will be found in the separate technical divisions.

For the estimator, the costs for testing are often placed in the job overhead portion of the estimate. Testing for concrete is best judged after all quantities are entered on the estimate. Like supervision or other job overhead expenses, it is easier to finish the main part of the estimate and then estimate the job overhead, including testing, tools and expendables, temporary utilities, etc.

Much testing is required for earthwork and concrete. This can be a part of a subcontractor's price or in the scope of the prime contractor. Some earthwork testing is to make sure that the ground (the site) is compacted sufficiently to hold up the structure. The concrete tests simply determine that the strength of the in-place concrete matches or exceeds the strength used to design with, in psi (pounds per square inch).

The jobsite testing for concrete includes "taking cylinders", which is placing wet concrete from jobsite concrete trucks into plastic canisters, hence the use of the term cylinders. Typical concrete specifications require four cylinders be taken from each pour (not each truck), with the first cylinder being tested. The cylinders are crushed in a lab at 7 days, the second at 14 days, the third at 28 days, and the fourth saved for possible later use in case something is wrong.

Borings, which are made with cylindrical pipes hammered into the ground to gather earth samples, allow the study of earth layers back at the lab. The owner often completes this testing before the bid and shares the information with bidders. After all, the ground is the owner's property! Often called a "geotechnical report", it concerns the strata, or layers, of earth brought up from drilled holes in the ground. The test report will state that there is, say, sand to 4′ deep, clay from 4′ to 6′, grandma's remains under that, then lime rock on down to 20′. The borings also reveal the depth that water is found, known as the "water table", an important statistic for the estimator.

The depth of the water table is the subject for risky estimating. The testing language in the specifications will often state that the owner has provided a geotechnical report dated months ago, but that it may be inaccurate at the time of construction. *However, any risk of water removal required to meet the testing goals of earthwork and concrete lies with the contractor.*

Construction waste management and control

This is found in Division 01 General Requirements.

The cost to remove debris at a construction site can be more accurately estimated than might be imagined. Estimators can predict the amount of debris in a methodical manner. The first consideration is that there are two kinds of debris – demolition debris and the debris resulting from general construction.

The hauling away of demolition debris can be considerable in both labor and material (dumpster expense). Subcontractors bidding demolition usually include the "hauling away" of their own work – the debris resulting from demolition. They load their own dumpsters and trucks and haul it away at their expense, and sometimes own the dumpsters and trucks.

General cleanup (labor hours) is a function of job duration, and there can be considerable labor but only a small amount of debris going into dumpsters. The debris accumulated each week or month on a construction site, due to regular construction activities not associated with demolition, will accumulate to a quantity of dumpster pulls the estimator will relate to on prior jobs. Perhaps that experience is that small jobs require one dumpster per month be hauled away, medium-sized jobs require one dumpster per week, and larger jobs a dumpster per day. It is plugged into the estimate as a material expense.

Cutting and patching

This is found in Division 01 General Requirements.

This is on everybody's list of General Requirements, the item of cutting and patching is a common topic for all project manuals. Simple as it sounds, owners and architects must consider it extremely important! Cutting and patching is covered in new projects, renovations, government jobs, and private ones.

Must of this subject is about simple instructions to repair existing surfaces when installing pipe penetrations or openings required for mechanical work. There is a lot of repetitive verbiage about restoring to the same condition as before, using new materials, and matching existing finishes.

It is hard to address cutting and patching in full on the plans, so it is described in much detail within the General Requirements. Some of the language is:

A Restore work with new products.
B Fit work tight to pipes, sleeves, ducts, conduit, and other penetrations through surfaces.
C Provide openings in elements of work for penetrations of mechanical and electrical work.
D Refinish surfaces to match adjacent finishes.
E Use materials identical to existing materials.
F Where removing floors, walls, or ceilings that extend one finish area into another, patch and repair surfaces into the new space.
G Restore surfaces to original conditions after installation of other work.

Cleanup

This is found in Division 01 General Requirements.

The prime contractor will have, at a minimum, cost categories for interim cleanup, final cleaning, and dumpsters. Cleanup is a major job overhead expense. It can also be a lot of work for a subcontractor working for a construction manager, who may require that all subcontractors provide a certain number of hours per week cleaning up at the direction of the CM's superintendent.

Interim cleaning, the hours per week it takes to keep the building and site clean, is largely determined by job duration. The estimator will be a better judge of cleanup needs after the rest of the job is studied. Interim cleanup is usually estimated to be so many hours per week times the duration of the job. The job size and the estimator's experience with similar jobs will determine what is estimated for cleanup.

A one-million-dollar job completed in eight months or a year might require two days per week of cleanup. A ten-million-dollar project completed in eighteen months might need one or two people working full time to keep the project clean. A thirty-million-dollar building completed in two years might require four or five people working full time.

The cleaning of floors is sometimes covered in the technical specifications, where the products of vinyl tile, carpet, and hard tile are covered, as well as in Division 01 General Requirements. It is usually the technical specifications where waxing and cleaning of floors is specified. Flooring subcontractors do not clean floors! That is a job overhead expense, done at the end of the job.

Final cleaning can have several components and require cleaning subcontractors instead of the general labor used for interim cleanup. Some general requirement language concerning site cleanup is as follows:

A Clean project site, yard, and grounds in areas disturbed by construction activities, including landscape development areas, of rubbish, waste material, letter, and other foreign substances.
B Sweep paved areas broom clean. Remove petrochemical spills, stains, and other foreign deposits. Rake grounds that are neither planted nor paved to a smooth, even-textured surface.
C Remove tools, construction equipment, machinery, and surplus material from project site.
D Clean debris from site, roofs, gutters, downspouts, and drainage systems.
E Clean exposed finishes, including windows. Touch up minor finish damage. Repair and restore marred, exposed finishes and surfaces. Replace finishes and surfaces that cannot be satisfactorily repaired or restored or that already show evidence of repair or restoration.

PART 12

Ethics

Section 7 Payment

Backcharges
Timely payment
Payment to vendors invested in the project (VIPs)
Payment to uninvested vendors

Section 8 Performance

Leadership by the GC or CM
Benevolence of the prime contractor

Section 1 Introduction

Codes of ethics

The codes of ethics of two national estimating organizations are recognized: the Professional Construction Estimators Association of America Inc. and the American Society of Professional Estimators (ASPE).

Many of the subjects discussed in these codes are discussed in this chapter. The organizations point out that estimators must be qualified by education or experience and should not stray from their area of expertise. A high standard is placed on the estimating profession, and the estimator should not accept quotations from unqualified companies or suppliers.

The preparation of detailed estimates from a full set of completed plans is paramount. Note that this precludes bidding some of the private jobs described in this chapter.

These organizations link low prices to inferior work. Their advice is to not undertake work believed to be unprofitable.

Bids should not be divulged to competing subcontractors or suppliers, prior to bid time or after, in an effort to drive down prices.

Subjects covered in this chapter

Ethical behavior discussed in this chapter concerns estimating, pricing, and bidding a construction project.

The steps taken to arrive at a bid are retraced by estimators repeatedly, and many well-worn paths are laden with transgressions. These ethical lapses have their own language, such as "shopping", which is not going out to buy groceries, and "last looks and final offers". Estimating ethics defines the correct and right moral action, or behavior, to take with decisions that are, while legal, questionable.

This chapter doesn't cover the illegal part – bid rigging and gift giving. The subject here is bidding ethics, and the topic of payment is only briefly covered and limited to bidding-related payment issues. On the topic of performance, since construction of a good job built on time is the most important thing that a prime contractor can do, a few comments are made about leadership, as principled team leadership of subcontractors helps to ensure a good performance.

Various bidding environments

There are several bidding "arenas" in which contractors operate. What is an ethical consideration in one environment may not apply in others. These different arenas allow room to argue ethical topics, because construction is an extensive industry. Bidders work for both public and private owners – there are strict competitive bids, and there are negotiated bids. Contractors can face a wide range of circumstances.

Construction projects include homebuilding, room additions, private jobs for bar and restaurant owners, large private jobs like hospitals, and small and large government work for cities, counties, and state and federal governments. There are jobs in a city with subcontractors all around, and then jobs in rural areas with only repair folks nearby. There are good owners and there are scoundrels. There are good and complete plans and there are plans with insufficient information. These factors can create quite disparate bidding environments.

Consider a community of owners and repeat bidders, in a city of some size, where most participants have worked together before. There is a bid for a new public library with a full set of plans and specifications already reviewed by the building department, ready for a permit. Four or five prime bidders attend a pre-bid meeting, presided over by the city, the owner. These contractors generate lots of activity in the local subcontracting and supplier community, not only within the city but also in the surrounding county and beyond. The bidders turn in bid bonds (guaranteeing that they will not change their bid and will sign a contract for that amount) and will furnish a performance and payment bond as a contract requirement if they are the low bidder. The city has set a firm bid time and date, and a bid one minute late will not be considered. The owner requires a "list of subs" to be turned in with the bid, naming who they are going to use (subcontractors cannot be changed without reason or owner request). There is a lot of pressure on all of the contractors to estimate the job carefully and accurately. The subcontractors and suppliers all work with the bid deadline, wanting their prices to be considered before the bid, not after.

Contrast the strict protocols of the above library bid with a shopping mall project for a developer known for building simple and inexpensive projects. The developer has some plans done, which consist of a floor plan and elevations but not much else, certainly not enough information to get a building permit. Several general contractors are asked, over a period of several months, to bid the project. There is no bid date and the plans lack detail. This is a rather loose bidding environment.

The ethical standards discussed in this chapter can be effected in the city bid, but not the mall project. The developer cannot expect contractors to give him or her the same considerations as those providing bids for the city. These considerations include a prompt bid price with guarantees for both honoring the bid and performing the construction. Indeed, a contractor providing a price to the developer might be better off providing a ballpark number with a host of exclusions, limitations, and qualifications. Or, no bid at all.

The bidding environment most applicable to the ethical practices discussed in this chapter is one in which repeat owners, who have hired architects and engineers to prepare full and complete plans and specifications, require strict bid dates and bid bonds and performance and payment bonds. The General Conditions of their contracts are most likely from AIA 201 or the federal counterpart, and construction law is understood by all of the parties. A full submittal process will be conducted by the contractor with full review by the A/E. This is a high-end structured construction environment. Ethical behavior is more applicable in this environment than in the mall project.

Section 2 Bad practice

Shopping

When the prime contractor shares one subcontractor's price with another for the purpose of inducing a lower price, it is called shopping and occurs downline. A huge ethical standard is broken. "Bid shopping" can happen on bid day or after.

Suppose on bid day there are four roofing bids from $250k to $325k, and they've all been turned in for an hour. It is 1 pm and the bid is at 2 pm. Then a roofing contractor, who was at $300k, revises their bid to $249k, which makes them one thousand low. Now isn't that magic? It is obvious that this guy has been told the low bid. Now, the only people in the world at the time who know the amount of those bids are the subs calling them in and the prime bidders. You can bet the subs have probably not talked to each other. It is obvious one of the prime bidders talked to the new low bidder, and the (formerly) low bidder at $250k was just robbed.

The sub at $249k might reward Loose Lips (the prime with the shopping basket) with a bid of $240k, but still make sure of being lower than the others by bidding $249k to them. This $9k spread is a reward to Loose Lips for "helping" him or her. These kinds of players benefit for a day, but they subvert the competitive bidding system. They also take a risk with a low price for the work.

Before the occurrence of the next bid, estimators will probably have figured out who Loose Lips is, or will figure it out over a period of several bids. On subsequent bids, honest prime bidders hope that the subs sabotage Loose Lips. But this doesn't always happen, because subs don't want to be left out – everybody wants a job. Our industry is not tough enough, or consistent enough, in excluding shoppers. This is the clear stance of the ASPE, that no bid should be given to known bid shoppers, and is admirable. The more common occurrence, at least observed by the author, is that many subcontractors still bid known bid shoppers, but do so with a higher price.

"Peddling" after a bid

Bid peddling occurs upline, when the prime contractor contacts the owner after a bid and attempts to negotiate a previously given price downward (the contractor lowers his/her price). It is also when a subcontractor initiates negotiations and offers to lower a price to the prime contractor.

As stated in Canon 7 by the ASPE,

> Bid peddling occurs when a subcontractor approaches a general contractor with the intent of voluntarily lowering the original price below the price established on bid day. This action implies that the subcontractor's original price was either padded or incorrect. This practice undermines the credibility of the professional estimator and is not acceptable.

Last looks and final offers

Last looks and final offers do not have any place in public bidding. Last looks are what you do when grandma dies, and final offers are perhaps made to car salesmen but not to building owners.

A last look is one in which an owner, at the end of a bidding or negotiation period, allows one prime contractor to know what the low bid is and the opportunity to beat it and get the job (there are a lot of variations of this situation; see the following paragraph). Sometimes it is a friend or someone who has worked long and hard to get the job, but sometimes a look is given to an undeserving company. A look can also be a prime contractor revealing a low bid to a subcontractor, sometimes at the end of arduous sessions of revisions and value engineering. An owner or prime contractor may feel beholden to the contractor who has helped get the project to this stage. The only good thing about a last look is, presumably, the owner or prime contractor will not continue the process of bidding and is saying, "I won't keep this up, I am through negotiating." Of course, this cannot be counted on.

There are exceptions to a high-minded stance concerning last looks. An exception to last looks can occur when there are alternates to a base bid, and there is confusion among the bidders as to what should be included or excluded in the alternate. This usually results from a poorly written alternate(s) by the architect, which is a common occurrence. There are perhaps other exceptions to the general rule that last looks are unfair to the other bidders.

Private jobs can be a dogfight with Wild West bidding. The owner or prime bidder who puts others through these situations has no credibility and is not interested in a fair bid. If the contractors would insist on a real *rebid*, it would help put the owner on an honest track. What helps keep bids honest and straightforward is the insistence on a bid date and time, and an official bid opening where all the bids are read aloud. Good luck.

Sometimes an owner will ask a contractor for a "best and final offer". This would not be happening, of course, in a straight-up competitive bid opened and read aloud. The best and final offer constitutes a *rebid*, and it sometimes occurs after many changes and revisions. If one bids private work, there will occur those times of over budgets and rebidding, where an overworked contractor may feel like it is his/her due to get a final chance for the job through a last look.

Suppose a bidder works long and hard through repeated revisions of the plans, even offering value engineering suggestions. This bidder may be partially responsible for getting a project within budget. He or she has worked hard. Then the owner, showing no allegiance to the hard worker, allows a new contractor to enter negotiations, or continues negotiations with original bidders. The owner shares the cost savings ideas and revisions all around.

The bidders feel like staggering boxers in the 12th round. There they are, bloody and barely standing, having rebid the job in eleven rounds already. Then the owner calls and says, I'm out of revisions, there will be no more changes, I know I have your latest price, but hey you two guys are close together, give me your *best and final offer.*

These paragraphs apply to the relationship between owners and contractors, as well as contractors and subcontractors/suppliers.

The best thing a contractor can do when he or she foresees such "endless rebidding" is to bow out of the game, or to charge for estimating and value engineering services.

Section 3 Estimating

In-house estimates

If the contractor has the resources (the in-house labor and supervision), and the estimator has the education and know-how, estimates are prepared for a trade that can be accomplished by company employees. If the contractor has carpenters, some framing may be estimated, labor and material. This may only be true for small jobs, and the contractor might seek subcontract prices for a large framing job. Or conversely, the estimator may accept a subcontract price for the framing of a small job if it is close to the in-house price. After all, there is always risk when it comes to labor, and sometimes going with a subcontractor's guaranteed price is the best way to go, even if it is a little higher than what an estimator thinks their company can do it for.

What if the GC figures his/her own price but does *not* have the capability to do the work? Perhaps the estimator knows the trade well enough to come up with a ballpark number of $50,000 for metal studs and drywall work. However, bids of $60,000 and $70,000 are received. Can the GC ethically use $50,000 in the bid? And then get more bids later when there is time? No! The contractor should not seek more bids after the bid date if there were two or more qualified bidders on bid day.

If the GC cannot accomplish the work with qualified employees, or the estimator is not qualified to bid the trade, the estimator should not use his/her own estimate on bid day. However, if qualified, there is nothing wrong with the prime contractor using an in-house estimate.

Having figured an estimate, the prime should not use an in-house quote to reduce a subcontractor's bid.

Bid mistakes and bids that are toooo low

The 5% bid bond penalty raises the stakes of bidding to such a high level that arithmetic mistakes are uncommon. A trained and experienced estimator handles each bid the same way, follows a format, and makes a major investment of time. Shortcuts are not taken. The serious nature of the money involved is a strong deterrent to bid mistakes.

Subcontractors do not usually "put up" a bid bond. However, mistakes are uncommon because they are figuring costs for their own trade and do not deal with the diverse components of a GC or CM bid. However, arithmetic mistakes do happen. If a subcontractor, whether or not prompted or questioned, acknowledges a mistake *before the bid*, the bid is of course corrected.

If the prime bidder receives three or more bids (it is harder to tell with just two!) and one of them is "way low", the estimator should evaluate the situation. Sometimes a discussion of scope with the subcontractor can quickly determine what the problem is. If this doesn't turn anything up, the estimator considers how well the players are known, whether any of them are erratic or infrequent bidders, and other relevant factors.

When there are three or more good bidders, and one is very low, the prime bidder should not be silent. If a mistake is not revealed until after a bid, chances are the subcontractor will not honor it, so it is better to find out about it sooner than later. It is the prime bidder that has to have a bid bond, not the subcontractor. Nobody wants to win an award based on a mistake.

If the prime bidder suspects a mistake in a bid, the best thing to do is flush it out and have it exposed. There should be a conversation like this, "Rick, just checking your arithmetic. I've got your price here and I want to use it, but I'm just making sure that it is correct. I would like for you to confirm it to me. And, by the way, what is your bond rate?" Rick will know exactly what's going on, having been told so in estimating language, and will start to sweat profusely. He knows he is low by probably a wide margin, and knows a bond may be required. In any event, to avoid a misunderstanding, there is nothing wrong with telling a subcontractor that they are "way low".

The author has been in this situation several times. After questioning a low bid, a mistake is often acknowledged within minutes, as arithmetic errors are found quickly, and it usually places the bidder out of contention. The subcontractor does not use the situation to raise their price just enough to remain the low bidder, and it would be unethical for them to ask, "Where do I need to be?" Their interest is in getting out of a mistake and not being awarded a bad job. There really was a mistake. The prime bidder has averted a catastrophe and prevented Rick from having a heart attack.

If the prime bidder suspects that a bid is too low, *it should be questioned. This is both smart business management and the ethical thing to do.* This may seem odd, because of the rules of bidding about being silent. However, when a bid mistake by a sub is suspected, it is a special case due to the seriousness of the situation for the prime. The prime bidder, the leader, must protect his/her price – it is guaranteed, the sub price is not!

In analyzing the ranking of subcontractor bids, if the low bid is not used, it should be for a serious reason. These include poor past performance and/or financial problems. The ethical test for these reasons being valid is whether the prime contractor, if challenged, would put them in writing.

Section 4 The bid price

Low, best, and responsible bids

What are the goals and purposes of the competitive bid system? How does it help enforce ethical behavior? Consider the terms that follow in this section.

The bid system should deliver low, best, and responsible bids to the owner. Government purchasing agents and facility directors for major businesses need "low, best, and responsible bids". The owners have the money! For them to keep spending it, over and over, on a system that determines the price paid for construction projects, the process should repeatedly deliver several good bids. The owner wants a stable of reliable contractors to line up for every project. Each contractor wants to pick up a job now and then, and that means there can't be too many bidders. A healthy American economy provides the supply and demand for this. However, for it to all work out equitably for all parties, there are ethical responsibilities on both sides, contractor and owner.

For the large client that needs construction projects completed often, the purpose of the competitive bid system is to deliver low and responsible bids *over a period of time*. It is best, when considering the rules of bidding, to picture the owner as a repeat owner such as the government or big business, not a private individual with one job. Like a card game with many hands or rounds, the rules need to be set up to work over and over for both sides, bid after bid, even if every single one isn't perfect.

The policies, procedures, and ethical considerations of the players (the contractors, designers and owners, cities, counties, school boards, states, and the federal government) are tailored towards the purpose of serving the owner through a fair bidding system. The owner is a repeat client that awards bids on a recurring basis. The contractors are repeat bidders, bidding the plans by architects and engineers that do business over and over with large clients.

Competitive bids

A competitive bid is not so low that it would foster nonperformance, nor does it take advantage of the owner by being too high. A good bid is not too high or too low, it is "competitive". But more importantly, having three or four bids close together, or maybe a couple of them close together and then two more 10% or 20% higher, or a spread of four bids without one being too low, defines good competitive bid results. This is reassuring to the owner, and the competitive bidding system has delivered.

The owner will only do business with the low bidder, but it wants several qualified firms to line up and compete again and again to provide competitive bids. That means that all bidders must show up at the same time at the same place to turn in a bid in an "open setting", meaning that bids will be read aloud for all to hear. Otherwise, unfairness can reign, and honest bidders will stay away, leaving the owner with a gallery of rogues.

Firm bids

A good bid is "firm", meaning that it won't change, that it will be "honored" by the contractor, and a contract will be signed for the bid amount. This sounds straightforward enough, but it isn't an easy matter. It takes a lot of work for a responsible estimate to be prepared, one that the contractor is confident of and will not waver from.

The incentive to do all of this work is that the owner is going to award a job to somebody, no ifs, ands, or buts. That mall owner may be looking for years, but the public owner is serious. The tax collector needs a larger office now!

A firm bid is a major foundation of the competitive bid system. In the bidding of public projects, "firm and timely" bids are rigidly controlled. When bid preparation falls short, it is often later revised or retracted. And when prices move around, the bidding system collapses and becomes a free-for-all. Except for perhaps the failure to perform and complete a project, there is no greater ethical principle than for a contractor to turn in one and only one price that remains the same and is a responsible bid price that the contractor will sign a contract for. This is enforced through bid bonds, and it is covered in another chapter.

The word "firm" means the price won't wiggle, it will not budge, it will stay the same all the way to the contract signing. It is extremely important that a bid does not change by even one dollar. Why does this have to be such an ironclad rule?

Consider a card game of one hand with rules made for fair play, a betting game with real money and high stakes. Contractors, like card players, sit around a table, at least in many open bid situations (some public owners now allow electronic bids, which are still "open bids" so long as the owner publishes the bids the next day). The owner's representative, say the architect or purchasing agent, like a dealer but not dressed as well, opens the bids.

While betting on cards depends on the card player's hand, the "bet" that the contractor makes depends on the risk inherent in the costs, the estimate. Is the contractor's staff familiar with this kind of work? Can that out-of-town plumber deliver on their low bid? Why is the low masonry bidder on COD (cash on delivery) at the concrete and block plant? Are the costs, as the estimator has figured it, a house of cards? Or are they a good representation of the eventual costs? That's why the card game analogy is apt – bidders do gamble in that risks are taken, and estimators are studious evaluators of risk.

The estimator has an estimate completed by the day or night before a bid with "plug numbers" used for those costs that are not going to be received until the next day. The cost of major products and subcontracts may move around all morning on bid day as actual quotes and sub prices replace plugs. The contractor is only going to add a few percent for overhead and profit, and this amount can certainly look small in comparison to the total costs. When the costs quit jumping around (hopefully) shortly before the bid is due, the contractor hopes he or she has them "covered". That is the estimator's job, projecting the eventual costs, just barely covering them.

The contractors study their hand. They consider how long the job will last, compare it to the owner's completion date, and evaluate the probability of completing the job on time for the estimated cost. Management will often dictate the markup. The job profit will go towards home office overhead – salaries, rent, and insurance, as do all of the jobs, and if it is a good year there will some left over for a profit. With a slim margin for victory, a firm bid is turned in.

The low bidder, committed to a firm bid, is content after the bid even if the other bidders are somewhat higher, because of confidence in the bid. Of course, "leaving money on the table" (the spread between the first- and second-ranked bids) is not fun, but it happens. Hopefully, the drama of the bid will not continue, for example a low subcontractor or supplier finding a mistake and withdrawing. Usually, this does not happen and all is well.

Regardless, the price to the owner will remain the same and will be honored. A firm bid, together with a timely bid, are bedrock principles that make the bidding system work, and if uprooted destroys the integrity of the system for the owner. For the contractor, a bid award will be lived with for months on end based on an amount provided on bid day, a onetime shot at the job.

Every single estimate should be completed with the best accuracy possible. If there is any compromising of detail or shortcuts made, if there is anything less than a fully detailed estimate, that is the job the contractor is likely to get! All of the serious study that a company is going to do on an estimate for a project, 100% of it, must occur *before* that bid is completed!

Some estimators consider that honoring a price rises to the highest level of good practice.

Fair bids

Bidding procedures should not allow advantage or favor to one bidder over another. As in a game of cards with bidding or betting, strict guidelines must govern the behavior of players or the game breaks down and becomes dysfunctional.

The large owner, besides wanting good and responsible bids, feels an obligation, realizing its enormous impact on the construction industry, to give an equal chance to all qualified bidders to get an award. The federal and state governments, more so than any other owners, are charged with considering the rights of all bidders and treating everyone equally. For the market to provide a good stable of bidders, it is important for the owner to treat bidders fairly and give them all an equal chance to get a job. The government has more experience at creating a level playing field than anyone! This level playing field is created by industry and owner criteria that make the competitive bidding system work best for owners and contractors.

These factors include:

1 A set bid date and time.
2 Bidders must honor their bid by going to contract for their bid amount, or suffer a penalty.
3 No shopping or peddling.
4 The owner must present the project through well-defined plans/specs by professional A/Es.
5 The owner should report clarifications to the documents at the same time to all bidders.
6 The owner should use the low bid.

Contrast this with private bids. The private owner may only be concerned about getting a price one time and have no concern about having a fair bid. If the owner does not have a specific time for a bid opening and collects them over a period of days or months, so be it. If someone wants to change their price the next day, have at it. If the owner has a buddy in the hunt and tells him or her what the other bids are, the other bidders are cheated. And, after all this happens, if the owner decides to have a "rebid" and start all over, everyone's time has been wasted.

Bidding multiple prices

There are many reasons that a subcontractor might quote various prices to the prime bidders on bid day. After all, good payment and good supervision are factors that make a good job for all. So, one can imagine a range of bids. However, a known bid shopper should not be given a bid at all. The best way to treat these rogues is to completely ignore them with a zero. This is an admirable stance; if the bidder upline is going to reveal your price to others (shop), then do not play ball with them at all.

However, what is seen in practice is that subcontractors and suppliers do in fact bid the unscrupulous bidder. But the price may be inflated, say by 5%–10%. Or, depending on how bad the sub bidder needs a job, they may bid the rogue the same price as others.

It is entirely up to each subcontractor or supplier to decide whether to provide the same price to everyone. Usually there is one price or quote to all. However, there is no obligation, legally or ethically, to give the same bid to everyone or even to furnish all of the bidders a price, even if they are not shoppers.

Loyalty is created after successful projects are completed. Good project management and field leadership makes jobs run on time and allows budgets to be met. A good relationship is why a subcontractor will give a lower price to an old business associate.

However, there is always pressure on bid day to get a job. The best odds for getting a job are to bid everyone the best competitive price that can be offered. Consider a sub or supplier bidding an important job and there is one GC that is non-local. Is the community of subs and suppliers going to favor the locals? Each job has its own dynamics, and if the out-of-towner has made themselves known to bidders, has visited the site, and has conversed with the subs and suppliers several times about job details and knows the job inside and out, why not give them a good price? This is not an ethical consideration – business is business.

Often the best company to get the job is the one that studies it more than anyone else does. If they are from out of town, so be it. Each job is a one-shot deal. Whoever gets to know the job best, and discusses the details of it multiple times with the bidding community, may be the best contractor for the job.

It is not unethical for a subcontractor to bid various prices to multiple bidders, even if they are not shoppers.

Minor pricing disputes

The participants of minor pricing disputes often face ethical considerations. These disputes can arise because of confusion (conflicts) in the documents, more than one interpretation of the documents, or an interpretation that has been made incorrectly. For example, consider project specifications that do not specify the grades of finish wood to be used in a remodeled courtroom. The material for a 4′ high wainscot is called "boards" on the plans in one place and "beadboard paneling" in another. What is it – 1 × boards or 4′ × 8′ sheets? The individual boards will take a lot longer to install.

A submittal for beadboard sheets was rejected by the architect and instructions were given to provide 1 × 6 tongue and groove V joint pine. The individual boards would cost the framing subcontractor a couple of thousand dollars more than figured, and their quote for the whole finish carpentry job was only $20,000, so they were not about to do it for free. They had relied on the paneling note.

The GC talked to the architect about a change order, but met reluctance. If there was confusion, why hadn't clarification been asked for during the bid? Besides, the general contract was for $3 million, and the architect didn't want to go to the owner for a couple of thousand.

In this case, the GC was being treated unfairly by the architect, but ethically, the GC should not pass this treatment on to the subcontractor. The GC and sub had an agreement, and the sub had used the words "beadboard paneling" in the quotation, just what it said on the plans (in one of two places). The right thing for the contractor to do is pay the sub the additional two thousand, regardless of the whether the prime contractor is paid.

There are circumstances where smallish disputes are handled between the GC and sub without the GC going to the owner for payment. Sometimes a contractor finds that a hard line is taken by the A/E in every instance, without any representation of the matter being made to the owner. A possible course for the GC to take is to bypass the architect and present a claim directly to the owner, perhaps combining several small issues together.

Section 5 The bid date

Timely bids

Consider a contractor traveling around the county doing repeat projects for a hotel or restaurant chain, building similar projects over and over. The contractor, job in hand, may roll into town and apply for a permit. The superintendent has been charged with getting local bids and meeting a budget already prepared by the estimator from historical data. The superintendent finally finds the desired site contractor for the clearing two weeks before the job is supposed to start, and will be collecting subcontractor bids for weeks to come. *There is no bid date.*

Public bids are opened precisely when they are due, to the minute. The official watching the clock and receiving bids, often a purchasing agent, facilities director, or architect, will say, "It's two o'clock and no more bids will be accepted." The bids will be on a table in front of him or her, typically in large manila envelopes, sealed with tape. They are opened and read aloud, often in no particular order.

First, the inclusion of a bid bond is verified. Next, the architect will look on the bid proposal to see if the addendums are acknowledged, numbered, and dated. If the bid package contains a bid bond, and all the addendums are acknowledged, the prices are then read aloud, including the base bid price, alternate prices, and unit prices, if applicable. On to the next bid. Last, the list of subs provided by the low bidder are usually read. The whole thing is over in ten minutes unless there are a lot of alternates to be read.

Consider a late bid at 3:10 pm when all of the bids for the 3:00 pm opening have been opened. Into the room rushes a bid runner, red faced and sweating, throwing a bid on the table and swearing at the traffic, stating his boss is going to kill him if that bid isn't opened. Let's say the owner agrees that the bid should be opened, and asks, "What's wrong with ten minutes?" After all, the owner knows that the contractor has been working hard on the bid and has been to the site several times, and what if the lowest price is in that unopened envelope!

The problem is there is no end to this if allowed. Late bids give the chance for contractors to come up with shenanigans of all sorts, conniving and colluding all the way. As with good business practices, policies, and procedures, *exceptions are not good.* Stick to the rules and procedures, do it the same way every time, and things work better in the long run. An exception to the practice of firm and timely bids allows a clever person to gain an advantage over someone else. And estimators are very, very clever.

Having bids on a certain day and time works towards a fair bid, one that unsuccessful bidders will not feel cheated or disadvantaged. They will come back next time, which is what the owner needs to happen!

All the participants of the bidding system, from owners to subcontractors and suppliers, should make best efforts to support the competitive bidding system by providing a firm bid on time. A bid should be like a footrace – an all-out effort, the best man or woman wins, and then it is over. It is amazing how much drama can be prevented and the bidding process benefit by simply insisting on a firm and unmovable time of day for bids to be received. Estimators can be mischievous. There would be no end to the games that would be played if contractors had a flexible time that a bid was due. Ingenious estimators could also think up reasons for changing a price just turned in. If there are any loopholes to a firm bid or a specific time, somebody will figure out how to game the system.

The owner should require that all bids be received at one time. No flexibility can be allowed in the timing of bids turned in. Bidding contractors should not accept late subcontractor bids.

Announcement of bids

An open bid is one in which the results are read aloud by the owner or architect, or nowadays with electronic bids, shared the next day. This is the fair way, everything is out in the open, all the bids are read. The low prime bidder should announce the ranking of sub bidders, or at least announce the low and successful bidder for each division of work. However, many prime bidders do not announce (at least to everybody) what the sub bid amounts are, because the sub may not be bidding the same price to everyone.

No announcement is due to non-bidders or late bidders.

There is one situation in which the subcontractor ranking should not be divulged. *If the owner, upon receipt of prime bids, determines that they are too high and the project will rebid at a later date, the sub list should be withheld.* Of course, the bids of the prime contractors are already "out on the street", known to all. However, the prime bidders should, if a project is rebid, withhold the ranking of subcontractor pricing.

When the owner receives one bid

The public owner prefers several bids to be received all in the same range. This gives confidence that the low bid is a good deal. If a city or school board receives only one bid, they have a problem. A purchasing agent or contracting officer has a hard time recommending (to the mayor, or the city commission, or the school board, etc.) an award when there is only one bid and can easily hide behind the safe course of rebidding the project. However, this will cost the owner more time and expense. Neither is a rebid good for the lone prime contractor who turned in a good price.

Many government agencies can accept a single bid *if it is in their best interest.* Some will accept a single bid *if* there is a budget by the architect or engineer supporting the bid amount (about the same or higher).

When an owner receives only one bid, there are no competitors to compare the lone bid to. The bidding contractor finds that the competition is a budget. Budgets are not usually very good bidders, because they are only the best guesses of designers and often not accurate. The owner prefers a market price bid to a budget prepared by the architect/engineer. Sometimes there is not even a budget estimate; the budget is simply the amount the owner has to spend, which often is not enough.

However, *if* there is an A/E budget that is in line with the bid amount, *and* there is enough money in the budget, *and* the owner really needs the job done, *or* there is a grant paying for part of the project that is running out of time . . . all of these things help get the job awarded. The law is, "if it is in the best interest of the owner". If there has only been one bidder, the owner may even "negotiate" with the low bidder – changes may be made, specifications changed, whatever is in the owner's best interest.

When the prime contractor receives one bid

There is a strong obligation by all involved in the competitive bid system to use the low bid. But, what if the low bid is also the only bid?

This is one of the thornier and complicated ethical issues.

What if the prime contractor only receives one painting or one electrical price? The prime may feel this "coverage" is spotty, not a competitive situation. Some contractors will "make up" their own number for, say, painting, when they receive only one bid. The prime bidder receiving only one painting bid of $100,000 may put $90,000 in the estimate. When this happens, it means the prime is going to go shopping later! In this case, is it ok?

Let's examine a single painting bid that a prime contractor doesn't use. Before we feel too sorry for that painter ten times out of ten, consider what might happen a few times out of ten. The painting contractor may have heard they are the only bidder, either from a prime or from a paint supplier who informs the subcontractor that they are the only company inquiring about the price for the special elastomeric paint specified for the job. The painter then increases their bid, jacking it up like it was on steroids. They are the only bidder and it feels like payday! Is the prime, in this instance, instead of being a bad guy going shopping, being prudent by later seeking other painting bids? After all, in at least some instances, a sub price might be inflated.

Now, consider another instance of the same thing. Most bidders will be hesitant to ignore a lone mechanical or electrical bid, because they are much bigger numbers than a painting bid. Instead of $100,000, the prime may be looking at a bid of several million. The stakes are raised (for the prime) when a lone bidder is a major subcontractor. Because of the magnitude and complexity of these bids, the prime is less inclined to use a lower number. But, isn't the ethical consideration the same whether it is a small painting bid or a large electrical bid? Shouldn't the same thing be done in both situations?

Most bids are not trumped up; a lone bid is usually a best and responsible price from that vendor. If the sub price is not used, that part of the estimate becomes a bet (instead of an estimate), because the estimator is not qualified to be a painting estimator. Since an estimator should ethically, "only perform services in areas of their discipline and competence", this leaves little room for disagreement with a qualified bidder. The estimator should not make up a price.

While there is a strong obligation by all involved in the competitive bidding system to use the low bid, some owners and contractors disagree with this when only one, or perhaps even two, bids have been received.

When only one price is obtained for a *product* or *material*, the situation is nuanced. A subcontractor puts in time and effort to bid a job and is *invested* in that project. Time and money has been spent. However, a price for a product or material can be different, some would say. There is less of an obligation to use a quote that doesn't include labor; it is ok for it to be shopped later. Or is the obligation, upon the receipt of only one quote, the same as with subcontracts?

The answer depends on how the quote was obtained, and whether the vendor is "invested" or "uninvested" in the project. A door and hardware supplier quoting a project with one hundred doors is invested in the outcome of the project because of the time and effort it took to estimate the job. When the prime calls a lumber yard and obtains a quote for a wood door and a few products, and a salesperson at the counter provides a price in a few minutes, that lumber yard is uninvested in the project.

There is an obligation to use lone prices of invested suppliers and fabricators because of the time and effort they spent. There is no obligation to uninvested suppliers.

Rebids

The government will rebid a job if it is in their best interest. If only one bid was received, the owner may not make many changes before rebidding the project. However, the government owner, after receiving several bids all higher than what they want to pay, will direct their architect/engineer to make changes that will reduce the pricing.

If an owner rebids a project, changes should be made if a different result is wanted. To rebid the job without making changes is to say, "You contractors have got it wrong!" The private owner is more likely to do this than the government.

Low bidder

Recall the comments about there being several bidding environments, the range of strict public bids to Wild West private bids. The following observations about low bidders concern a fair and open competitive bid arena of good plans, good bonded contractors, and repeat owners.

Except perhaps during a depression, the *best* and *lowest* bid is usually the same bidder. The competitive bid system works remarkably well for the owner, and the market magically provides three or four and sometimes five prices for a job. Considering the amount of trouble and expense it takes to prepare a bid, it is a good and reliable functioning system for the owner. How many prices does an owner deserve to receive? Three or four is plenty; when there are six or seven bidders, the low bid will be too low; no money will be made and the owner may not like the job that is provided.

There is a strong obligation by all involved in the competitive bidding system to use the low bid. To do otherwise subverts and erodes the bid process, unless there are disqualifying financial or performance considerations. In a classic competitive bid situation, the low bidder is usually the best contractor for the job.

The main reason that a contractor gets a job is that his/her costs are lower, or at least their estimate shows them to be so! For a bidder, 90% of the job is cost and 10% is profit (to use an average), so it is usually cost that helps get a job more than a low markup! Having lower costs is a good thing, an envious position to be in. One cause for this is that the kind of work fits the contractor, i.e. some of the trades are done in-house, or the contractor is good at supervising and running them. Or, the job category, for instance schools, may be the contractor's specialty. Other reasons include good payment, good reputation, and running jobs on time, and subcontractors may favor them with better pricing than the next person who pays slow. These reasons make a GC have low costs. The low bidder is not making less money than the next person, nor are corners being cut – *the job is costing less!*

However, consider the depression starting in 2009. The author, after bidding for over thirty years against two to five bidders on any given job, bid against fifteen or twenty competitors! Some of them came from far away. It was a surreal bidding period, a complete breakdown of the competitive bid system. Having fifteen to twenty bidders means that several of the lowest bids are no good! On these bids, it is the author's opinion that the lowest half dozen were below any reasonable cost. But bidders from near and far away had families to feed and had bids made of hope instead of good sense. There simply were not enough bids to go around.

Section 6 Plans

Ill-defined plans

There are two high-risk estimating categories concerning plans, or lack of plans, mentioned at this time. One is ill-defined plans; the other is unforeseen conditions. Risk taking discussed here is not about the everyday judgment that estimators make in calculating labor and material costs. Rather, it is when the tools and conditions that estimators rely on are absent or lacking that situations arise that get contractors into unusual risk taking.

The first of these is work having an "indefinite scope", often occurring because of ill-defined plans. It happens most on private jobs in which a wide range in the quality of plans is seen. It happens least, but still occurs, in open bids with architects and engineers and owners all pursuing clarity through an RFI and addendum process. Sometimes the risk is spread among trades, sometimes it lands in one division, sometimes it is not understood at the time, and regardless of the situation, the consequences can have an ethical component.

An unclear view of the work can happen in many ways, one being catch-all plan language such as "rework as required" or "adjust as required." Such fuzzy language found in renovation work is hard for the estimator to quantify, and quantities are what an estimator needs. Estimators like specific instructions; they want to know length, width, and height – not "maybe" language. Vague language can result in an argument about what the architect thought he or she described but the estimator did not.

Consider this simple example where a subcontractor takes on the risk of an indefinite amount of acoustical ceiling work. It was foreseen ahead of time by the bidders, the risk was squarely assigned, and it is an example from a real case. As a reason for its occurrence, it falls into the category of ill-defined plans.

At a school renovation, the existing building had acoustical ceilings throughout, but a grid plan was not shown on the architectural portion of the plans. There were no reflected ceiling plans. Some lighting fixtures were shown to be replaced on the electrical plan, but the grid pattern of an acoustical ceiling was not shown there either. However, there was a note stating, "Adjust ceiling grid as required for new lighting fixtures". Without providing the existing acoustical grid layout, there was no way to tell if the new lights fit into the existing grid layout.

The general contractor and Bob the acoustical ceiling subcontractor discussed what to do before the bid . . . well, like two hours before. Too late for an RFI! The subcontractor is pressed by the GC to provide a bid, the GC saying, "There won't be much to it!" The sub does so, taking a risk. The bid of $30,000 was half the cost of a completely new ceiling of grid and acoustical tile. It was a guess at the work for a lot of rooms.

The job was awarded. The acoustical ceiling subcontractor met with the GC's superintendent and studied the first classroom. Without changing the grid layout, some tiles were removed for a few 2′ × 4′ lay-in light fixtures to be placed in that room. The lights were installed – it wasn't exactly the same as the electrical plan, but was close. Twelve lights went in that room and it lit up like a Christmas tree. New ceiling tiles were placed as needed in the former light fixture locations. The grid was unchanged.

The architect was happy with the room. The remaining classrooms were then completed using the existing grid layout and some new tiles. The acoustical ceiling subcontractor breathed a sigh of relief.

The general contractor saw the subcontractor complete a $30,000 job for around $10,000. Is the general contractor in a position to reap some of the money the subcontractor is making? The GC said to his subcontractor, "Hey, I told you this would be easy. And, didn't my superintendent help you? Let's split the savings, I'll bet you're making $20,000 on a $30,000 job!"

In this case the GC should have no problem paying the subcontractor $30,000 and is wrong to suggest splitting the savings. Consider a different outcome of the work – what if it had cost the subcontractor more than their bid? Would the GC offer to pay the sub extra?

These are ethical considerations, and *since contractors have to live with overages if they happen, if any savings occur they should go to the risk-taker.*

Addendums

An addendum clarifies, amends, or adds new information about the bid. It can address the plans or specifications or the bid proposal, including making an extension to the bid date. Addendum items include the answers to RFIs written by bidding contractors. Addendums are numbered, dated, and become a contract document. Sometimes the information in the addendum causes a bid extension, because the bidders may need time to process the information.

Addendums can have an ethical component, not for the contractor, but for the owner. The *public owner* is committed to ensuring that all bidders have the same information and receive it at the same time. If they decide that a sidewalk on the plans isn't needed, this information will be included in an addendum. It will be described with no ambiguity as to which sidewalk it is, perhaps even defined in a sketch. Smart owners require that their architects answer all RFIs and get the contractor's questions answered.

Information that changes a bid can be handled differently on *private jobs*. An owner might tell one contractor about not needing a sidewalk and forget to tell another bidder. The owner may receive various questions from multiple contractors and have several long-winded conversations with them. The contractors might come away with different versions of changes here and there.

A bungling owner such as the above is usually a one-time or small-time owner. Large private owners such as hospitals and insurance companies that have recurring projects are careful to provide a level playing field for the bidders. This responsibility rises to the level of ethical behavior because of the potential harm caused to bidders if they do not get the proper information.

The owner should present identical changes, revisions, and addendums to all bidders at the same time.

Unforeseen conditions

This is the second general category where contractors can stumble into risky situations that can have an ethical component. Unforeseen conditions often cause problems where the scope of work is not clearly in one camp or another. No one wants to take on misfortune, and if underground water is found to be higher than expected, the dewatering that is required (when it wasn't figured on) might result in finger-pointing between the owner, the prime, the site contractor, maybe the concrete subcontractor, and anybody else that might be around. In business, finger pointing often means the little guy, the one downline contractually, gets hurt, because the big guy gets paid first.

Consider this actual example in which two contractors appraised the situation and took risks on dewatering and they were both wrong. When the scope of work changes during the job, determining who should be left holding the bag is an ethical consideration.

Some bus lifts for a school board needed to be installed inside an existing large metal building. Much of the interior slab, at the same elevation as the asphalt outside the building, was to be demolished. Four-foot-thick concrete was to be poured there, supporting powerful lifts that would hoist buses into the air to allow work to occur underneath. The owner had provided a six-month-old geotechnical report that showed the water table to be about four feet deep.

A GC, and a site contractor who owned trucks and bulldozers and dumpsters, set out to get the job. Both had worked in the same area of town without groundwater trouble in the past few years. The site sub said, "No problem, if there is a little water, I'll dig a hole in the corner of the excavations and pump it off".

The GC concurred with this expectation; he had been to four and five feet deep on a nearby job and had no problems with water. There would be some water but not a lot. The sub would demolish the interior slab and excavate the earth to about four feet deep. A bit of water would be pumped out by the sub and then the GC would pour the supporting concrete for the lifts. They had done this kind of thing many times together. Water would not be a problem.

But water was a big problem from the outset: rains, continuous rains, a water table two feet deep instead of four feet ... blame it on global warming. The subcontractor couldn't just pump it away with one hole in the corner with a small rented pump as he had imagined – there was too much water. Well points needed to be placed surrounding the entire area and some serious dewatering needed to occur.

The GC, *seeing that the problem was larger than anticipated*, took over and quickly hired a dewatering subcontractor. The site was made dry so that construction could continue, but the GC spent $30,000 very quickly.

Both contractors had been wrong in their assessment of water being present. The GC did not expect the site sub to handle the dewatering problem because the GC knew the sub had only meant to bid minor dewatering, not major dewatering. Things had changed.

Then another issue came up, an unforeseen condition. During the demolition, the slab thickness was found to be different from the six-inch slab shown on the drawings. There was a continuous thickened slab at the center, several feet wide and a couple of feet deep. The demolition work suddenly became greater. The site contractor, having the equipment and dumpsters and knowing his friend had jumped in to take care of the dewatering, readily took on the extra demolition.

The GC took photos and documented the extra concrete demolition to the school board owner. He sent notice that a changed condition had been encountered. Everyone acknowledged it was unforeseen; the old plans that the architect and owner had used to base their current plans on did not show the deepened slab, so the bid set simply showed a six-inch slab. The additional demolition work took two weeks and a lot of concrete was hauled away.

The GC submitted a change order request, not for dewatering, which would have been denied, but for extra demolition. Because of the extra slab thickness, there was a lot of cubic yardage of concrete to demolish and haul away, and there was the additional time, which meant a bit extra for General Conditions. The owner and GC agreed to $20,000 for this work, which was a legitimate representation of the extra demolition.

In this case, the subcontractor eased the situation for the GC by not requesting extra payment for the demolition. The work was done by machine, it did take more time and extra dumpsters, but he was already there, he owned the dumpsters, the GC was a friend of his, and *he had not been forced to eat the expense for dewatering.* After all, he did demolition work in-house, but not well points and dewatering.

The subcontractor was relieved that he didn't have to do any major dewatering. After all, if he were working for anyone else, *he might have been charged for all $30,000 of the major dewatering.* He had been prepared to spend $5,000 in minor dewatering, which he mostly saved. He did spend $10,000 extra in demolition of the concrete, but he didn't charge the GC for it. All figured, he was only down $5,000. Not as bad as it could have been!

The GC acknowledged that both he and his subcontractor had made a mistake in dismissing the possible extent of the dewatering. While there was a clause in their contract for the subcontractor to handle dewatering, the GC was not about to make the subcontractor responsible for the major dewatering that had to be done. The subcontractor's contract was for about $50,000, the GCs was about $1 million. In these cases, the bigger guy has to step up. But, the GC did keep the change order money, which helped offset the dewatering he had paid for.

The GC carried the risk for major dewatering. The fact that $20,000 of the $30,000 dewatering cost was recouped was luck. Contracting is a risky business, and it is the estimator who sees it all.

Section 7 Payment

Backcharge(s)

The prime contractor should not have the stingy habit of charging subcontractors for every transgression that comes along. There is some "benevolence" that comes with being a leader. Being harsh and demanding does not create long-term relationships and impedes short-term results. The example of charging for dumpsters in the Wild West project is one version of the prime contractor taking advantage of those downline.

Backcharges are a touchy subject with subcontractors. A penny pinching or spiteful contractor can hand out backcharges like traffic tickets for cleanup or moving material out of the way. Of course, some of these may be valid. But whatever the case, a backcharge should be dealt with openly and promptly.

Some backcharges are legitimate. Among the many business transactions that occur during a construction project, one company may inadvertently or by nonperformance or any number of reasons harm (cost) another company. For example, a GC may have an agreement with subcontractor A to provide all of the scaffolding, or a crane, or some service like cleanup, for the benefit of all the subs to use and depend on. During the course of the job, if the obligation is not met by A, the prime contractor has to step in and provide the neglected service for subcontractor B. Otherwise, the schedule is impacted. In this case, there is a legitimate backcharge. The GC spent money on scaffolding because subcontractor A, whose job it was to provide scaffolding, was busy on another job.

Timely payment

The prime contractor has the responsibility to pay a large amount of money every month to scores of subcontractors, fabricators, supply houses, and other vendors, sometimes several states away. People are hurt and lives suffer when payment is not made. Slow payment is a bad practice, unwarranted back charges are unethical, and nonpayment can be illegal. Nevertheless, within the area of payment, there is a lot of room for monkey business.

There are a myriad of situations concerning payment that have an ethical component. One could think up scenarios all day long that could happen during the performance of a construction project. The good practices here are limited to the most important payment consideration, and that is being paid timely.

It is the prime contractor's job to keep everyone paid and to have a good bookkeeping system and people that can keep up. This is the part where the contractor has to have money – enough money to "front-end" a job, even front-end two or three of them at the same time. Prime contractors, GCs and CMs, must have an equity in the business that will take care of the initial cost of a job. The contractor may have to pay out more than is received, and it might take several months before a job is paying for itself. The contractor must have money, not borrowed money, and use it, or jobs will grind to a halt.

Timely payment to subs and suppliers is good practice. Over the course of a construction project, an entire community of subs and suppliers are affected by the payment practices of those at the top of the food chain. With a large amount of money flowing through the prime contractor from the owner, there are many opportunities for misuse of funds, nonpayment, slow payment, etc.

Payment to vendors invested in the project (VIPs)

There are major subs and minor subs, and major suppliers and minor suppliers. There are those companies that have actively bid or quoted parts and pieces of the project, and then there are minor materials (concrete, lumber, etc.) that the project manager and superintendent buy during the job. Sometimes, these two groups are not treated the same when it comes to payment, because the suppliers of minor materials are not invested in the project.

The largest subcontracts are usually the mechanical and electrical contractors. Other subs with large subcontracts can include sitework, structural steel, roofing, some of the finish trades like painting and flooring, and glass and storefront. Major suppliers of material and products to the GC or CM would include doors and hardware, and lockers and toilet partitions. These items are quoted job specific.

The General Conditions of the contract, Division 00, instructs that the prime contractor pay a subcontractor within seven days of receipt of the owner's draw. This is *good practice*, and it works for larger subcontractors, suppliers, and fabricators and is called "pay when paid".

Payment to uninvested vendors

The prime is at the head of the chain, having the contract with the owner. Being privy to information from the owner and having a bonding capacity, the prime should have the financial capability to pay the small and minor subcontractors before being paid. The prime should also take care of the local supplier who has no special interest in the job, whether the owner has paid or not.

If a supplier sells everyday materials to a contractor for some job they haven't pre-quoted and don't even know who the owner is, they are uninvested in the project. They just want to be paid once a month for what they have sold. The $10,000 of form lumber that the GC bought at the local lumber yard should be paid for on the first of the month.

Some (often small) subcontractors supply labor only, or labor and tools, or labor and some material but not major material. These are the unit price subcontractors who work by the piece. For instance, the prime contractor buys the block and the mason only provides labor and scaffolding. This kind of subcontractor may want to turn in an invoice every week, to heck with every month.

The good prime contractor takes care of the smaller contractors that help, with their good pricing, get a job and get it done. The prime should provide good supervision and timely payment in advance of owner payment. The "pay when paid" practice should not extend to minor subs and suppliers. They don't care if the contractor has received an owner draw or not. The local lumber yard and the small subcontractor should be paid promptly.

The customary term for payment, and a good practice to maintain, is 30 days.

Section 8 Performance

Leadership by the GC or CM

Performance and leadership are not bidding topics, but they are important, so some *good practices* are mentioned here at the end of this book.

Completing a job on time, without issues of quality and payment, is the good practice goal for all GCs and CMs. This book has covered specific duties, obligations, and matters of ethical concern for contractors, but there is no question that in the big picture performing a good job on time is the most important thing the contractor can do for the owner. To do this takes a team effort, and undoubtedly *leadership* is the most significant ingredient the GC or CM, the prime, can contribute.

The construction industry benefits when the prime contractor exercises good and strong leadership skills. The prime, having the money to bond the job and being able to pay minor subcontractors and suppliers prior to receipt of the owners draw, is the boss. And the boss has the responsibility for the operation of the entire team. The prime contractor should have experienced people to provide proper supervision and project management.

The prime bidder should always stand ready to help a subcontractor. They are on the same team, and the prime should be the biggest fan of the folks working for him or her. They both want the other to succeed. It is unfair for the prime to trump up charges against the sub in order to facilitate a termination. The prime should look to the group of subcontractors like a team that has a range of talent, like any team has. Some team members need more help than others do. That's ok; they are on the same team. What is not appropriate is for the prime to insist on an A+ performance by everybody and dispassionately fire anybody that gets in the way of an all-out schedule.

A strong nonperformance clause is always contained in the contract authored by the GC or CM that the subcontractor signs. Nonperformance of the contract is cause for termination, and GCs and CMs are adept at writing letters outlining the subcontractor's failings, small and large. This documentation can pile up, hammering home a story of failed duties and obligations that makes the sub look sorrier than dirt.

There can be many reasons for nonperformance, and causing delays to a project schedule is a powerful reason for the prime to complain. The owner has to be answered to; there are liquidated damages, reputation, and future work to consider. The prime will point out sub deficiencies such as not having enough onsite labor (not "staffing" the job), poor or lack of supervision of employees, sloppy non-skilled employees, poor quality, mistakes, and failed inspections. But, simply getting behind is usually the main problem.

Getting thrown off a job is a big deal, and having it happen just one time can ruin a subcontractor. When this happens, the prime contractor is probably going to take a big bite. Slow performance will not be tolerated. During the time leading up to termination, the GC or CM may be cutting the sub short on payment, building up a nest egg to hire the next person.

It is in the prime's best interests, and certainly is good practice, to help the sub perform. A construction project is made up of various companies working together for one project, and then it is over. A separate team is made for each job, and teammates sometimes perform at different levels.

Benevolence of the prime contractor

If the prime has to help one sub more than the others, spending more time in management, so be it as a first step. They are on the same team. Sometimes a second step is supplementing the subcontractor's crew with additional labor. Often this extra management and labor will turn the tide. The prime should not be like a wolf ready to pounce on a non-performing subcontractor without the above measures tried first.

In the course of completing a ten, thirty, or a hundred million dollar project, the contractor, as these pages have shown, is responsible for all of the wide-ranging circumstances that can befall a schedule, as the owner and architect have shouldered many burdens upon the contractor, including to catch the architect's mistakes, being in charge of and responsible for all construction techniques and scheduling, tracking the submittal process, and notifying the owner and architect of any potential delays. With all of this to do, some things will go wrong and there will be mistakes, and it will rain.

While these abundant duties can easily breed harsh management and leadership in the contractor, it is wise to remember the fate of the mouse in Robert Burns' poem written in 1786. Seems the mouse, in choosing the location for his residence, thought an empty field was far from any misfortune and dug out his new abode underground. It was a good plan, the mouse figured, but his foresight (planning ahead) was in vain. Burns came along and, not knowing of the recent construction, plowed him up in preparation for planting. Burns apologized in his poem *To a Mouse*, which included these words:

> In proving foresight may be vain: The best laid schemes o'mice an' men gang aft a-gley, an' lea'e us nought but grief an' pain.

These lines are remembered as, "The best laid plans of mice and men often go awry." Which is to say, even when everything is all planned out, part of it may go wrong, and it will rain again.

In the course of the business relationships resulting from construction projects great and small, there are times when the general contractor can use a club and dagger in the treatment of subcontractors, or provide some help and leadership. A good leader, whether parent or king, in order to provide the most consistent good results over a long period of time, should first offer guidance.

GLOSSARIES

Concrete glossary
Masonry glossary
Steel joist and steel deck glossary
Carpentry glossary
Doors and hardware glossary

CONCRETE GLOSSARY

AFF Above finished floor.

Anchor bolt template A form, often horizontal and placed flat at the top elevation of an impending pour, with holes drilled through for the placement of anchor bolts in a precise location. Anchor bolts are often used to tie down a structural steel column base plate. Also see "Rebar templates".

Angle of repose Slope of earth embankment, maximum 90 degrees.

Backfill Earth excavated and reused on the same site. Placed back into a foundation and recompacted (it is loose fill after excavation). Earth placed against one side of a wall.

Beam bottom The horizontal form at the bottom of a concrete beam. Beam bottoms are counted separately from the sides because they take more labor per square foot (than the sideforms do) to build. A beam bottom form may be extended into a platform for staging. Because of the heavy load of concrete, they are carefully supported with posts, post jacks, and sometimes scaffolding.

Blockout A form used to create an opening or indentation in concrete. The form "blocks out", or displaces, the concrete to provide the later passage of a pipe or duct, window or door. The form used to create the opening is called a blockout form. On the takeoff of concrete, a blockout is shown as an "out", a reduction in the volume of concrete.

BOF Bottom of footing.

Bond beam Horizontal block filled with concrete.

Brick ledge A ledge for vertical brick to be supported on. In formwork, a "blockout" installed to the side of a wall form can create a brick ledge.

Building pad Compacted earth at the elevation of a slab pour. Building pads are commonly machine graded to a tolerance of plus or minus 1/10th of a foot, or about 1-1/4″ (survey rods are graduated in tenths of a foot).

Chamfer strip A "blockout" type of form often used to prevent 90-degree concrete edges (for example vertical concrete columns). Chamfer strips are sometimes triangular in shape and an inch or so on all three sides. Attaching them at the corners of a column creates 45-degree edges instead of 90-degree edges, which looks better and keeps the sharp edges from being chipped off.

Cold joint The point of contact between separate concrete pours. Sometimes used to define the limits of a pour, for example, "Put a cold joint down at the end of 400 LF of sidewalk and pick up there the next day."

Concrete pad A square or rectilinear footprint of concrete.

Continuous footing A footing usually a few feet wide or less that has a "running length" and typically carries evenly distributed loads. Sometimes called a "spread footing".

Continuous pour A single pour without cold joints. Some pours are specified by engineers to be one pour. The author made a "mat pour" of 5,400 cy for the foundation of a building (42″ deep) which took 24 hours and required an MOT plan covering a square mile, concrete deliveries from three plants, and 540 trucks of concrete.

Core-bore, verb. Core-bore machine, noun. To saw a hole into or through concrete. The holes in concrete slabs for handrail posts are often core-bored by a machine with cylindrical saw blades. The holes are drilled wider than the soon-to-be-embedded pipe to allow an installation tolerance and to provide an area for grout. "A small crew is going to core-bore all day."

Curb and gutter (also see "Header curb") The concrete at the side of many streets is known as a curb (vertical) and gutter (horizontal) and is "L" shaped.

Curing compound A liquid product applied to concrete after a pour. Its purpose is to minimize water evaporation by sealing the concrete surface, retaining moisture in the concrete for best curing strength. Sold by the gallon or by 55-gallon barrels and sprayed or brushed on. Some curing compound products are not placed on slabs at ceramic tile locations because they reduce bonding. Also known as a concrete accessory.

Diagrammatic An architectural term meaning that the plans are not to be taken too literally! Lines and intersecting lines on a plan are in two-dimensional plan form, not 3D. Overlaps and depths can only be shown diagrammatically. The dotted lines of a footing may not accurately depict, because of depth, one footing above another.

Diagrammatic is a word used to describe the schematic nature of plans as opposed to shop drawings, which are more specific. Also see "Interpretation (plan interpretation)".

Dowel, smooth dowel Several meanings. Vertical rebar in filled block cells can be termed dowels even if extending several stories. Shorter pieces of rebar of a few feet in length and bent 90 degrees at the bottom are referred to as dowels and are "doweled" into a footing. Short pieces of rebar a foot or two long are dowels, sometimes used to tie separate slabs together. The slab poured first will be drilled for half the length of the dowel, and a smooth dowel (no serrations) often inserted, then the second slab poured. In this way the slabs are tied together, but there can still be some slight movement without causing cracking.

Earthform Using the ground as an edgeform for concrete.

Edgeform The form at the edge of a slab pour; sizes of less than 12″ in height are called edgeforms. They are held in place with vertical stakes. When the height of forms increases beyond a foot in height, they are usually referred to as a "sideforms". Also see "Sideform" and "Mono edgeform".

Elevator sill angle Steel angle embedded in the slab adjacent to an elevator.

Embed item Steel embedded flush to the surface of fresh concrete, later to be welded to bar joists or other steel.

Equipment pads Small concrete slabs underneath mechanical or electrical equipment, interior or exterior.

Excavation required for concrete (also see "Overexcavation") A term used to describe earthwork removal required for the placement of concrete. This term is used to divide responsibility between the sitework and concrete trades.

Expansion joint The product placed at a joint in concrete. Four- and six-inch-high fibrous expansion joints are sold in lengths of 10′, ten in a bundle. Also known as a concrete accessory.

Fine grading Leveling, by hand and flat point shovel, required for slabs and the bottom of excavations after rough grading. Many estimators do not figure fine grading for the bottom of continuous narrow footings, but would when the footing or pad is wider than two or three feet. Asked if he knew what fine grading was, the student answered, "What you do at test time!"

Format, formatting Labeling the vertical columns at the head of the takeoff or estimate sheet.

Geotechnical engineering The owner often employs a testing company to make onsite borings (drilled holes into the ground with pipes that collect soil samples). The testing company supplies written results (usually provided to the bidders) called a "geotechnical report", which describes the type of soil in the foundation and the depth that water was encountered at the various borings. This kind of testing is called geotechnical engineering.

Grade beam A concrete beam on the ground shaped with a height greater than width, with the width remaining constant. A grade beam can exist at the perimeter of a slab or have slabs or both sides and can be poured with the slab or separately. A "monolithic edge" or "thickened edge" is usually not a grade beam; only when the width and height become similar to the rectangular shape of a "header curb" is it a grade beam.

Header curb (also see "Curb and gutter" and "Grade beam") Concrete on grade with a running length, the top half or third often exposed to view on one side, approximately 16″ high and 8″ wide often at the edge of a street.

Imported fill Fill dirt delivered, or "imported", from offsite.

Interpretation (plan interpretation) Architectural term used to describe how plans are "read". With related information in various places on the plans and separate written specifications, the contract documents are said to be interpreted (read as a whole), with emphasis on:

1 Specifications govern over drawings.
2 Specific governs over general.

Also see "Diagrammatic". See the chapter on General Conditions for an in-depth explanation of plan interpretation and "design intent".

Item An estimating unit of measure! Sometimes used as a quantity when the normally used unit of measure is very small and doesn't make sense. Or, "item" can be used to count the quantity of a single activity or a collection of small quantities. It is similar to "each" as a unit of measure.

Keyway The concrete equivalent of a tongue and groove wood joint, which ties separate concrete pours together by the shape of their connection.

Like kind A group of similar tasks sharing the same labor and material pricing. Like kind tasks are grouped together on the takeoff. For example, ten differently sized footing pads may require ten lines on the takeoff but then are all added together, with one total quantity forwarded to the estimate.

Maintenance of traffic (MOT) Maintenance of traffic is a requirement of highway or roadside jobs such as a streetscape project of bricks, lights, and trees. An MOT plan is often a requirement of the specifications, a written plan showing the location of warning lights and barricades, which may change as the job progresses. This work is often subcontracted to a company in this business.

Mono edge The perimeter, or edge, of a monolithic slab.

Mono edgeform The vertical form at the perimeter of a monolithic slab. Also see "Edgeform" and "Sideform".

Monolithic slab A concrete slab poured with an integral thickened edge at the perimeter. Also called a "floating slab" because it acts structurally like a tabletop that may shift but stay intact.

NTS Not to scale.

Overexcavation In Division 3 terms, not Division 2 language, that amount of earth removal required beyond the flat area of the pour (excavation straight down is just called excavation). Overexcavation is usually done to provide a work area; it is a part of the excavation needed to "complete the concrete work". Overexcavation and other earthwork terms (explained here from the viewpoint of the concrete estimator) have different meanings for those in the earthwork trade.

Piers In residential construction, concrete or block columns used in a foundation are called piers. They are often just a couple of feet high to elevate frame construction above grade. When this support is taller, the word "pier" is used less and the word "column" is used more; in marine construction, however, they're more apt to be termed piers whether made out of wood or concrete.

Place and finish concrete Term used to describe all of the operations of a single concrete pour. These include handling the delivery of concrete in ready mix trucks, depositing the concrete, screeding it level, edging, bull floating, and metal troweling. Once the concrete has been deposited (often very early in the morning), the finishing crew is locked into a string of operations that time step with the ongoing hardness of the concrete surface. The curing of the concrete creates a deadline for later in the day.

Placing and finishing concrete usually is priced by the square foot.

Plyform Name of a plywood product used as a form. The typical thickness is 3/4″ for a 4′ × 8′ sheet with one smooth side covered with a thin layer of wax.

Rake beam A continuous beam on incline, often matching the roof slope, hence the phrase, "follow the rake of the roof".

Ready mix concrete Concrete ingredients mixed by truck and delivered to the jobsite.

Rebar cage Refers to a section of reinforcement consisting of longitudinal rebar tied together with a series of stirrups.

Rebar template A form for rebar extending from one that provides accurate placement of the rebar into the next adjacent pour. Wood or steel forms with holes for rebar. Templates are "hung" or otherwise supported by other forms. Also see "Anchor bolt template".

Rough grade Has more than one meaning. To the concrete trade, rough grade often means the ground level that the Division 2 subcontractor is providing. This could be the grade for the main building pad (plus or minus 1/10th of an inch) or the grade provided at the exterior sidewalks. The term is also used to describe the lay of the land of a site before construction. In this case, rough grade is the same as existing grade. Also see "Building pad" and "Subgrade".

Screed Used as a noun, a straight piece of lumber, aluminum, or steel used to "strike off" or rake and level concrete. Also used as a verb, "Herb, go screed the concrete flat".

Shop drawing A drawing showing precise dimensions for the fabrication of reinforcement, structural steel, wood cabinets, etc. The detail shown is more precise than that of the plans, and the drafter of shop drawings "interprets" the plans to create the shop drawing. Subcontractors and suppliers send shop drawings to the general contractor or construction manager, sometimes requesting them to confirm dimensions from the site or from another trade. Shop drawings reviewed by the prime contractor require "coordination between the trades". It is the responsibility of the prime to ensure that the skylight opening dimensions (from the manufacturer's product data sheets) match the supporting steel drawn on the steel shop drawings.

Shoring, reshoring Shoring is the vertical support for elevated slab formwork. The posts and bracing erected prior to an elevated slab pour. When posts remain (after the slab is poured) while another slab is being formed above, the posts are called reshoring. Depending on the construction time between slab pours, reshoring can remain on several floors.

Sideform The forms on one or both sides of a beam or wall. Not generally used to describe slab edgeforms unless they are higher than a foot. Also see "Edgeform" and "Mono edgeform".

Similar In architectural drawing, when Section 2 only differs from Section 1 in an unimportant way, or an obvious way, but is much the same otherwise, Section 2 is said to be similar to 1. The second location is similar but not exactly the same. Often denoted by the abbreviation "sim" adjacent to a section or detail name.

Slab on deck A concrete slab poured on top of a steel deck that has corrugations, or ribs.

Slab on grade A concrete slab poured on top of the ground.

Staging Several definitions. Can mean a staging compound area to work within, or a scaffolding/platform structure to work on top of.

Stair nosings Metal or rubber protection extending the width of the front nosing of a step to protect the concrete from being chipped. These nosings are often abrasive in texture to provide good footing. Stair nosings are attached to a form and embedded in the concrete or installed after forms are removed.

Stem wall A short concrete or masonry wall typically used in a foundation, the top often being the same elevation as the floor.

Stepdowns, footing stepdowns Stepdowns are elevation changes to the top of footings, often due to a sloping grade.

Stirrups Sometimes called "hoops" because they can resemble a basketball goal, stirrups are rebar "ties" that hold longitudinal rebar in position within a beam or column.

Subgrade Grade established (often to pour concrete) below eventual higher grade, and sometimes referred to as a "temporary" grade. There can be many levels of subgrades in one foundation.

Template A form used to "position" rebar, anchor bolts, or other embed.

Thickened edge (TE) Linear edge of a slab thicker than the main slab, usually a few feet or less in width.

Thickened slab (TS) Linear thickened portion of a slab interior.

Tie beam Horizontal concrete beam on top of block or concrete walls continuing around the perimeter of a structure.

Top of concrete (TOC) Usually referring to the elevation of the top surface of concrete, such as a slab or other concrete.

Top of footing (TOF) The elevation at the top of a footing.

Top of wall (TOW) Usually referring to the elevation at the top of a wall.

Unit of measure How labor or material is measured, for example, each, LF, pounds, cubic yards, etc.

Visqueen The word Visqueen is a brand made by British Polythene Industries Limited for thin polyethylene film typically 4, 6, and 8 mils thick (thinner than a dime) used as a vapor barrier. As a concrete accessory, it is placed on top of grade and under the reinforcement. It prevents water in the concrete from seeping into the ground, allowing for better curing strength. It is sold in rolls from one to several thousand square feet. Visqueen is used as a vapor barrier in wall construction and in the temporary protection of other materials. Also known as a concrete accessory.

Waste factor concrete The percentage added to net concrete on a takeoff to account for enough to do the job. The reasons for needing a margin of error include:

1. Earthforming provides an irregular shape.
2. Small pours; see number 4.
3. Waste and spillage, but this is usually not much.
4. Cannot afford to be short! Being a half yard short can cause additional labor, potential quality issues, and extra cost for the material because of an undersized truckload.

Waste factor earthwork The percentage added to loose earth quantities to account for compaction.

Waste factor forms The percentage added to the linear feet of exact form lengths to account for enough to actually do the job. Random lengths of concrete edges, corners, and turns all send carpenters off to cut lumber. When a carpenter cuts form lumber, it is waste. A waste factor is not added when "contact surface area" is used as a unit of measure.

Waterstop Flexible rubber-like material, usually an inch or less in thickness, installed at concrete joints to control water seepage.

Weld plate A typical weld plate is a small 4″ square piece of flat steel that the ends of bar joists are welded to. They are embedded in concrete (or the grout in a block wall) and often have a "stud" on the bottom (steel shaped like a bolt with a head on it). Weld plates can be embedded in vertical or horizontal concrete for welding to other steel besides bar joists. To the concrete trade, a weld plate is an "embed".

MASONRY GLOSSARY

Ashlar masonry Masonry units of a large size, and sometimes multiple unit sizes in the same wall.

Backup Masonry behind the exterior facing.
Batter Recessing masonry back in successive courses from the face of a wall; the opposite of corbel.
Bed joint The horizontal layer of mortar on which a masonry unit is laid.
Belt course A horizontal course made to stand out; sometimes projected.
Block fill insulation Insulation "poured" or dropped or injected into empty block cells.
Blockwork Refers to a horizontal or vertical measurement that corresponds to modular block construction. For wall lengths, this is whole feet and multiples of 4 inches. Any wall length ending in an odd number of inches (10′-1″, 11′-5″, etc.) is not blockwork. Elevations, or heights, that correspond to modular block coursing would also be in increments of 4″.
Bond A pattern such as the various brick patterns. Also used to describe the adhesion or grip between mortar and masonry units.
Buttering Using a trowel to place mortar on a masonry unit.
Cavity wall A wall built of multiple wythes that contains an air space. Wythes are bonded together with metal ties. Also see "Hollow wall".
Closer The last masonry unit (whole or piece) to be placed in a course.
CMU (concrete masonry unit) A concrete block.
Coping The finish or cap to the top of a wall, sometimes precast.
Corbel In masonry wall construction, when a block or brick (or often an entire course) extends beyond the face of the wall, often for decorative purpose.

See the online resources for diagram 4G.1

Course One course of block or brick consisting of one horizontal row.
Drill and dowel An abbreviation of instructions on plans. First a hole is drilled into a masonry or concrete surface, then epoxy may be inserted, then a dowel (a foot or two long).
Fire wall A wall that resists the spread of fire.
Flashing Thin bent metal (or other manufactured product) placed in the mortar of hollow and cavity walls to capture moisture and divert it to the exterior through weep holes. The "high side" of flashing is the backup wall; the "low side" is the mortar joint of the facing wall. Flashing is placed at all locations where the air space is interrupted.
Header A masonry unit laid perpendicular to the face brick, extending into a backup wythe to tie them together.
Hollow wall A wall built of multiple wythes that contains an air space. The facing and backing wythes are bonded with masonry units. Also see "Cavity wall".
Horizontal joint reinforcement Galvanized wire (slightly thicker than clothes hangers) that is placed horizontally in block coursing. It consists of two strands of wire approximately 6″ apart (for 8″ block) and crisscrossed with ladder-like strands.

Lead walls The first portions of a wall built, at wall ends or corners, so that guide strings can be pulled to facilitate the straight construction of the remainder of the wall.

Lifts When block cells are poured with concrete, the "dropping" or depositing of concrete (and the subsequent vibrating operation) can cause the aggregate to be dispersed unevenly throughout the concrete. For this reason, building codes limit the number of vertical feet that cells can be filled in any single pour. This height that can be poured at one time is called a lift.

Parapet wall A wall above the roof line.

Parging A troweled operation with mortar, typically on a vertical masonry surface. Parged surfaces aid in moisture penetration, and are often on foundations and coated with waterproofing.

Screen wall A masonry wall built to visually shield one side from the other but to allow some sight. Sometimes used to hide HVAC or electrical equipment or to build a semi-transparent patio wall.

Shelf angle Steel angle supporting masonry placed across an opening.

Story pole A layout rod with marks on it used for laying out masonry courses.

Toothing Block or brick repair one at a time, for example at a door or window jamb created by a new opening in an existing wall.

TOW Top of wall.

Truss straps Metal straps embedded into concrete tie beams or block bond beams to hold down wood trusses.

Tuckpointing Either the repair of mortar to missing areas or the removal of mortar and its replacement. With age, masonry buildings may need "pointing up", the repair of small cracks and crevasses in the mortar.

Veneer An unbounded facing wythe of masonry.

Weep holes Openings in mortar of facing material at flashing levels to allow moisture to escape.

STEEL JOIST AND STEEL DECK GLOSSARY

The following definitions are provided by the Steel Joist Institute and the Steel Deck Institute.

ASD Allowable strength design. Method of proportioning structural components such that the allowable strength equals or exceeds the required strength of the component under the action of the ASD load combinations. Also see "LRFD".

Bar joists Another name for steel joists when the webs are made from round steel bars.

Bay The distance between the main structural frames or walls of a building.

Bearing The distance that the bearing shoe or seat of a joist or joist girder extends over its masonry.

Bearing plate The steel plate used for a joist or joist girder to bear on when it is supported by masonry or concrete supports. The plate is designed by the specifying professional to carry the joist reaction to the supporting structure.

Bottom chord extension (BCX) The two angle extended part of a joist bottom chord from the first bottom chord panel point towards the end of the joist.

Bridging A member connected to a joist to brace it from lateral movement. Also see "Diagonal bridging" and "Horizontal bridging".

Bridging terminus point A wall, beam, tandem joists (with all bridging installed and a horizontal truss in the plane of the top chord) or other element at an end or intermediate point(s) of a line of bridging that provides an anchor point for the steel joist bridging.

Buckling Limit state of sudden change in the geometry of a structure or any of its elements under a critical loading condition.

Bundle The banding together of joist products, bridging, and decking into certain sizes, weights, pieces, lengths, etc. to expedite shipping, unloading and storage, and erection at a job site.

Camber An upward curvature of the chords of a joist or joist girder induced during shop fabrication. This is in addition to the pitch of the top chord.

Cantilever The portion of a joist product that extends beyond its structural support. A lateral brace may need to be provided at the end of the cantilever to ensure that it is stable during erection and under load.

Ceiling extension A bottom chord extension of a steel joist where one of the two chord angles is lengthened.

Choker A wire rope or synthetic fiber rigging that is used with hoisting equipment to attach a load.

Chords The top and bottom members of a joist or joist girder.

Clear span The actual clear distance or opening between supports from one end of a joist to the other; the distance that the joist spans between walls or beams.

Cold-formed steel structural member Bending without heat.

Connection Combination of structural elements and joints used to transmit forces between two or more members.

Deck A floor or roof covering made out of gauge metal attached by welding or mechanical means to joists, beams, purlins, or other structural members and can be galvanized, painted, or unpainted.

Diagonal bridging Two angles or other structural shapes connected from the top chord of one joist to the bottom chord of the next joist to form an "X" shape. These members are almost always connected at their point of intersection.

Diaphragm Roof, floor, or other membrane or bracing system that transfers in-plane forces to the lateral force resisting system.

Erection bridging The bolted diagonal bridging that is required to be installed prior to releasing the hoisting cables from the steel joists.

Extended end The extended part of a joist top chord with the seat angles also being extended from the end of the joist extension back into the joist and maintaining the standard end bearing depth over the entire length of the extension.

Fall restraint system A safety belt system that prevents the user from falling beyond a certain distance. It consists of a body belt or body harness, anchorage, connectors, and other necessary equipment such as a lanyard, lifeline, and other devices.

Header A structural member located between two joists or between a joist and a wall that carries another joist or joists. It is usually made up of an angle, channel, or beam with saddle angle connections on each end for bearing.

Hoisting cable A chain, strap, or cable that is attached at each end that is used to facilitate the moving and lifting of joist products, bridging, decking, etc.

Hoisting equipment Commercially manufactured lifting equipment designed to lift and position a load of known weight to a location at some known elevation and horizontal distance from the equipment's center of rotation. Hoisting equipment includes cranes, derricks, tower cranes, gin poles, and gantry hoist systems.

Horizontal bridging A continuous angle or other structural shape connected to the top and bottom chord of a joist.

Joint Location where two or more ends are attached.

Joist A structural load-carrying member with an open web system that supports floors and roofs utilizing hot-rolled or cold-formed steel and is designed as a simple span member. Currently, the SJI has the following joist designations: K-Series including KCS, LH-Series and DLH-Series, and CJ-Series.

Joist girder A primary structural load-carrying member with an open web system designed as a simple span supporting equally spaced concentrated loads of a floor or roof system acting at the panel points of the member and utilizing hot-rolled or cold-formed steel.

Joist substitute A structural member intended for use with for very short spans (10 feet or less) where open web steel joists are impractical. They are usually used for short spans in skewed bays, over corridors, or for outriggers. It can be made up of two or four angles to form channel sections or box sections.

LRFD Load and resistance factor design. Method of proportioning structural components such that the design strength equals or exceeds the required strength of the component under the action of the LRFD load combinations. Also see "ASD".

Placement plans Drawings that are prepared depicting the interpretation of the contract documents requirements for the material to be supplied by the seller. These floor and/or roof plans are approved by the specifying professional, buyer, or owner for conformance with the design requirements. A unique piece mark number is typically shown for the individual placement of joists, joist girders, and accessories along with sections that describe the end bearing conditions and minimum attachment required so that material is placed in the proper location in the field.

Shear stud Headed shear connector specifically designed for use on a steel joist or beam, which permits composite action between concrete slab and joist or beam.

Span The centerline-to-centerline distance between structural steel supports such as a beam, column, or joist girder or the clear span distance plus four inches onto a masonry or concrete wall.

Splice Connection between two structural members joined at their ends by either bolting or welding to form a single, longer member.

Tagged end The end of a joist or joist girder where an identification or piece mark is shown by a metal tag. The member must be erected with this tagged end in the same position as the tagged end noted on the placement plan.

Tie joist A joist that is bolted at a column.

Top chord extension (TCX) The extended part of a joist top chord. The type of extension only has the two top chord angles extended past the joist seat.

Webs The vertical or diagonal members joined at the top and bottom chords of a joist or joist girder to form triangular patterns.

CARPENTRY GLOSSARY

Anchor bolt A bolt partially embedded in concrete often used to "hold down" a wood base plate. In the anchorage of frame walls, bolts are typically 1/2″ to 3/4″ in diameter, 8″–12″ long and spaced 4′ o.c.

Beam, floor beams or roof beams One or more boards used to span an opening and support a load.

Floor beams or roof beams One or more boards used to span an opening and support a floor or roof load.

Ridge beam A board, usually horizontal, that roof rafters lean against at the ridge of a roof. See Sections A (diagram 645.2) and B (diagram 645.4) of the Church roof framing plan (diagram 645.1).

Bearing, bearing height Structural term meaning a "build to" elevation. The bearing height of trusses is the elevation of the top of the wall the trusses sit on.

Bird-mouth cut The notch made in a rafter where it bears on a wall or other support.

See the online resources for diagram 6G.1

Blocking, bridging Boards that are perpendicular to the run of studs, joists, or rafters. The purpose of bridging is to keep lumber from "bowing", or twisting at the bottom. While bridging is usually the same size as the lumber it is perpendicular to, it can be a smaller size, even consisting of two small pieces in an X brace. Bridging is often required at mid or third points to a run of joists or rafters. If the joists are 24″ o.c., then the bridging pieces are 22-1/2″ long. At the ridge of a roof, the boards perpendicular to the trusses or rafters are called either blocking or bridging.

Board feet A unit of measure no longer in much use. Lumber used to be priced by the board foot. One board foot is 12″ square and 1″ thick, nominal size. Nominal sizes of lumber are used to count board footage, i.e. a 1 × 12 that is 1′ long is one board foot. An easy way to remember how to count board footage is to take the size of the board, say a 2 × 4 or 2 × 6, and multiply the numbers together, getting 8 or 12, and dividing them by 12. A 2 × 4 is 8/12 or 0.67 board feet per linear foot, and a 2 × 6 is 12/12 or 1 board foot per linear foot.

Let-in braces To laterally brace a wall, the building code allows a single 1 × 4 diagonal (45 degrees, one half of an X) board to be nailed from the top plate to bottom plate at the end of a wall. The let-in brace is let into the wall, meaning the wall construction (studs and plates) are notched where the brace occurs, allowing a flush exterior wall surface.

Temporary bracing Used for walls and trusses to hold them plumb and in position so that other framing work can proceed.

Truss bracing Bracing added in the field to wood trusses as required by the manufacturer for the trusses. See the truss shop drawings below on the XYZ job for an example of three braces that the manufacturer is instructing the contractor to install so that the trusses carry loads as engineered.

See the online resources for diagram 6G.2

Cull, culled To discard or reject. In a lumber pile, the crooked piece is "culled" and not used.

Decking The horizontal surface of a floor, walk deck, or boardwalk. Also refers to roof sheathing (but not wall sheathing, which is vertical). Decking can be boards or 4′ × 8′ panels.

Drip A board, often a 1 x placed at the top of a fascia, used to allow water to drip down from. As a protective measure, to cover or shed water away from an opening such as a door or window.

Fascia Perimeter board at the edge of a roof. The fascia is "plumb cut" if it is vertical. If the fascia is at a 90-degree angle to the slope of the roof, it is "square cut".

See the online resources for diagram 6G.3

Firestopping Blocking within a concealed space, usually a wall, to inhibit fire passage. Division 10 blocking lumber, sometimes called "deadwood", installed for bathroom accessories to be nailed to. See the Introduction in Part 6 Carpentry.

Five quarter board A board1-1/4″ thick. A 5/4 board four or six inches wide makes a good decking (walking) surface that allows the joists underneath to span 24″ to 30″, whereas a 1 x piece, at 3/4″ thick, would require support at 16″ o.c.

Gable Used to describe a wall or roof at the end of a house or building where the wall extends to the bottom of the sloped roof structure. The roof "end" consists of two planes sloping up to a ridge and shaped like the letter A. See the sketch below.

See the online resources for diagram 6G.4

Girder A term usually used in floor framing to describe a beam supporting floor joists.

Glu-laminated (glu-lam) Board lumber of any size glued together, usually to create a beam, though arches can also be made with glu-lam lumber.

Gusset plate Thin metal connection plate, often galvanized and consisting of a serrated surface, pressed against adjacent members of a truss (web to top or bottom chord, for example) or other structural intersections of wood to wood.

Header In frame construction, the supporting members above an opening. In 4″ wall construction, headers are typically two pieces of lumber with a plywood spacer sandwiched in between. In 2 × 6 wall construction, there can be three header pieces separated by two 1/2″ plywood spacers.

Hip A roof end that consist of three planes. See diagram 6G.6 for a drawing of hip rafters. Hip trusses would have the same layout.

Hip rafter Beginning at the end of a roof at the corners and extending to the middle ridge. See the sketch in the glossary under "Truss bracing".

Jamb The sides and top of a door or window are called the side jambs and head jamb. When the side and head jamb of a door is an integral unit, it is called a door frame. The jamb is the vertical component consisting of wood construction between two windows.

> **Band joist** A horizontal joist, often doubled, at the perimeter of floor framing, often directly underneath exterior wall framing.
>
> **Floor joist** Horizontal structural framing to support floor loads, usually 16″ or 24″ on center.

Joist hanger Three-sided metal (usually prefabricated) device for holding the end of a board against a beam or structure.

Knee wall, pony wall Short wall.

Lattice Thin, narrow piece of trim lumber, often used to cover joints in other wood, such as plywood.

Ledger A board placed against and secured to a surface (usually a wall). The ledger supports floor, ceiling, or roof framing, whether boards or trusses. Unlike a beam, a ledger is fixed in place against a support.

Let-in brace The building code allows a single 1 × 4 diagonal (45 degree) board to laterally brace a frame wall at wall ends, fastened to both base plate and top plate.

See the online resources for diagram 6G.5

Mullion (also see "Jamb") The vertical separation in-between two (or more) doors or windows. When two windows are connected side by side at the factory, they are "mulled" together. Individual glass panes are separated into smaller units of glass with mullions.

Nailer A board installed for the purpose of supporting another product. A nailer is installed against other wood members already in place.

Gypsum board nailer The 2 × 4 boards installed in wall or ceiling framing to provide nailing for gypsum boards. The amount of nailers provided for gypsum board can be considerable and is not shown on the plans.

OSB (oriented strand board) 4′ × 8′ panel sheets made of wood fibers and glued for interior or exterior use. Used for wall and roof sheathing. An alternative to plywood. Although it is accepted by building codes, it is inferior to plywood.

Outlook, outlooker Short roof rafters turned 90 degrees to the direction of other rafters and resting on top of an exterior wall, often extending past the wall on the exterior side of a building and creating an overhang.

Pitch Rise over span.

Base plate The horizontal board at the bottom of a stud wall. Vertical wall studs stand on top of the base plate. Treated with chemicals if the base plate is in contact with concrete.

Sill plate The horizontal board, usually on top of walls, that joists and rafters bear on. Sill plates typically are placed on top of masonry or concrete. Sill plates are also another word for base plates.

Top plate(s) The horizontal board(s) on top of wall studding. Non-load-bearing walls can have a single top plate, while bearing walls require two plates.

Platform framing A "platform" is floor framing (first, second, third, etc.) that includes the subflooring on which frame walls are built.

Plumb Several meanings all involving "vertical". Walls are built "plumb", and a carpenter checks a wall "for plumb". A "plumb bob" is a weight tied to a string. A "plumb cut" is vertical, e.g. a fascia is plumb cut when it is vertical (also see "Square cut").

Plywood spacer (also see "Header") The strip of plywood installed between two headers. It has two purposes, the primary reason being structural, the second being to take up the "space" difference (width) between a 2 × 4, which is 3-1/2″, and double 2 by lumber, which is 3″. In addition, with 2 × 6 wall framing and its 5-1/2″ width, it takes three headers and two pieces of 1/2″ plywood to equal the wall width.

Post and beam construction Refers to a structural wall system of concentrated loads using posts and beams as opposed to ordinary frame walls that are continuously loaded. The downward wall loads are typically transferred to a foundation composed of piers or columns on center, not a continuous foundation wall.

Rafter Generic term meaning many types of bearing members of roof framing. Horizontal or inclined, the structural component of roof framing usually extending from wall to wall or from wall to ridge.

Hip jack rafter The rafters, parallel to full-length rafters, extending from the fascia over the bearing wall and to the hip rafter.

Hip rafter The rafter (often at 45 degrees) bisecting a turn (typically 90 degrees) in a fascia, extending up and over the corner of a wall on towards the ridge. The hip rafter forms a ridge since it is the intersection of two planes.

Sheathing Wall and roof sheathing, typically plywood or OSB, is installed against wall and roof framing and provides a structural membrane. Sheathing can be 4′ × 8′ panels or individual boards.

Shim A small "sliver" of wood used to make up a gap in framing members.

Soffit The horizontal "belly", or underside, of a roof overhang.

Stringer The inclined (typically a 2 × 12) board used in the frame construction of stairs. Typical stair widths of 3′–4′ have three stringers (one on each side and the middle) or four stringers (two on each side). Treads and risers attach to the stringers.

Square cut (also see "Plumb cut") The end cut of a rafter or truss being perpendicular to its rise. Often used to describe the fascia condition.

Studs Vertical members of a wall, usually 2 × 4 or 2 × 6.

Cripple studs The studs *under and on top* (not the jamb) of an opening, such as a window opening.

Cripple studs are placed on center to a layout beginning with one end of the wall.

Jack studs Wall studs to the left and right of an opening, extending vertically up to and supporting a header.

King studs The stud beside an opening that extends up to the top plate and does not support a header.

Precut studs 2 × 4 or 2 × 6 studs cut to a length less than 8′, so that when bottom and top plates are added, a wall height of approximately 8′-1″ is achieved.

Subflooring In frame construction, the boards or plywood directly on top of floor joists or floor trusses that receive finish flooring.

Slope Rise over run.

Timber truss Truss built with large pieces and often exposed to view. The "gussets" of a timber truss are typically fabricated with structural plate steel (say 1/4″ or thicker) and drilled with holes to receive bolts (or all thread) that may extend completely through the timbers.

Top plate (also see "Base plate" and "Sill plate") Horizontal wood at the top of a wall and the same width as the wall studs. In some non-load-bearing conditions, a single top plate is used. For bearing walls, double top plates are used.

Truss A floor or roof structural unit made of multiple boards bracing each other to make a load-bearing unit.

Drop truss A nonbearing truss usually sitting on top of an end wall. A drop truss is shorter than the regular trusses adjacent to it by the depth of the outlooker.

Floor truss A horizontal truss (say 2′ high) made of common lumber in a manufacturing facility that provides the structural engineering drawings for the truss. A floor truss is typically made with the 2 × 4 top and bottom chords being flat, not turned on edge.

Hip truss A truss forming a ridge at the intersection of two planes.

See the online resources for diagram 6G.6

Mono Truss A half truss.

See the online resources for diagram 6G.7

Scissor Truss A truss with a sloping bottom chord.

See the online resources for diagram 6G.8

Truncated truss In truss configurations, a geometric shape that is stopped short at one end; see truss "a". Truncated also describes trusses that "step down" at a hip end truss set, with a horizontal top chord; see truss "b".

See the online resources for diagram 6G.9

Valley In roof construction, the intersection of two roof planes at "inside corners", providing a path for water to flow.

Vapor barrier A thin membrane made of non-permeable material placed on top of wall and roof sheathing as a "moisture" barrier.

DOORS AND HARDWARE GLOSSARY

Some of the following definitions are from WDMA I.S.1A-11, *Industry Standard for Architectural Wood Flush* Doors and WDMA I.S.6A-11, *Industry Standard for Architectural Stile and Rail Doors*.

This glossary contains the following definitions that are *not* used by the DHI. They are architectural "styles", not door types. However, these terms might find their way onto an occasional architectural door schedule. These styles are:

Bifold door
Bypass door
Double acting door
Dutch door
French door
Louver door
Panel door
Pocket door
Sliding door

Active door The door in a double-door set that is opened first and to which a functional lockset is applied.

Adjustable frame (or wraparound frame) A frame with two legs and a head made of two pieces so that they can slide past each other and be used for various wall thicknesses. They can be wood or metal and have different attachment methods, and are often used to wrap around an existing wall.

Armor plates Usually located on the "push side" of the door, metal (usually stainless steel) sheet protection installed on the door for use when feet, carts, or wheelchairs are used to open the door. An armor plate usually covers the entire bottom half of a door, but can be more or less.

Astragal Often the shape of a "T" and called a "T" astragal, a piece installed at the edge of a door from top to bottom of the slab. Used with double doors to hide the space between them, the T astragal provides for a close fit and prevents air, sound, or light passage. Can also be used horizontally on the bottom edge of a flush transom panel or the bottom edge of the top half of a Dutch door. Shown below is an astragal attached to one leaf of a double door – the other leaf closes against it. Astragals can be made from metal, wood, or plastic.

See the online resources for diagram 8G.1

Backset The distance from the centerline of a lock bore (the drilled opening) and the edge of a door.

See the online resources for diagram 8G.2

Bifold door A door "style" featuring two halves folding vertically in accordion fashion against the door jamb. They often have small cabinet-sized inoperable fixed knobs on the exterior side. Bifold doors are usually two- or four-leaf units and are not hinged to a side jamb. Individual panels are hinged together but not at the jamb. At the top and bottom edges of these units, on both the left and right sides, small "pivots" fit into channel shaped "guide tracks" at the head jamb and at the floor. Bifold doors are often seen in residences. When they occur in commercial buildings, the hardware (pivots and guide tracks) are usually specified to be thicker and stronger than those sold in lumber yards for houses.

Bore The drilled opening (hole through the door) that allows insertion of a lockset.

Bumper threshold The door closes against and touches a bumper threshold, the bumper portion of the threshold rising up higher than the bottom of a door. Typically used with exterior doors to prevent water or insect intrusion.

See the online resources for diagram 8G.3

Certified wood Wood products that have been qualified by an independent third party agency as satisfying their proprietary requirements for responsible environmental practices.

Core The innermost layer or section in component construction. For typical door construction these are used: particleboard core, medium density fiberboard core (MDF), structural composite lumber core (SCL), staved lumber core, laminated veneer lumber core (LVL), fire resistant composite core, and other special core types.

Cylinder The keyway part of a lock that provides the ability to lock or unlock a door with a key.

Cylindrical lockset The entire assembly of a door latching device – latchbolt, cylinder, and knob or lever.

See the online resources for diagram 8G.4

Deadbolt A lock in which the latchbolt is engaged by using a key or a thumb turn. There is no spring and the latchbolt stays in place, either open or closed, "deadlocked" in position.

Delamination Separation of plies or layers of wood or other material through failure of the adhesive bond.

Detention door and hardware Name given to assemblies designed, manufactured, and tested for the containment of people in correctional facilities.

Door bottom Attachment to the bottom of a door slab to minimize the transmission of sound, wind, or weather under the door. Sometimes this material resembles the "brushes" of a broom, and door bottoms are also called door "sweeps" or "brushes".

See the online resources for diagram 8G.5

Door closers, automatic door closers A device typically attached at the top of both the door slab and frame, "pulling" the door slab shut after being opened. A standard closer with no hold-open capability is called "self-closing". Closers that automatically shut by way of a fire alarm are "auto closing". Door closers can be adjusted to pull *or push* with more, or less, force. This is important sometimes for not only fire door requirements but also handicap accessibility. The number of seconds that it takes the closer to shut and latch the door is an important fire code consideration and the pounds of force (for example 5 lbs maximum opening force) to open a door is an important consideration for the handicap code.

Door lite, also see "Factory glazed door" A piece of glass in a door. "Lite" is spelled properly here, the industry differentiating it from vision light. The metal stop that surrounds the glass is called a "lite kit". The glass may be furnished by a storefront and glass subcontractor while the lite kit is often supplied by the hollow metal supplier/fabricator. Both of these items can be "furnished only" to a general contractor or CM. The size of lites is restricted in fire rated doors, and lites are not permitted in three-hour rated doors.

Door stop A device to stop the swing of a door being opened, often placed 90 degrees from the closed position of the door. They can be placed on a floor or wall.

See the online resources for diagram 8G.6

Double acting door (or double egress) A door "style" that hinges on one side and swings in both directions.

See the online resources for diagram 8G.7

Dutch door A door "style" featuring two halves, a top half and lower half, often with a shelf on the bottom leaf.

Engineered materials A general term used to describe any wood or plant fiber composite panel. Such products as particleboard, MDF, SCL, and LVL are described as engineered fiber. Typically they are made from wood or plant fiber or wood pieces and have specific quality requirements.

Face plate Located on the edge of a door, covers the latchbolt area.

See the online resources for diagram 8G.8

Face veneer The outermost exposed wood veneer surface of a veneered wood door, usually only 1/50th of an inch thickness!

Factory glazed door A wood door manufactured with a wood or metal lite kit and glass.

Fire rated doors A door that has been constructed in such a manner that when installed in an assembly will pass a fire test as described by NFPA 80. The door must carry an identifying label on the edge of the door from a qualified inspection agency. Fire doors must be inspected once per year!

Fixed door One or more *non-operable* assembled leaves or panel within a door frame or sliding door track.

Flush bolt A device that secures (holds) an inactive leaf of a pair of doors to a jamb. The "bolt" slides back and forth or up and down into a slot in the jamb. Flush bolts are usually mortised into the face or edge of a door.

See the online resources for diagram 8G.9

Flush door A flat smooth door having no glass lites, panels, louvers, or grilles.

French door A door "style" with an assembly of stiles and rails with muntins and bars surrounding multiple glazed openings often called "divided lites" that extend for the full height and width of the door slab.

Glazed lite A noun; the opening in a door that is to receive glazing material.

Glazing A verb; the process of installing glazing materials.

Glazing material A transparent or translucent material installed in the glazed lite of door assemblies and windows. If a contractor asks a door supplier quoting a job, "Did you include glazing?" he or she is asking if the supplier is getting the doors from the manufacturer with lites and kit already in the door.

Hardware group or hardware set The pieces of hardware for a door that may apply to multiple doors on a door schedule. The compilation of hanging, locking, and closing devices, installed on multiple doors with the same hardware.

Hinges The connection piece between a door and a frame. There are usually two moving pieces of a standard hinge, connected by a hinge pin. Both the frame and door are mortised (recessed) to receive the hinge. For 6′-8″ and 7′-0″ doors, there are usually three separate hinges, a pair and a half, per door. Residential hinges are typically 3-1/2″ × 3-1/2″ and commercial hinges are 4-1/2″ × 4-1/2″. Other types of hinges include piano hinges and spring hinges.

See the online resources for diagram 8G.10

Hospital stop When the stop on a metal door frame does not extend to the floor, the door frame being straight at the bottom with fewer edges and turns at the floor. This allows for easier cleaning of the floor. See Sketch 1, Door and Frame Profiles, in Part 8, Section 4 Door Frames.

Jamb depth or width The overall width of a metal door frame, out to out. See Sketch 1, Door and Frame Profiles, in Part 8, Section 4 Door Frames.

Keyed lockset For classroom and entry functions, a lockset on a door that requires a key to retract the latchbolt, thereby unlocking the door. For a storeroom function, the key first retracts the latchbolt, but the latchbolt returns to the locking position when the key is removed.

Keypad A digital keypad used to unlock a door given the proper number sequence.

Kick plate Located on the "push side" of the door, metal (usually stainless steel) sheet protection about six to twelve inches high installed near the bottom of a door. Protects the door from foot scuffmarks.

Latchbolt The metal bolt that secures the door when closed. Usually operated by a knob or lever handle. The latchbolt closes against a strike plate. A positive latching latchbolt is spring operated. The latchbolt of a deadbolt is not spring operated.

Latch guard A security plate that covers the latch area to make it inaccessible for tampering.

See the online resources for diagram 8G.11

Lead-lined A door or frame lined with lead to prevent radiation penetration. For use near X-ray equipment in hospitals or doctor's offices.

Leaf A fixed or moving portion of a side-hinged door system, glazed or unglazed, surrounded by a fixed frame. Leaves are also known as "slabs".

Lite Door glass. Four versions are drawn below.

See the online resources for diagram 8G.12

Louver door A door "style" that is an assembly of stiles and rails where part of the slab surface is filled with slat or chevron (inverted "V") louvers.

Louvered A panel constructed of wood or metal slats installed in an opening to allow light, air, and noise. Common types are slat and chevron (inverted "V") louver.

Mop plate Located on the "pull side" of the door, metal (usually stainless steel) sheet protection, typically 4″ to 12″ high, installed on the door for protection against water and mop buckets.

Mortise A recess.

Mortise lockset Lock type that installs into a pocket (typically about 5-1/2″ high and 4″ deep) cut into the edge of door.

See the online resources for diagram 8G.13

Mullion A vertical member used to separate doors or side-lights in an HM frame. Also known as a "mull".

Muntin Horizontal or vertical bar used to separate individual pieces of glazing material in a door slab. Muntins can also be surfaced applied on top of glass, giving the appearance of individual panes.

Narrow lite door A door slab that contains a narrow piece of view glass, typically 3″–8″ wide. Also see "Door lite".

Panel door A door "style" made by applying components to a door. These pieces are vertical stiles, horizontal rails, and one or more panels. Intermediate rails or mullions are used to separate panels. Panels can be raised, flat, or recessed.

Panel, raised Where indentations in the door face give the appearance of raised panels; the raised areas match the full thickness of the door and the areas surrounding the panels are less than door width. The panel edges are shaped to fit into grooves in the stiles, rails.

Peephole A small vision cylindrical (bullet shaped 1/2″ to 9/16″ dia.) piece, 1-3/4″ long when installed in a solid core door, allowing sight through a door. Usually positioned in the middle of the door at eye-level height.

Pivot hinge A door having a dowel shape extend from the bottom of the door into a hole in the floor, allowing the door to turn or pivot. Another short dowel is at the top of the door extending into the jamb head, see diagram 8G.14.

See the online resources for diagram 8G.14

Ply There are three plies in a door – outer, underlayment, and core – assembled into industry standard 3 ply, 5 ply, or 7 ply. Each ply can consist of more than one layer, for example, a 5-ply door consists of two each outer layers, two each underlayment layers, and a core. A single sheet of veneer or several strips laid with adjoining edges that may or may not be glued, which forms one veneer lamination in a glued panel. In some constructions, a ply is used to refer to other wood components such as particleboard or MDF.

Pocket door A door "style" that slides on a "track" parallel to the wall. When the door is recessed into the wall, the opening is clear for passage. The pocket door slides out of the wall cavity to close the opening.

Push/pull plates For use on doors without locksets. One plate goes on one side of the door, the other one is placed on the opposite side.

See the online resources for diagram 8G.15

Rail The horizontal structural member of a stile and rail door. Fits between the vertical stiles.
Rail, bottom The bottom rail of a stile and rail door
Rail, intermediate A rail, other than the top and bottom rail, used to separate panels or to separate panels from glazing materials in a combination door. Also referred to as "cross rail".
Rail, top The uppermost rail of a stile and rail door.
Rain drip Rain drips are located above doors or windows or at the bottom of a door leaf to divert water.

See the online resources for diagram 8G.16

Revolving door An exterior door consisting of two or more leaves that pivot about a common vertical axis within a cylindrically shaped vestibule.
Rim lock Lockset type that is surface mounted, usually on the inside face of a door slab.

See the online resources for diagram 8G.17

Rough opening The size of the wall opening, width and height, into which a door and frame is to be installed.
Saddle threshold The middle portion of these thresholds rest directly beneath a closed door, hence the name, but can be used to transition between floor heights on each side of the door; see threshold shown on the right.

See the online resources for diagram 8G.18

Side lite Operable or non-operable glass product installed on one or both sides of an operable door or a fixed door. Side lites often have their own separate frame or are contained within the frame of a composite assembly.
Silencer A small piece of resilient material attached to the stop of a door frame to cushion the closing of a door, minimizing sound and the wear and tear of the door and frame. Sometimes called a mute.
Sliding door, also called bypass doors Door that consists of manually operated door panels, one or more of which slide or roll horizontally within a common frame, and can also contain fixed lites/panels. Note: Typically, operating panels are identified with an (X) and fixed lites or fixed panels are identified with an (O).
Sound transmission class (STC) A single number rating system derived from measured values of sound transmission loss or the acoustical performance of a building element, such as a door, window, or wall. The higher the STC value, the better the rating and the better the acoustical performance value. Tested in accordance with ASTM E413 and E90.
Spring hinge A type of hinge, sometimes adjustable, that closes the door by spring force when the door leaf is open.
Stile The vertical members on the left and right sides of a door or fixed panel.
Stop The element of a door frame that the door slab closes against.
Strike plate The metal attachment that is mounted onto the edge of a door frame, covering the recess that accepts a latchbolt.

See the online resources for diagram 8G.19

T astragal see "Astragal".
Throat The width of the opening on the back (open) side of a door frame, usually the same as the wall thickness. See Sketch 1, Door and Frame Profiles, in Part 8, Section 4 Door Frames.

Throw With the latchbolt engaged (locking position), the measurement of the projection beyond the face plate or edge of the door. The length of the throw into the edge of a frame can be an important consideration with fire doors and hurricane rated doors if the door leaf opens too easily.

Transom The opening above a door or set of doors that may contain fixed glazing material, an operating sash, panel, or other filler.

Undercut As seen by the door supplier, the distance between the bottom of a door slab and the bottom of the frame, which the door supplier can control. As viewed by the architect and contractor, the undercut is the distance to the finished floor from the bottom of the door slab. For the contractor, this subject is affected by two variables, the floor levelness and the thickness of the flooring. The architect, perhaps concerned about allowing a lot (or a little) amount of air to move from room to room, may specify that certain doors have a 1-1/2″ undercut, which is a good amount of space. Or concerned about sound transmission, the specification might be for a 3/4″ undercut. Without instructions to the contrary, a carpenter will set a door frame against the concrete floor. The floor finish might be 1/8″ thick vinyl tile, or 1/2″ carpet, or 3/4″ wood. For the contractor to achieve specific door undercuts, some preplanning needs to occur with the finish flooring thickness conveyed by project management to door supplier to superintendent to carpenter.

See the online resources for diagram 8G.20

Vision light A glazed opening in a door.

Weatherstrip Usually attached to the stop of a door frame, a thick resilient or cushion material that a door edge closes to. Helps prevent water infiltration and the passage of air, smoke, and sound.

INDEX